Lecture Notes in Geoinformation and Cartography

Series Editors: William Cartwright, Georg Gartner, Liqiu Meng, Michael P. Peterson

Lars Bernard · Anders Friis-Christensen ·
Hardy Pundt (Eds.)

# The European Information Society

## Taking Geoinformation Science One Step Further

Springer

*Editors*

Prof. Lars Bernard
Technische Universität Dresden
Helmholtzstrasse 10
01069 Dresden
Germany
lars.bernard@tu-dresden.de

Dr. Anders Friis-Christensen
European Commission
Joint Research Centre
Institute for Environment
and Sustainability
TP 262, 21027 Ispra(VA)
Italy
anders.friis@jrc.it

Prof. Hardy Pundt
University of Applied Studies
and Research Harz
Friedrich Str. 57-59
38855 Wernigerode
Germany
hpundt@hs-harz.de

ISBN: 978-3-540-78945-1 e-ISBN: 978-3-540-78946-8

Lecture Notes in Geoinformation and Cartography ISSN: 1863-2246

Library of Congress Control Number: 2008924326



Typesetting: Camera-ready by the editors

*Cover design:* deblik, Berlin

Printed on acid-free paper

9 8 7 6 5 4 3 2 1

springer.com

# Preface

The Association of Geographic Information Laboratories for Europe (AGILE) was established in early 1998 to promote academic teaching and research on GIS at the European level. AGILE seeks to ensure that the views of the geographic information teaching and research community are fully represented in the discussions that take place on future European research agendas and it also provides a permanent scientific forum where geographic information researchers can meet and exchange ideas and experiences at the European level.

In 2007 AGILE provided - for the first time since its existence - a book constituting a collection of scientific papers that were submitted as full-papers to the annual AGILE conference and went through a competitive and thorough review process. Published in the *Springer Lecture Notes in Geoinformation and Cartography* this first edition was well received within AGILE and within the European Geoinformation Science community as a whole. Thus, the decision was easily made to establish a Springer Volume for the 11$^{th}$ AGILE conference held 2008 in Girona, Spain, and led to what you now hold in your hands.

The 11$^{th}$ AGILE call for full-papers of original and unpublished fundamental scientific research in all fields of Geoinformation Science resulted in 54 submissions, of which 23 were accepted for publication in this volume (acceptance rate 43%). These figures are similar to those of the 2007 volume and indicate that having full-paper submissions leading to an annual high-quality scientific edition is a promising model for following AGILE conferences.

The scientific papers published here, cover a number of basic topics within Geoinformation Science. The papers included in this book span fundamental aspects of geoinformation processing: Measuring spatiotemporal phenomena, quality and semantics of geoinformation, spatiotemporal analysis, spatiotemporal modelling and decision support, and spatial information infrastructures. We believe that the papers comprise innovative research and take Geoinformation Science one step further.

Organising the programme of an International Conference and simultaneously editing a volume of scientific papers necessarily requires time and effort. We therefore would like to gratefully acknowledge the efforts of the authors and reviewers of this book, who in adhering to a strict timetable, helped to finalise this book so that it could be delivered at the AGILE 2008 conference. We thank the local chair Irene Compte (University of Girona) and her team in Girona for giving all kind of local support to make the AGILE 2008 conference happen. Veronica Schemien (TU Dresden) put a lot of effort by supporting us with the editing of this book. Additionally, we would like to thank Definiens AG, ESRI Inc., and Intergraph Corporation who for years have sponsored the annual AGILE conferences. Last but not least, our thanks go to Agata Oelschläger (Springer) who helped us to finalise this book in time.

*Lars Bernard, Anders Friis-Christensen, Hardy Pundt*

February 2008

## Programme Committee

Programme Chair Lars Bernard,
TU Dresden (Germany)

Programme Co-Chair Anders Friis-Christensen,
European Commission, Joint Research Centre (Italy)

Programme Co-Chair Hardy Pundt,
University of Applied Sciences Harz (Germany)

## Local Committee

Local Chair Irene Compte
University of Girona (Spain)

## Scientific Committee

Itzak Benenson, Tel Aviv University (Israel)
Michela Bertolotto, University College Dublin (Ireland)
Lars Bodum, University of Aalborg (Denmark)
Arnold Bregt, Wageningen University (Netherlands)
Christoph Brox, University of Münster (Germany)
Gilberto Camara, INPE (Brazil)
Nicholas Chrisman, Laval University (Canada)
Christophe Claramunt, Naval Academy Research Institute (France)
Arzu Çöltekin, University of Zurich (Switzerland)
Helen Couclelis, University of California - Santa Barbara (USA)
Max Craglia, European Commission, Joint Research Centre (Italy)
Isabel Cruz, University of Illinois - Chicago (USA)
Leila De Floriani, University of Genova (Italy)
Pasquale diDonato, University of Rome - La Sapienza (Italy)
Sara Fabrikant, University of Zurich (Switzerland)
Peter Fisher, University of Leicester (United Kingdom)
Anders Friis-Christensen, European Commission, Joint Research Centre (Italy)

Michael Gould, University Jaume I - Castellón (Spain)
Francis Harvey, University of Leicester (United Kingdom)
Gerard Heuvelink, Wageningen University (Netherlands)
Hartwig Hochmair, University of Florida (USA)
Kathleen Hornsby-Steward, University of Iowa (USA)
Gary Hunter, University of Melbourne (Australia)
Bin Jiang, University of Gävle (Sweden)
Didier Josselin, Avignon University (France)
Eva Klien, University of Münster (Germany)
Thomas Kolbe, TU Berlin (Germany)
Menno Jan Kraak, ITC (Netherlands)
Lars Kulik, University of Melbourne (Australia)
Rob Lemmens, ITC (Netherlands)
Roger Longhorn, GEO:connexion (United Kingdom)
Michael Lutz, European Commission, Joint Research Centre (Italy)
Bela Markus, University of West Hungary (Hungary)
Gabor Mezosi, University of Szeged (Hungary)
Javier Nogueras-Iso, University of Zaragoza (Spain)
Atsuyuki Okabe, University of Tokyo (Japan)
Henk Ottens, University of Utrecht (Netherlands)
Volker Paelke, University of Hannover (Germany)
Marco Painho, Universidade Nova de Lisboa (Portugal)
Dieter Pfoser, Research Academic Computer Techn. Hellas (Greece)
Poulicos Prastacos, Foundation for Research and Techn. Hellas (Greece)
Florian Probst, University of Münster (Germany)
Hardy Pundt, University of Applied Sciences Harz (Germany)
Ross Purves, University of Zurich (Switzerland)
Martin Raubal, University of California - Santa Barbara (USA)
Tumasch Reichenbacher, University of Zurich (Switzerland)
Wolfgang Reinhardt, UniBwMünchen (Germany)
Femke Reitsma, Edinburgh University (United Kingdom)
Jochen Renz, University of New South Wales Sydney (Australia)
Claus Rinner, Ryerson University (Canada)
Andrea Rodríguez, Universidad de Concepción (Spain)
Mauro Salvemini, University of Rome - La Sapienza (Italy)
Monika Sester, University of Hannover (Germany)
Takeshi Shirabe, Vienna University of Technology (Austria)
Spiros Skiadopoulos, National Technical University of Athens (Greece)
Bettina Speckmann, TU Eindhoven (Netherlands)
Thérèse Steenberghen, Katholieke Universiteit Leuven (Belgium)
John Stell, University of Leeds (United Kingdom)
Juan Suárez, Forestry Commission (United Kingdom)

# Contributing Authors


**Alexander Almer**
Joanneum Research, Austria

**Arda Alp**
PhD Student: Artificial inteligence Laboratory - Ecole Polytechnique Federale Lausanne, Switzerland

**Peter Bachhiesl**
School of Telematics/Network Engineering, Carinthia University of Applied Sciences, Austria

**Alberto Belussi**
University of Verona, Italy

**Michela Bertolotto**
School of Computer Science & Informatics, University College Dublin, Ireland

**Elena Camossi**
School of Computer Science & Informatics, University College Dublin, Ireland

**Claudio Carneiro**
PhD Student: GIS LAB - Ecole Polytechnique Federale Lausanne, Switzerland

**Martin Charlton**
National Centre for Geocomputation, National University of Ireland Maynooth, Co Kildare, Ireland

**Diego Díaz Doce**
The University of Edinburgh, United Kingdom

**Jürgen Döllner**
Hasso-Plattner-Institute, University of Potsdam, Germany

**Martin Espeter**
University of Münster, Germany

**Peter Foley**
National Centre for Geocomputation, National University of Ireland Maynooth, Co Kildare, Ireland

**A Stewart Fotheringham**
National Centre for Geocomputation, National University of Ireland Maynooth, Co Kildare, Ireland

**Andrew U. Frank**
Vienna University of Technology, Institute for Geoinformation and Cartography, Austria

**Mauro Gaio**
LIUPPA, France

**François Golay**
Swiss Federal Institute of Technology, Switzerland

**Klaus Granica**
Joanneum Research, Austria

**Gerald Gruber**
Fachhochschule Kaernten, Austria

**Alex Hagen-Zanker**
Urban Planning Group, Technische Universiteit Eindhoven, Netherlands

**Henning Sten Hansen**
Aalborg University, Denmark

**Brent Hecht**
University of California, Santa Barbara, United States

**Maarten Hilferink**
Object Vision BV, CIMO-Vrije Universiteit Amsterdam, Netherlands

**Manuela Hirschmugl**
Joanneum Research, Austria

**Hartwig Hochmair**
University of Florida, United States

**Jens Ingensand**
Swiss Federal Institute of Technology, Switzerland

**Krzysztof Janowicz**
University of Münster, Germany

**Markus Jobst**
Vienna University of Technology, Austria

**Farid Karimipour**
Vienna University of Technology, Institute for Geoinformation and Cartography, Austria

**Tahar Kechadi**
School of Computer Science & Informatics, University College Dublin, Ireland

**Carsten Kessler**
University of Muenster, Germany

**Eric Koomen**
Vrije Universiteit Amsterdam FEWEB/SPINlab, Netherlands

**Martin Krch**
School of Geoinformation, Carinthia University of Applied Sciences, Austria

**Federica Liguori**
GIS Expert, Italy

**Willem Loonen**
Netherlands Environmental Assessment Agency (MNP), Netherlands

**Haik Lorenz**
Hasso-Plattner-Institute, University of Potsdam, Germany

**Pierre Loustau**
LIUPPA, France

**Jose Macedo**
PosDoc researcher: Database Laboratory - Ecole Polytechnique Federale Lausanne, Switzerland

**Jody Marca**
Politecnico di Milano, Italy

**Christian Menard**
Fachhochschule Kaernten, Austria

**Hossein Mohammadi**
The University of Melbourne, Geomatics Department, Australia

**Gerhard Navratil**
Vienna University of Technology, Institute for Geoinformation and Cartography, Austria

**Mauro Negri**
Politecnico di Milano, Italy

**Thierry Nodenot**
LIUPPA, France

**Tonny Oyana**
Southern Illinois University, United States

**Ilija Panov**
University of Münster, Germany

**Genevieve Patenaude**
The University of Edinburgh, United Kingdom

**Gernot Paulus**
School of Geoinformation, Carinthia University of Applied Sciences, Austria

**Giuseppe Pelagatti**
Politecnico di Milano, Italy

**Michal Petr**
Forest Research, United Kingdom

**Johann Raggam**
Joanneum Research, Austria

**Abbas Rajabifard**
The University of Melbourne, Geomatics Department, Australia

**Martin Raubal**
University of California, Santa Barbara, United States

**Bernhard Schachinger**
Fachhochschule Kaernten, Austria

**Thomas Schnabel**
Joanneum Research, Austria

**Johannes Scholz**
School of Geoinformation, Carinthia University of Applied Sciences, Austria

**Mirco Schwarz**
University of Münster, Germany

**Kara Scott**
Southern Illinois University, United States

**Stefano Spaccapietra**
Senior researcher (LAB's responsible) and full professor: Database Laboratory - Ecole Polytechnique Federale Lausanne, Switzerland

**Emmanuel Stefanakis**
Harokopio University of Athens, Greece

**Juan Suárez**
Forestry Commission, United Kingdom

**Harry Timmermans**
Urban Planning Group, Technische Universiteit Eindhoven, Netherlands

**Matthias Trapp**
Hasso-Plattner-Institute, University of Potsdam, Germany

**Michael van Dahl**
VCS Aktiengesellschaft, Austria

**Alexander C. Walkowski**
Institute for Geoinformatics, University of Münster, Germany

**Marc Wilkes**
University of Münster, Germany

**Ian Williamson**
The University of Melbourne, Geomatics Department, Australia

# Table of Contents

## Measuring Spatiotemporal Phenomena



## Quality and Semantics of Geoinformation

## Spatiotemporal Analysis



## Spatiotemporal Modelling and Decision Support

## Spatial Information Infrastructures

# Forest Stand Volume of Sitka Spruce Plantations in Britain: Can Existing Laser Scanning Methods Based on the Conventional One Provide Better Results, a Comparison of Two Approaches

Michal Petr[1,2], Genevieve Patenaude[1], Juan Suárez[2]

[1] Institute of Geography, School of Geosciences, The University of Edinburgh, Edinburgh EH8 9XP, UK, M.Petr@sms.ed.ac.uk, genevieve.patenaude@ed.ac.uk
[2] Forest Research, Northern Research Station, Roslin, Midlothian EH25 9SY, UK, michal.petr@forestry.gsi.gov.uk, juan.suarez@forestry.gsi.gov.uk

**Abstract.** This paper looks at different datasets obtained from an airborne Light Detection And Ranging (LiDAR) system and compares the reliability of two contemporary analysis approaches. Estimates of different stand parameters, such as top tree height, were derived using regression analysis and a segmentation approach on data obtained from small-footprint laser scan were contrasted with the field measurements in 7 plots, specifically volume and basal area. Plots of 2,500 $m^2$ containing plantations of Sitka spruce (*Picea sitchensis Bong. Carr.*) were scanned with two different point densities in years 2003 and 2004. These plots were divided into training and test regions of 625 $m^2$ each. Regression analysis was performed using percentiles corresponding to the canopy tree height at different vertical levels and a segmentation method was used to delineate individual tree crowns where tree metrics can be determined. The bias of the estimated values for the stand volume and basal area ranged from 1.21 to 6.49 $m^3ha^{-1}$ (0.17 to 0.92 %) and - 2.69 to 1.23 $m^2ha^{-1}$ (- 3.9 to 1.7 %), respectively; and the bias calculated from the segmentation using 0.5 and 1m dataset ranged between - 349.77 to - 434.76 $m^3ha^{-1}$ (- 49.7 to - 61.8 %) for the stand volume and - 33.36 to - 42.24 $m^2ha^{-1}$ (- 48.5 to - 61.4 %) for the basal area. The results showed that the regression models estimated stand volume and basal more accurately compared with values calculated from the segmentation. Furthermore, it is shown that there was no significant difference in the estimates from the regression model when using different point densities.

**Keywords:** forest, stand volume, basal area, LiDAR, segmentation

# 1 Introduction

The forest is a dynamic and complex ecosystem that represents an integral part of world's environment providing essential elements to different spheres of our society. In the United Kingdom, coniferous forest covers more than 57 % of the total woodland (*Forestry Statistics 2006.*), and the current growing stock is important for a sustainable forest management. The need for faster methods of forest inventory compared with the conventional field measurements has motivated the employment of remote sensing techniques that can cover large areas of woodland forest and with fewer constraints such as manpower. Except for the common techniques such as aerial photography, new systems such as Light Detection and Ranging and radar are currently being applied.

Many studies have used airborne laser scanning (ALS) for the estimation of various forest stand metrics, mainly mean tree height, stand basal area, and stand volume (Naesset 1997a; Holmgren 2004; Hyyppa et al. 2001; Maltamo, Eerikainen et al. 2004). The most significant advantage of LiDAR is its ability to provide almost three-dimensional information about forest canopy structures (Naesset 2004). Data gathered by ALS systems can provide complete coverage as opposed to the sampling normally achieved with the field data collection. The width of study area will depend on the flight altitude and scanning angle (Hyyppa et al. 2005).

LiDAR systems have been frequently employed in forest studies because of their ability to provide both horizontal and vertical information about trees, whereas most optical sensors are only capable of providing horizontal information (Lim, Flood et al. 2003; Donoghue and Watt 2006). Another significant advantage of LiDAR sensing is that it readily obtains specific information about trees, such as canopy height, very accurately (Naesset 2004). Even though the cost of obtaining LiDAR data is high (Tilley et al. 2004), it gives an essential and effective tool for the estimation of tree characteristics; and as this data can be also adopted to various other purposes, a reduction in the cost can be expected with wider application.

The use of airborne LiDAR systems in forestry focuses primarily on the determination of tree metrics in the stand level for inventory purposes (Naesset 2004; Holmgren 2004; Coops et al. 2007; Naesset 1997a) or at the tree level for measurement of single trees (Hyyppa et al. 2005; Maltamo, Yu et al. 2004; Peuhkurinen et al. forthcoming). Airborne LiDAR with small footprint and low point density (about 1 point/m$^2$) has proven to be an efficient tool at the stand level in the determination of tree metrics (Holmgren 2004). However, at the tree level a point density of at least 5 points/m$^2$ is required to achieve good results for tree segmentation and to provide sufficient information about individual trees (Maltamo et al. 2005).

In forest inventories, where information is related to the stand, several parameters such as mean tree height or basal area are necessary to determine. Previously airborne LiDAR has effectively and accurately determined mean tree height (Naesset 1997a; Holmgren, Nilsson, and Olsson 2003; Hyyppä et al. 2000), basal area (Hyyppä et al. 2000; Lim, Treitz et al. 2003), stand volume (Naesset 1997b; Means et al. 2000), and dominant tree height (Lim, Treitz et al. 2003). When airborne discrete LiDAR data was used for stand volume estimation it was shown that the accuracy varies for different tree species (Naesset 1997b). Reasons for this variation are mainly due to crown shape and the type of vegetative material (leaves or needles) of specific species. Furthermore, the scanning point density influences the capability of LiDAR to detect tree tops; however, in most cases it omitted the correct tree top (Suárez et al. 2005) and therefore underestimated tree height (Nilsson 1996; Naesset 1997b). Many studies also focused on the estimation of dominant tree height of young forests and proved that mature and even young trees can be accurately estimated by LiDAR ( Naesset and Bjerknes 2001).

Another method to determine forest metrics from LiDAR data is the single-tree level approach. This approach is based on tree delineation algorithms that identify, locate, and relate tree characteristics to a single tree. Several studies have suggested and used high scanning point densities from 10 - 20 points/m$^2$ for the detection of single trees (Maltamo, Yu et al. 2004; Hyyppa et al. 2001). The canopy height model (CHM) which gives the canopy height as the difference between the Digital Elevation Model (DEM) and the Digital Surface Model (DSM) is derived from LiDAR data and has been used for tree segmentation. Segmentation algorithms have been primarily used for the detection of trees from the CHM where local maxima were given by tree tops in segments. The retrieved height using this method provided reliable results and a recent study has shown an underestimation of only 0.97 m (Maltamo, Yu et al. 2004). The stem volume of an individual tree has been calculated from volume equations that used predetermined variables, tree height and crown diameter (assumed to be a circular shape) from the CHM and predicted stem diameter as well (Naesset et al. 2004). The study of (Persson, Holmgren, and Soderman 2002) presented estimates of stem volume with RMSE of 0.21 m$^2$, corresponding to 22 % of the field stem volume. The main advantage of the tree segmentation compared with the stand approach is its capability of identifying spatial heterogeneity within a stand where mean variables are estimated. However, these algorithms are developing and still need improvement.

The aim of this study is to estimate the stand volume of coniferous trees with the use of airborne laser scanning data and to compare the accuracy of this estimate with ground measurements. The main objectives pursued in this study are: 1) to determine whether a relationship between the stand volume and canopy heights obtained from LiDAR data exists for our study site; 2) to determine how precisely it is possible to estimate the stand volume of coniferous trees from LiDAR data in the study area; 3) to evaluate the effect of different hit densities on the estimation of stand volume; and 4) to determine the differences of volume estimates between a regression based method and a segmentation method at the stand.

This paper is divided into 5 sections, each focusing on a different part of the research. The following section describes available field and LiDAR datasets, and also deals with an applied segmentation algorithm. Section 3 then suggests an appropriate methodology which should be used whereas section 4 will presents the final results of the study. Finally, section 5 will provide discussion of the results and overall conclusions.

# 2 Materials

## 2.1 Study Site

The study area of this project is the Kielder Forest District located in northern England (Fig. 1) (55° 14` N 2° 35` W), and is managed and owned by the UK Forestry Commission (FC). The area is characterised mainly by low hills with altitudes between 30 m and 600 m, and a mean slope angle of 6°. Sitka spruce (*Picea sitchensis Bong. Carr.*) is the dominant species in the area followed by other species like Norway spruce (*Picea abies (L.) H. Karst*) and Lodgepole pine (*Pinus contorta Douglas).* This study focuses on the Sitka spruce plantations located within the study plots.

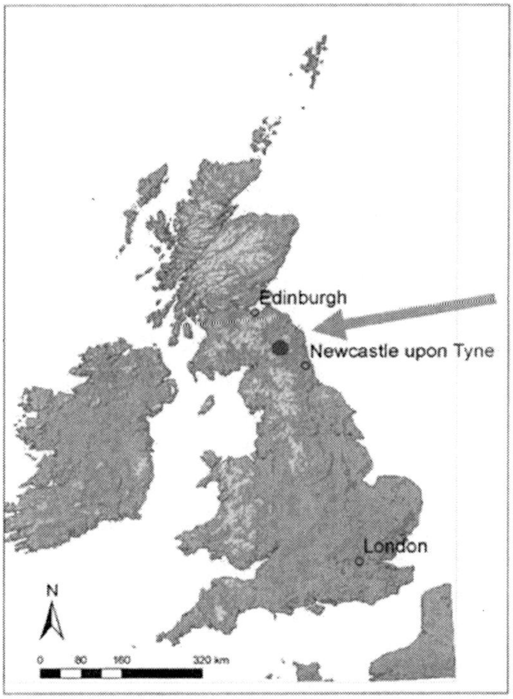


**Fig. 1:** Localization of the Kielder forest in the UK (dot) with example of sample plots 8 and 9

## 2.2 Field Data

The field work in the study area was carried out by the Forestry Commission between February and April 2003. Field data was obtained from 7 study plots with dimensions 50 m x 50 m. These plots were located separately in three different sites with similar climate conditions. In the first location the trees were 64 years old, in the second location trees of 33 prevailed and in the third location the dominant tree age was 36 (Table 1). Each study plot and an individual tree within each plot were accurately located with the differential GPS and a Total station. Tree parameters collected consisted of tree height, diameter at breast height (DBH) and dominance type. Furthermore, for selected trees, additional information about tree crown dimensions in N-S and E-W directions and height to the first live whorl was registered.

Then, the reference data, which is presented in Table 1, was calculated. First, top height ($t_h$) was calculated as the average of the 100 largest trees with the largest DBH per hectare as in (Philip 1994); however, in this study only 25 trees per plot were used. Second, the mean diameter at breast height ($_m$DBH) was computed as an average of all living trees. Next, the basal area of a plot (G) was computed as the sum of all individual basal areas. The Crop form coefficients were derived from the top height of the study plots. Finally, the stand volume (V) was computed as the sum of stem volumes of all living trees located within a study plot. The stem volume for each tree was calculated as basal area multiplied by a crop form coefficient for Sitka spruce from (Hamilton 1975).

**Table 1** Summary of the field plot reference data (50 m x 50 m)

| | Plot area [$m^2$] | $t_h$ [m] | N [$ha^{-1}$] | $_m$DBH [cm] | G [$m^2\ ha^{-1}$] | V [$m^3\ ha^{-1}$] | Age |
|---|---|---|---|---|---|---|---|
| Plot1 | 2590 | 32,5 | 948 | 32,5 | 82,6 | 1209,6 | 64 |
| Plot2 | 2541 | 27,5 | 1168 | 25,0 | 61,7 | 748,0 | 64 |
| Plot3 | 2384 | 19,8 | 2024 | 18,2 | 62,8 | 509,3 | 38 |
| Plot4 | 2456 | 19,9 | 2040 | 18,1 | 62,0 | 521,6 | 38 |
| Plot5 | 2709 | 17,3 | 1980 | 19,9 | 59,4 | 476,2 | 33 |
| Plot6 | 2386 | 20,6 | 2792 | 17,3 | 72,6 | 611,9 | 36 |
| Plot7 | 2494 | 21,0 | 2196 | 20,0 | 72,4 | 651,6 | 33 |

$t_h$ top height, N stem number, $_m$DBH mean diameter at breast height, G stand basal area, V stand volume.

Due to the small number of reference plots available for this study, each 50 m x 50 m study plot was split into four sub-plots of 25 m x 25 m (625 $m^2$). These 28 sub-plots were randomly divided into two same size groups of training and test plots, where each group always contained two sub-plots from the original plot. The stand volume and the basal area were the only forest metrics computed in this study with the same methods as above. For the calculation, just the trees registered within a sub-plot were used. A summary of training and test sub-plots for the basal area and the stand volume is presented in Table 2.

**Table 2** Summary of the sub-plots reference data (25 m x 25 m)

| Characteristics | Range | Mean |
|---|---|---|
| Training plots (n = 14) | | |
| G [$m^2 ha^{-1}$] | 51.64 - 85.63 | 66.46 |
| V [$m^3 ha^{-1}$] | 465.02 - 1210.81 | 678.21 |
| Test plots (n = 14) | | |
| G [$m^2 ha^{-1}$] | 54.85 - 92.76 | 68.79 |
| V [$m^3 ha^{-1}$] | 466.72 - 1311.65 | 703.79 |

G stand basal area, V stand volume.

## 2.3 Laser Scanner Data

LiDAR data for the study area was acquired in two years, 2003 (26th March) and 2004 (July). The contractor for data from year 2003 was UK Environmental Agency using an Optech ALTM 2033 LiDAR system which was the same system for 2004. Operating at 1064 nm the Optech 2033 is a discrete return system, recording only the first and the last laser returns. Detailed information about ALTM 2033 is presented in Table 3. The LiDAR data from 2003 was acquired with a 305 m wide swath and 70 % side overlap, and the flight altitude was 905 m. The scanning angle at nadir was 20°, and produced distances of 1.8 m between scanning lines and 0.3 m between points. The provided LiDAR data was in ASCII format with the X, Y, Z coordinates in the British National Grid (BNG); the intensity of the laser pulse was included in this information. Height was stored in Z coordinate and referenced to the Ordnance Survey of Great Britain OSGB 1936 Datum (Donoghue and Watt 2006).

**Table 3** Summary of the laser scanner

| Parameter | Performance |
|---|---|
| Sensor | ALTM Optech 2033 |
| Laser pulse frequency | 33 kHz |
| Field of view | 10° (20°)[a] |
| Beam divergence | 0.3 Mrad |
| Horizontal accuracy | ± 0.15 m |
| Vertical accuracy | 0.60 m |
| Laser classification | Class IV laser product (FDA CFR 21) |

[a] for LiDAR data acquired in 2003.

The LiDAR data from 2003 and 2004 have different point densities. The density of LiDAR data from year 2003 is lower (above 2 points per $m^2$) when compared with the LiDAR 2004 data with point density of about 7 points per $m^2$ (Table 4). Nevertheless, there were almost no differences in point densities between training and test plots.

**Table 4** Sampling density of laser scanner data

| | No. of observations | No. of transmitted pulses[a] ($ha^{-1}$) | Mean point density (points/$m^2$) |
|---|---|---|---|
| **LiDAR 2003** | | | |
| Training plots | 14 | 12192 – 35584 | 2.29 |
| Test plots | 14 | 11680 – 43808 | 2.38 |
| **LiDAR 2004** | | | |
| Training plots | 14 | 41440 – 96960 | 7.01 |
| Test plots | 14 | 39456 – 98976 | 6.89 |

[a] refers to first pulse data

The last laser points of the LiDAR 2004 data were filtered and classified using Terrascan software (Terrasolid, Finland) to determine and separate the ground points from the vegetation hits. Only the last laser points from 2004 data assumed to represent ground were used to generate two grids with 0.5 m and 1 m spacing using the Surfer software (Goldensoftware, Golden, Colorado, USA). Then, these grids were interpolated by method of Delaunay triangulation with linear interpolation. The final DEM was created from previous grids in ArcGIS using the Nearest Neighbor interpolation algorithm with predefined 0.5 m and 1.0 m pixel sizes. However, the DEM from 2003 was delivered already complete using Treesvis software (Weinacker, 2004) with 1.0 m pixel size.

From the original 2003 and 2004 LiDAR datasets using only the first laser returns, new data was derived for further analysis. All LiDAR points in the study area were registered to a specific DEM based on their coordinates in the OS BNG. The relative height of each registered LiDAR point was calculated as a difference between the DEM and elevation value (Z). The same approach for the determination of relative height has been employed in many studies (Naesset 2004; Naesset and Okland 2002). In some cases the relative height of several LiDAR points was low caused mainly by ground vegetation, such as grasses and shrubs. Hence, the values of relative height below 2 m, which represented low shrubs or stones were excluded from further analysis (Naesset 1997a; Nilsson 1996).

## 2.4 Single-Tree Segmentation

The purpose of the segmentation is to delineate individual tree crowns and to derive information about tree height and other important parameters. The input data necessary for the segmentation is the CHM consisting of the DEM and the DSM. The DSM was assumed to represent the vegetation cover which was created only from the first laser returns (the first point reflected from the top of objects) of the LiDAR 2004 data. Similar steps were followed as in the creation of DEM for the DSM, with emphasis on vegetation using Surfer software operating with pixel sizes of 0.5 m and 1.0 m. The surface layers, corresponding to the ground as the DEM and the vegetation cover as the DSM, were afterwards employed for the calculation of the CHM. This model giving the relative height of vegetation was calculated on a per pixel basis where each pixel from the DEM was subtracted from the DSM (Fig 2.).

The segmentation algorithm that was applied to the high-resolution CHM was previously introduced by (Pitkänen et al. 2004; Pitkänen 2005) for the segmentation of coniferous trees. The high point density LiDAR 2004 dataset (about 7 points/m$^2$) was exclusively used as it had shown success in recent studies (Hyyppa et al. 2001; Peuhkurinen et al. forthcoming).

The first step in the segmentation process was the filtering of the CHM which utilized a height-based filtering method dependent on the degree of low-pass Gaussian filter with used input value of 0.8 to reduce the effects of noise caused by branches (Peuhkurinen et al. forthcoming). Afterwards, the local maxima were detected and assumed to be the tree tops in the filtered image. The individual tree crowns were then discriminated using a watershed algorithm with a drainage direction (Pitkänen 2005); segments with local maxima (i.e. tree height) below 2 m were discarded. Each segment or individual tree contains information, for example, tree height (local maxima), coordinates of the local maxima and centroid, the crown area, and the maximum crown diameter.

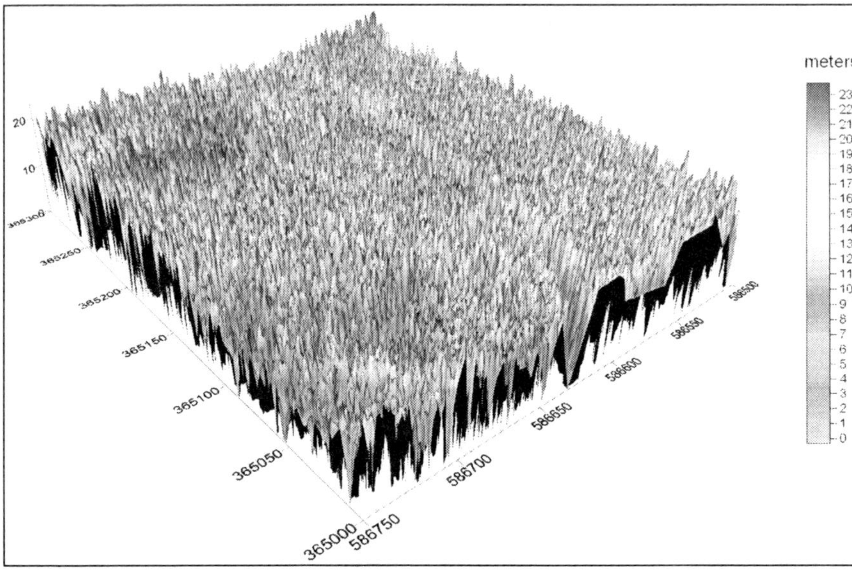


**Fig. 2:** LiDAR 2004 canopy height model of the plots 8 and 9 with 0.5 m pixel

# 3 Methods

## 3.1 Estimation of the Stand Volume and the Basal Area with Percentiles

The relative height values of the vegetation, calculated from derived LiDAR data and the DEM, was spatially registered to 14 training and 14 test plots (25 m x 25 m); data outside these plots was excluded from further analysis.

A multiple regression analysis was used to determine the relationship between field and laser derived data, with a final analysis applied for the estimation of the stand volume and basal area. First, the creation of the regression models employed independent variables represented by the mean height value, percentiles of 10, 20, ..,100 % and also two additional of 95 and 99 % of the relative height values. Similar statistics for the laser derived relative height data were introduced in many previous studies (Naesset 2004; Holmgren 2004; Nelson, Krabill, and Tonelli 1988; Naesset 1997a; Naesset 2002). An additional independent variable for the age of trees, as it varies among plots, was also tested. So in total, this study employed and evaluated 13 independent candidates to find the best predictor variable or variables using multiple linear regression with the first model, Model (1), formulated as:

$$Y = \beta_0 + \beta_1 h_{10} + \beta_2 h_{20} + \ldots + \beta_{10} h_{100} + \beta_{11} h_{95} + \beta_{12} h_{99} + \beta_{13} h_{mean} \quad (1)$$

Including the additional variable for age gives Model (2) as:

$$Y = \beta_0 + \beta_1 h_{10} + \beta_2 h_{20} + \ldots + \beta_{10} h_{100} + \beta_{11} h_{95} + \beta_{12} h_{99} + \beta_{13} h_{mean} + \beta_{14} v_{age} \quad (2)$$

where Y represents stand volume V ($m^3ha^{-1}$) or stand basal area G ($m^2ha^{-1}$); $h_{10}$, $h_{20, \ldots,}$ $h_{100}$, $h_{95}$ , $h_{99}$ represent percentiles of relative height for 10%, 20%,...,100%, 95% and 99%; $h_{\mathrm{mean}}$ represents mean value for relative height; and $v_{\mathrm{age}}$ represents age of trees in the stand.

Selection of the most suitable independent variables for (1) and (2) regression models was based on the stepwise multiple selection similar to that used by (Naesset 2004). To determine whether they should be retained in a model, these variables were tested with a partial F statistic at significance level greater than 0.05; only variables below this significance level were included in the final prediction. In equation (2), age was excluded from testing with F statistics and entered as a second independent variable into a regression model.

Residuals of both models were studied which assisted to decide and improve the final regression models. Additionally, an analysis of variance of these models was considered with the emphasis on the principle of the least squares method using a minimum distance between predicted and observed values. Then the validation of the 2 previously created models (1) and (2) was done on 14 test plots. The limitation of these 2 models was that the testing plots were not spatially independent from the training plots. However, in this study it was the best available scenario which proved the usefulness of the studied models. Furthermore, these two models were used to estimate the stand volume and basal area, with the results compared against the field data.

In the end, the differences between estimated and observed values for the stand volume and basal area in the test plots were evaluated with a t-test for a significance. A paired two-tailed t-test was employed to compare if there was a significant difference between calculated means of estimated and observed values. Identical testing of this significance was previously carried out by (Naesset 2004).

## 3.2 Estimation of the Stand Volume and the Basal Area with Single Tree Segmentation

### 3.2.1 Individual Tree Parameters

The segmented high-resolution LiDAR images from 2004 with 0.5 m and 1.0 m spatial resolution were used for the determination of tree parameters. First, the created segments were spatially registered to the test plots based on the location of tree tops (local maxima), and only those segments within these plots were recorded for later analysis. The tree height for each segment was obtained as a maximum pixel value corresponding to a tree top. The crown diameter (CD) was calculated from every segment's area (A) corresponding to the area of tree crown, using the following formula:

$$CD = \sqrt{\frac{4A}{\pi}} \tag{3}$$

Another important parameter for the estimation of stem volume is the basal area and the DBH that was calculated using Forest Research Environment Database (FRED), which is a model based on the linear relationship between the DBH, tree height and crown diameter. Then, the calculated DBH was employed to compute basal area (BA) for each tree using the formula:

$$BA = \frac{\pi \cdot DBH^2}{40{,}000} \tag{4}$$

where the value of 40,000 was used to relate the calculated BA to square meters. Finally, the stem volume for single tree was determined by the basal area and the crop form coefficient for Sitka spruce (Hamilton 1975).

Additionally, a linking process was made in ArcGIS (ESRI, Redlands, USA) for those trees detected by the segmentation algorithm. They were compared based on the location of their tree tops with the dominant and the co-dominant trees in each plot and then matched only with the closest one in the field. The detected trees with the distance less than 1.5 m to the field trees were included in further analysis. Metrics derived from the detected trees were compared against those in the field.

### 3.2.2 Stand Parameters

The estimates of stand parameters were determined by individual segmented trees located within the test plots. The stand volume ($m^3ha^{-1}$) and basal area ($m^2ha^{-1}$) from segmented trees were calculated as the sum of all stem volumes and the sum of all basal areas for detected trees, respectively.

# 4 Results

## 4.1 Stand Level

The results of the regression analysis showed that only one independent variable derived from LiDAR datasets was suitable for both estimation of the stand volume and also of the stand basal area; therefore linear regression models were employed. The final regression models for the estimation of the stand volume were based on the 30th percentile of relative height ($h_{30}$) for both LiDAR datasets. However, when additional information about tree age was available in a training plot, the regression models containing age as a predictive variable ($v_{age}$) were used. The coefficient of determination ($R^2$) for Model (1) was 0.94 for the LiDAR 2003 dataset and 0.91 for 2004 (Table 5). When age was used as an additional predictive variable in Model (2), the coefficient of determination ($R^2$) increased for both datasets to 0.95 (LiDAR 2003) and 0.93 (LiDAR 2004). The accuracy of the predictive models stated by root mean squared error (RMSE) for the stand volume showed for Model (1) results of 75.84 and 78.40 in $m^3ha^{-1}$ for LiDAR 2003 and LiDAR 2004 respectively, and for Model (2) showed higher accuracy with results of 73.44 and 76.48 respectively (Table 5).

The regression models were also employed to determine the stand basal area using the stepwise selection to find the best predictor. The 10% percentile of the relative height ($h_{10}$) was used as the most suitable independent predictor in all final models. A similar procedure for age as an additional variable was used as in the estimation of the stand volume. The range of the coefficients of determination ($R^2$) for both datasets and regression models varied from 0.64 to 0.82 (Table 5). The RMSE for the stand basal area in ($m^2ha^{-1}$) for Model (1) was 5.60 (LiDAR 2003) and 6.08 (LiDAR 2004); and for Model (2) was 4.48 and 5.44 respectively (Table 5).

**Table 5** Representing relationship between ground values from training plots (25x25 m) and laser derived metrics from stepwise multiple regression analysis

| Dependent variable[a] | Predictive model[b] | $R^2$ | RMSE [e] |
|---|---|---|---|
| **LiDAR 2003** | | | |
| V (1) [c] | $-11.079 + 3.459 * h_{30}$ | 0.94 | 75.84 |
| V (2) [d] | $-10.517 + 3.883 * h_{30} + 0.163 * v_{age}$ | 0.95 | 73.44 |
| G (1) [c] | $2.546 + 0.111 * h_{10}$ | 0.64 | 5.60 |
| G (2) [d] | $2.763 + 0.166 * h_{10} -0.023 * v_{age}$ | 0.80 | 4.48 |
| **LiDAR 2004** | | | |
| V (1) [c] | $-12.927 + 3.415 * h_{30}$ | 0.91 | 78.40 |
| V (2) [d] | $-12.6 + 3.726 * h_{30} -0.123 * v_{age}$ | 0.93 | 76.48 |
| G (1) [c] | $2.595 + 0.113 * h_{10}$ | 0.72 | 6.08 |
| G (2) [d] | $2.842 + 0.174 * h_{10} -0.025 * v_{age}$ | 0.82 | 5.44 |

[a] V stand volume [$m^3ha^{-1}$], G stand basal area [$m^2ha^{-1}$].
[b] $h_{10}$ ,$h_{30}$ percentiles of relative height for 10, 30 % (m), $v_{age}$ representing age of trees.
[c] Model (1) without age variable.
[d] Model (2) with age variable.
[e] RMSE for V [$m^3ha^{-1}$] and for G [$m^2ha^{-1}$].

The created regression models of (Table 5) were used afterwards for the estimation of the stand volume and the basal area on 14 separate test plots. The estimated values from the test plots were then compared with the field data. For the comparison of the difference between the estimated and field data, the mean difference (Mean) representing bias was used (Table 6). The stand volume derived from both LiDAR datasets overestimated the field values in Model (1) by 1.21 and 4.00 ($m^3ha^{-1}$), and in Model (2) by 4.35 and 6.49 ($m^3ha^{-1}$) (Table 6). However, values for the stand basal area also underestimated the field values in Model (1) from LiDAR 2003 dataset by 2.66 ($m^2ha^{-1}$) but overestimated them by LiDAR 2004 dataset by 0.03 ($m^2ha^{-1}$). Similarly, the results for G in Model (2) underestimated field values from LiDAR 2003 dataset by 2.69 ($m^2ha^{-1}$) but overestimated in the LiDAR 2004 dataset by 1.23 ($m^2ha^{-1}$). Overall, the estimated mean values of V and G were not significant on the predefined level of significance (p=0.05). The results of the estimation for both stand parameters are presented in Fig. 3 and Fig. 4.

**Table 6** Difference (D) between estimated values from LiDAR derived metrics and ground measurement values for test plots (25x25 m)

| | | D | | | |
|---|---|---|---|---|---|
| Variable [a] | Observed mean | Range | | Mean | |
| **LiDAR 2003** | | (1) | (2) | (1) | (2) |
| V | 703.79 | -236.8 to 97.6 | - 241.6 to 105.6 | 1.21 [ns] | 4.35 [ns] |
| G | 68.79 | -15.20 to 8.32 | - 16.96 to 4.16 | - 2.66 [ns] | - 2.69 [ns] |
| **LiDAR 2004** | | | | | |
| V | 703.79 | - 219.2 to 116.8 | - 220.8 to 123.2 | 4.00 [ns] | 6.49 [ns] |
| G | 68.79 | - 11.84 to 11.68 | - 12.32 to 8.64 | 0.03 [ns] | 1.23 [ns] |

(1) Model (1) without age variable.

(2) Model (2) with age variable.

[ns] not significant ($p > 0.05$).

[a] V stand volume [$m^3ha^{-1}$], G stand basal area [$m^2ha^{-1}$].

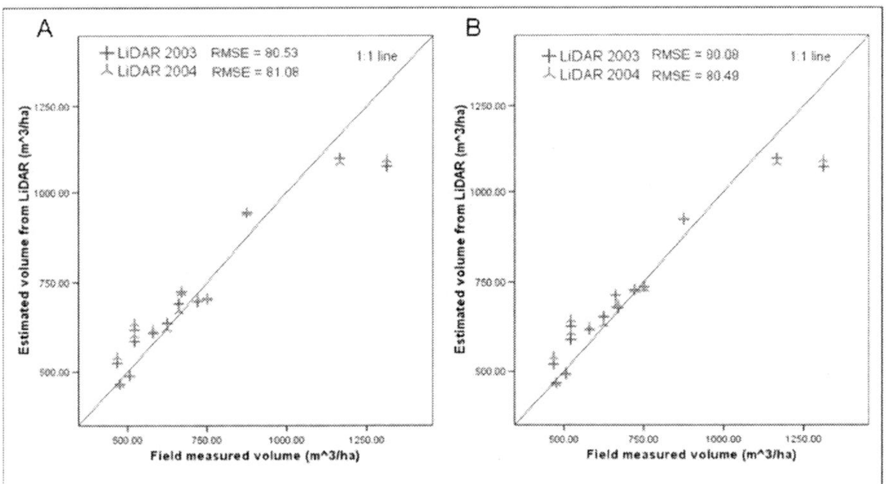


**Fig. 3:** Field observed volume is plotted against stand volume estimated from LiDAR derived data: (*A*) representing Model (1) without additional independent variable of tree age; (*B*) representing Model (2) with additional independent variable of tree age

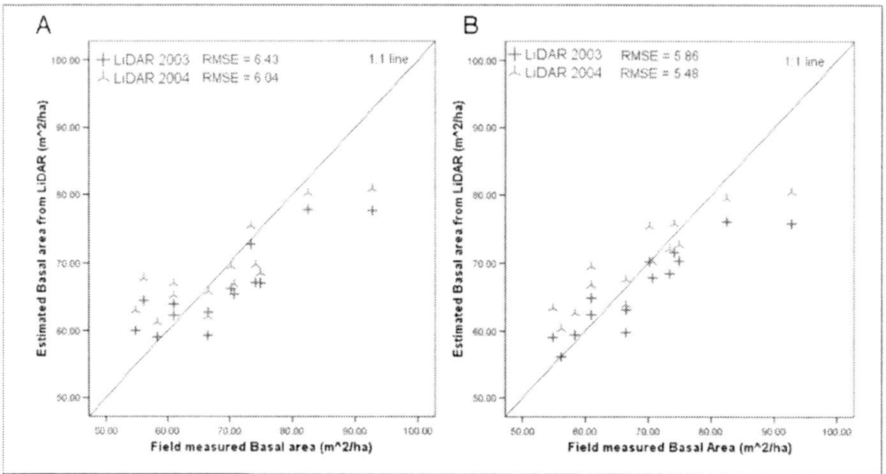


**Fig. 4:** Field observed volume is plotted against stand basal area estimated from LiDAR derived data: (*A*) representing Model (1) without additional independent variable of tree age; (*B*) representing Model (2) with additional independent variable of tree age

## 4.2 Tree Level

Using the segmentation algorithm on LiDAR 2004 dataset with 0.5 m spatial resolution, 496 trees in total (without test plot no. 7 and 8) were detected in 14 test plots by local maxima (Table 7). Unfortunately, in test plots 7 and 8 coordinates of measured trees in the field were not available and so they were not used for the comparison of the detected trees. The segmentation was able to detect only 35.5 % of all trees within test plots; however, it achieved a higher number (51.4 %) of the dominant and the co-dominant detected trees (DCD). Furthermore, in the test plots 1 and 2, the number of detected DCD was over 70 % (71.8 % and 70.3 %, respectively) whereas in test plot 12 only 35.2 % of DCD (Table 7). Additionally, the distribution of various dbh classes for DCD trees from the segmentation and field measurements is presented in Fig. 5 where the level of tree detection using segmentation was higher at classes with dbh above 30 cm.

**Table 7** Number of trees measured in the field and detected by segmentation algorithm from LiDAR dataset with 0.5 m resolution in test plots

| | Field | | LiDAR |
|---|---|---|---|
| Test plot | All trees | Dominant and co-dominant trees | Segmented trees[a] |
| 1 | 62 | 39 | 28 |
| 2 | 62 | 37 | 26 |
| 3 | 62 | 37 | 25 |
| 4 | 77 | 45 | 26 |
| 5 | 139 | 75 | 46 |
| 6 | 119 | 77 | 39 |
| 7 | - | - | 49 |
| 8 | - | - | 41 |
| 9 | 118 | 91 | 52 |
| 10 | 110 | 94 | 53 |
| 11 | 184 | 123 | 51 |
| 12 | 191 | 125 | 44 |
| 13 | 135 | 110 | 57 |
| 14 | 138 | 112 | 49 |

[a] segmented trees from 2004 LiDAR dataset with spatial resolution 0.5 m, - missing value.

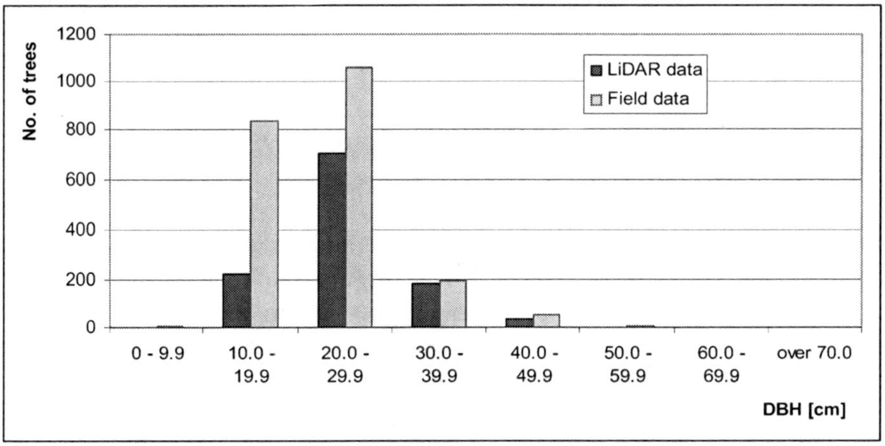


**Fig. 5:** Comparison of the number of dominant and co-dominant trees obtained from the segmentation (LiDAR data) and field dataset (Field data)

An example of the segmentation analysis is presented in Fig. 6 with both detected and measured trees in the test plot 2. When Fig. 6 was visually examined, in many cases the detected trees were identical to the measured trees in the field.

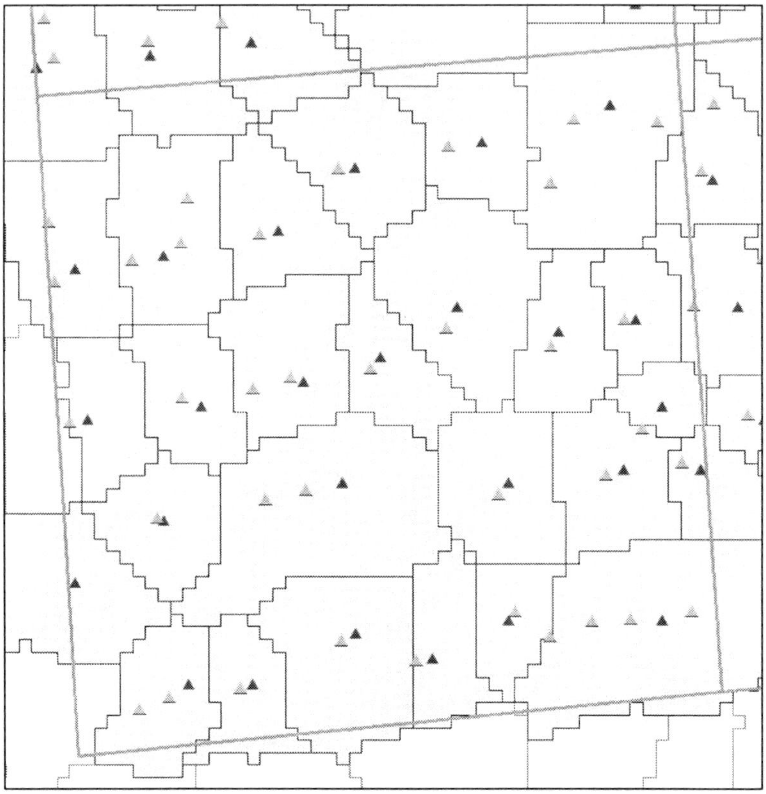

**Fig. 6:** The result of segmentation process in test plot no. 2 with trees measured in field (green triangles) and detected trees by segmentation algorithm (red triangles)

The results of the calculated stand volume and stand basal area for all 14 test plots are presented in Table 8. To determine both metrics two different datasets, each with various spatial resolutions (0.5 m and 1.0 m), were employed. The calculated stand volume was lower and underestimated in both datasets; a mean difference (bias) for 0.5 m dataset was 349.77 ($m^3ha^{-1}$) and was even higher for 1.0 m dataset with 434.76 ($m^3ha^{-1}$). The difference between measured and calculated V was the highest for the 1.0 m dataset with the underestimation of 874.9 ($m^3ha^{-1}$). However, the lowest underestimation was 213.7 ($m^3ha^{-1}$) representing 54.2 % of the field V.

The stand basal area in all 14 test plots was calculated as a sum of all basal areas of segmented trees within each plot and in all cases was lower than the field values. The achieved mean difference (bias) between observed and calculated G was underestimated by 33.36 ($m^2ha^{-1}$) for the 0.5

m dataset and 42.24 ($m^2ha^{-1}$) for the 1.0 m dataset. The highest absolute difference between observed and determined G was 61.87 ($m^2ha^{-1}$) and the smallest was 24.10 ($m^2ha^{-1}$), corresponding to 57.1 % of the field G. More details can be seen in Table 8.

**Table 8** Difference (D) between estimated values of segmented trees and ground measurement values for test plots (25x25 m)

| Variable [a] | Observed mean | D: Range | | D: Mean | |
|---|---|---|---|---|---|
| LiDAR 2004 | | 0.5 m | 1.0 m | 0.5 m | 1.0 m |
| V | 703.79 | - 686.0 to - 213.7 | - 874.9 to - 293.1 | - 349.77 | - 434.76 |
| G | 68.79 | - 48.52 to - 24.10 | - 61.87 to - 29.12 | - 33.36 | - 42.24 |

[a] V stand volume [$m^3ha^{-1}$], G stand basal area [$m^2ha^{-1}$].

## 5 Discussion

This study explored two methods for estimating two forest parameters using LIDAR datasets: the stand volume and the stand basal area. The first method focused on the stand level using percentiles as independent variables while the latter one focused on the tree level, delineating individual trees using a segmentation algorithm. The field data applied in both methods had high level of autocorrelation as test plots were adjoined to the training plots. Furthermore, variations in studied metrics V and G between all plots were observed and considered in the final results.

The first method applied both LiDAR datasets in a regression analysis. However, it used only the 30$^{th}$ height percentile which proved to be highly correlated, compared to the other percentiles, with the stand volume and was used as the best predictive variable in the regression models. Whereas other studies used for the estimation of the stand volume different percentiles (Holmgren 2004; Naesset and Okland 2002; Naesset 2004). For example, the study of (Holmgren 2004) used regression models that were based on the 90$^{th}$ height percentile derived from LiDAR data for estimating the stand volume.

Two LiDAR datasets with different point densities (approx. 2 and 7 points/$m^2$) were used for the estimation of the stand volume and demonstrated no significant difference. Nevertheless, the dataset with lower point densities showed better results using Model (1) which estimated volume more accurately (mean of 0.17 % of observed values) when compared with the LiDAR 2004 dataset (mean of 0.57 %). Applying Model (2), the mean increased to 0.62 % and 0.92 % of observed average values, respectively. Additionally, coefficients of determination, $R^2$, for both models were between 0.93 and 0.94. The study by (Gobakken and Naesset 2005) and (Naesset 2004) reported similar results to this study with a mean ranged between -5.1 % to 1.9 % and 2.6 % to 5.6 %, respectively for young and mature forests. However, differences between their methodology and forest parameters should be considered when compared with the results achieved in this study. Another study by (Bollandsas and Naesset 2007) based their method on the distribution of diameter classes and registered a significant bias on the independent dataset with an underestimation of 8.3 %. They found that bias increased with increasing observed volume possibly due to the stand variables representing the above-ground biomass. The results showed that the stand volume was more reliably estimated with the dataset of lower point densities even though the difference was small. Furthermore, the estimated results were close to the average field values which was mostly due to the fact that both, training and test plots were spatially highly correlated.

The RMSE of the estimated stand volume achieved in this study was approximately 11.4 % of average stand volume for both datasets and even for both regression models (Fig. 3). Similar results of the RMSE for the stand volume was achieved by (Holmgren 2004) and also by (Naesset 2002) with the RMSE between 11 % and 14 %. This study demonstrated an insignificant difference between observed and estimated stand volume with low and high point densities. Similar conclusions for the estimated stand volume were attained by (Holmgren 2004) with point densities from 0.1 to 4.3 points/m$^2$ and a footprint of 1.8 m. The additional age variable included in the regression models did not improve the final result and hence the stand volume can be estimated only by height metrics derived from LiDAR data.

The best single prediction variable for the regression models derived from the LiDAR datasets for the stand basal area was revealed to be the 10$^{th}$ percentile of the relative height that was highly correlated with the field values. On the other hand, previous studies employed mainly higher percentiles e.g. 80$^{th}$ or 90$^{th}$ (Naesset 2004; Holmgren 2004). The Model (1) estimates was lower than the field values by 2.66 ($m^2ha^{-1}$) for the LiDAR 2003 dataset contrary to the overestimation by 0.03 ($m^2ha^{-1}$) determined by the LiDAR 2004 dataset. The Model (2) obtained almost the same mean for the LiDAR 2003 dataset as the previous one but increased to 1.23

($m^2ha^{-1}$) for the other dataset. The coefficient of determination for the estimated stand basal area ranged from 0.72 to 0.75 for Model (1) and increased to 0.82 and 0.83 for Model (2), respectively. A previous study by (Naesset 2004) reported that the bias of observed stand basal area for young and mature forests was between 1.5 % to 8.4 %; however, more accurate estimates were obtained for young forests even though the field G was similar. The mean of the stand basal area achieved in this study was from - 3.9 % to 1.8 % of the observed values for both models and showed better results than (Naesset 2004). The underestimation of the stand basal area for the LiDAR 2003 dataset was most likely caused by a lower point density and also by one test plot with high basal area that largely influenced both models (Fig. 4). The results for the stand basal area achieved with the regression models demonstrated that the estimates were more accurate for the dataset of higher point densities. Additionally, both estimates were close to the average field values which was probably caused by the high level of spatial autocorrelation of the training and the test plots.

Comparing the accuracy of the estimation, the RMSE was employed for the stand basal area using Model (1) and achieved 6.43 ($m^2ha^{-1}$) and 6.04 ($m^2ha^{-1}$) corresponding to 9.3 % and 8.8 % of the average field value respectively (Fig. 4). The estimated G value with Model (2) achieved a higher accuracy corresponding to 8.5 % and 8.0 %. At the stand level results reported by (Holmgren 2004) showed RMSE of 10 % for the average basal area and in a further study by (Naesset 2002) an accuracy of RMSE between 8.7 % and 11.7 % was obtained, similar to our results. Even though test plots of different age were used, the accuracy of the estimation was comparable to the other studies. The estimation of the basal area with proposed regression models might prove to be more accurate in cases where test plots are stratified according to tree age.

The second approach tested the segmentation method at the tree level for the estimation of the stand volume and basal area. It showed a significant underestimation compared with the field data. The stand volume was underestimated for both 0.5 m and 1 m datasets with bias corresponding to 49.7 % and 61.8 % of observed values, respectively. Similarly, the calculated RMSE for both estimates was 52.8 % and 65.4 % respectively. Other studies using segmentation approach on the laser derived metrics achieved better results of the estimated timber volume with RMSE below 30 % (Maltamo, Eerikainen et al. 2004) and improved accuracy up to 16 % when using parameter prediction for small trees. Furthermore, a study by (Hyyppa et al. 2001) achieved the bias of 10.5 % for the estimated stand volume which was higher than conventional methods used in forest inventory. A study by (Peuhkurinen et al. forthcoming) applied same segmentation algorithm and showed an underestimate of the volume for pulpwood by 23 % or 25 %. This study applied the basic settings of the segmentation algorithm and more testing of algorithm is deemed necessary.

The stand basal area was estimated with similar accuracies as the stand volume. The bias of the observed values indicated better results for 0.5 m dataset (48.5 % of observed values) compared with 1.0 m dataset (61.4 %). According to the calculated results, RMSE was 49.5 % for 0.5 m dataset and 62.5 % for 1.0 m dataset. However, basal area is an important parameter for volume estimation and calculated results presented identical values of bias and RMSE especially for the 0.5 m dataset. This indicates that the employed segmentation has a potential in the future.

The low accuracies of estimates of the volume and the stand basal area were potentially due to the following factors: low detection level of small trees in used datasets, underestimation of tree height, the influence of tree clumping and overlapping of tree crowns in dense forests. The tree height of linked dominant and co-dominant trees with field measurements showed an underestimation in the average by 7.3 % with a RMSE of 1.93 m. The second factor that delineated the crown area for dominant trees is usually underestimated in a dense forest while that for small trees remains undetected in many cases (Hyyppa et al. 2001). Additionally, the segmented crown can represent two or more suppressed trees together or, conversely, one dominant tree with a large crown from the field can be segmented by the algorithm into several smaller trees. The low number of detected trees which ranged from 23 % to 48 % compared with all trees within test plots of 0.5 m dataset may also be responsible for the inaccuracies obtained. However, when only dominant trees were considered, up to 72 % of trees were detected with a similar number of detected dominant trees of 83 % in (Maltamo, Yu et al. 2004). The segmentation method was capable of estimating volume more accurately in mature rather than in young forests. Additionally, some errors were caused by incorrect registration of trees and the omission of adjacent trees to test plots.

An additional parameter influencing estimates was the scanning angle of LiDAR data acquisition. This study used scanning angle of 20° and 10° for 2003 and 2004 datasets, respectively. Differences in the estimated values by these two angles may influence determined tree height and underestimate the true height as was remarked in (Naesset 1997a), however the differences in scanning angle should be studied more.

The final results of this study presented that there is no significant difference between the estimation of the stand volume and basal area with LiDAR datasets of various point densities. On the other hand, results achieved with two methods at the stand and the tree level presented diverse results under the same datasets. Furthermore, the results achieved with both regression models showed that age was not essential information for the estimation of the stand volume and basal area. The percentile approach in this study proved to be a better estimate of the stand volume and basal area with accuracy similar to conventional methods when compared with the segmentation method. The potential of the segmentation method is high and previous studies showed it to be a good estimate of volume (Peuhkurinen et al. forthcoming). However, in this study it further exploration is needs to be increase its usefulness.

## 6 Conclusions

This study presented two different methods at the stand and the tree levels of Sitka spruce for the estimation of the stand volume and basal area. We have shown diverse results achieved by both methods. Estimates by the percentile approach at the stand level for the volume and basal area were comparable with conventional methods, and proved to be more accurate for studied species. In contrast, at the tree level results of the estimation were unsatisfactory and showed that this method requires further testing. In the future, field samples should be considered with emphasis on the stratification of trees into same age groups with higher numbers of independent test plots. Furthermore, new research should be carried out to clarify and identify these differences between estimates of volume and basal area by percentile and tree segmentation methods, and as well as expand the application of the latter method. Finally, in forest management continuous cover of LiDAR survey will enable foresters to monitor woodlands and easily update information about trees at the stand level with lower costs and faster compared to the traditional field measurements.

## Acknowledgements

This paper was written as a master dissertation at the University of Edinburgh, UK. The author would like to thank my supervisors Genevieve Patenaude and Juan Suárez for their help. Furthermore, I would like to thank to Juho Pitkanen from the Finnish Forest Research Institute for providing me with his segmentation algorithm. Finally, thanks to people from Forest Research, Northern Research Station, especially Thomas Conolly for help with statistics.

## References


Bollandsas, O. M., and E. Naesset. (2007) Estimating percentile-based diameter distributions in uneven-sized Norway spruce stands using airborne laser scanner data. *Scandinavian Journal of Forest Research* 00022 (00001).

Coops, N. C., T. Hilker, M. A. Wulder, B. St-Onge, G. Newnham, A. Siggins, and J. A. Trofymow. (2007) Estimating canopy structure of Douglas-fir forest stands from discrete-return LiDAR. *Trees-Structure and Function* 21 (3):295-310.

Donoghue, D. N. M., and P. J. Watt. (2006). Using LiDAR to compare forest height estimates from IKONOS and Landsat ETM+ data in Sitka spruce plantation forests. *International Journal of Remote Sensing* 27 (11):2161-2175.

*Forestry Statistics 2006.* (2006.) [cited December 1 2006]. Available from http://www.forestry.gov.uk/pdf/fcfs206.pdf/$FILE/fcfs206.pdf.

Gobakken, T., and E. Naesset. (2005). Weibull and percentile models for lidar-based estimation of basal area distribution. *Scandinavian Journal of Forest Research* 20:490-502.

Hamilton, G. J. (1975). Forest mensuration handbook, Forestry Commission booklet ; 39. London: H.M.S.O.

Holmgren, J. (2004) Prediction of tree height, basal area and stem volume in fores stands using airborne laser scanning. *Scandinavian Journal of Forest Research* 19 (6):543-553.

Holmgren, J., M. Nilsson, and H. Olsson. (2003) Estimation of Tree Height and Stem Volume on Plots Using Airborne Laser Scanning. *Forest Science* 49:419-428.

Hyyppä, J., M. Engdahl, S. Linko, Y. H. Zhu, H. Hyyppä, and M. Inkinen. (2000) Accuracy comparison of various remote sensing data sources in the retrieval of forest stand attributes. *Forest Ecology and Management* 128 (1-2):109-120.

Hyyppa, J., O. Kelle, M. Lehikoinen, and M. Inkinen. (2001) A segmentation-based method to retrieve stem volume estimates from 3-D tree height models produced by laser scanners. *Ieee Transactions on Geoscience and Remote Sensing* 39 (5):969-975.

Hyyppa, J., T. Mielonen, H. Hyyppa, M. Maltamo, X. Yu, E. Honkavaara, and H. Kaartinen (2005) Using Individual Tree Crown Approach for Forest Volume Extraction with Aerial Images and Laser Point Clouds. Paper read at ISPRS WG III/3, III/4, V/3 Workshop, September 12-14, 2005, at Enschede, the Netherlands.

Lim, K., M. Flood, P. Treitz, M. Wulder, and B. St-Ongé. (2003) LiDAR remote sensing of forest structure. *Progress in Physical Geography* 27 (1):88-106.

Lim, K., P. Treitz, K. Baldwin, I. Morrison, and J. Green (2003) Lidar remote sensing of biophysical properties of tolerant northern hardwood forests. *Canadian Journal of Remote Sensing* 29 (5):658-678.

Maltamo, M., K. Eerikainen, J. Pitkanen, J. Hyyppa, and M. Vehmas. (2004) Estimation of timber volume and stem density based on scanning laser altimetry and expected tree size distribution functions. *Remote Sensing of Environment* 90 (3):319-330.

Maltamo, M., P. Packalen, X. Yu, K. Eerikainen, J. Hyyppa, and J. Pitkanen. (2005) Identifying and quantifying structural characteristics of heterogeneous boreal forests using laser scanner data. *Forest Ecology and Management* 216 (1-3):41-50.

Maltamo, M., X. Yu, K. Mustonen, J. Hyyppä, and J. Pitkänen. (2004) The accuracy of estimating individual tree variables with airborne laser scanning in a boreal nature reserve. *Canadian Journal of Forest Research* 34 (9):1791-1801.

Means, J. E., L. Emerson, C. J. Hendrix, S. A. Acker, B. J. Fitt, and M. Renslow (2000). Predicting forest stand characteristics with airborne scanning lidar. *Photogrammetric Engineering and Remote Sensing* 66 (11):1367-1371.

Naesset, E. (1997a). Determination of mean tree height of forest stands using airborne laser scanner data. *ISPRS Journal of Photogrammetry and Remote Sensing* 52 (2):49-56.

Naesset, E. (1997b). Estimating timber volume of forest stands using airborne laser scanner data. *Remote Sensing of Environment* 61 (2):246-253.

Naesset, E. (2002). Predicting forest stand characteristics with airborne scanning laser using a practical two-stage procedure and field data. *Remote Sensing of Environment* 80 (1):88-99.

Naesset, E. (2004). Practical large-scale forest stand inventory using a small-footprint airborne scanning laser. *Scandinavian Journal of Forest Research* 19 (2):164-179.

Naesset, E., and K.-O. Bjerknes. (2001). Estimating tree heights and number of stems in young forest stands using airborne laser scanner data. *Remote Sensing of Environment* 78 (3):328-340.

Naesset, E., T. Gobakken, J. Holmgren, H. Hyyppa, J. Hyyppa, M. Maltamo, M. Nilsson, H. Olsson, A. Persson, and U. Soderman. (2004). Laser scanning of forest resources: The Nordic experience. *Scandinavian Journal of Forest Research* 19 (6):482-499.

Naesset, E., and T. Okland. (2002). Estimating tree height and tree crown properties using airborne scanning laser in a boreal nature reserve. *Remote Sensing of Environment* 79 (1):105-115.

Nelson, R., W. Krabill, and J. Tonelli. (1988). Estimating forest biomass and volume using airborne laser data. *Remote Sensing of Environment* 24 (2):247-267.

Nilsson, M. 1996. Estimation of tree heights and stand volume using an airborne lidar system. *Remote Sensing of Environment* 56 (1):1-7.

Persson, A., J. Holmgren, and U. Soderman. (2002). Detecting and measuring individual trees using an airborne laser scanner. *Photogrammetric Engineering and Remote Sensing* 68 (9):925-932.

Peuhkurinen, J., M. Maltamo, J. Malinen, J. Pitkänen, and P. Packalen. forthcoming. Preharvest measurement of marked stands using airborne laser scanning. *Forest Science*.

Philip, M. S. (1994) *Measuring trees and forests*. 2nd ed. Wallingford: CAB International.

Pitkänen, J. (2005) A multi-scale method for segmentation of trees in aerial images. In *Forest Inventory and Planning in Nordic Countries, Proceedings of SNS Meeting at Sjusjoen, Norway*, ed. K. Hobbelstad, 207 - 216: Norwegian Institute of land Inventory.

Pitkänen, J., M. Maltamo, J. Hyyppa, and X. Yu. (2004) Adaptive methods for individual tree detection on airborne laser based canopy height model. *Laser scanners for forest and landscape assessment. Proceedings of the ISPRS working group VIII/2* XXXVI, part 8/W2:187-191.

Suárez, J. C., S. Snape, C. Ontiveros, and S. Smith. (2005) Use of airborne LiDAR and aerial photography in the estimation of individual tree heights in forestry. *Computers and Geosciences* 31 (2):253-262.

Tilley, B. K., I. A. Munn, D. L. Evans, R. C. Parker, and S. D. Roberts. (2007). *Cost Considerations of Using LiDAR for Timber Inventory* 2004 [cited January 15 2007]. Available from http://sofew.cfr.msstate.edu/papers/0504tilley.pdf.

Wehr, A., and U. Lohr. (1999). Airborne laser scanning--an introduction and overview. *ISPRS Journal of Photogrammetry and Remote Sensing* 54 (2-3):68-82.

Weinacker, H., Koch, B., Heyder, U., Weinacker, R., (2004) Development of filtering, segmentation and modelling modules for lidar and multispectral data as a fundament of an automatic forest inventory system; In: Proceedings of the international Conference. Laser-Scanners for Forest and Landscape Assessment - Instruments, processing Methods and Applications. Freiburg im Breisgau.Germany.

# Assessing Stand and Data Variability Using Airborne Laser Scanner

Diego D. Doce[1], Juan C. Suárez[2], Genevieve Patenaude[1]

[1]The University of Edinburgh, Geography B., Drummond Street, Edinburgh EH8 9XP, UK.
[2]Silviculture North, Forest Research, Northern Research Station, Roslin, Midlothian, EH25 9SY, UK.

**Abstract.** An efficient forest management requires accurate and cost-effective measurements of forest inventory parameters. The cost of LiDAR surveys are directly dependent on the number and size of validation plots as well as the sampling density of points needed to adequately estimate forest inventory parameters. This study investigates (i) the spatial variability of the forest stand, i.e., the effect of the area chosen on the prediction of forest parameters and (ii) the relation between prediction accuracy and sampling point density for the estimation of top height, basal area and volume at plot level. Assessment of the stand's spatial variability was accomplished by comparing the accuracy of the top height estimations, using the 99$^{th}$ percentile of a normalised distribution of points, over areas of different size. Original sampling density was synthetically reduced to 10, 5, 4, 3, 2, 1, 0.50, 0.33, 0.25 and 0.20 returns per m$^2$. Forest parameters were subsequently estimated for each point density by means of 99$^{th}$ percentile (top height) and linear regression models (basal area and volume). Predictions were validated using 11 stands, each containing one $50 \times 50$ m$^2$ plot. Results show that the optimum area for forest parameters prediction is 1600 m$^2$ with an average top height accuracy of 95.05% and a standard deviation of 3.41%. Larger sizes will merely increase the cost of field data collection without improving accuracy. Interestingly, top height predictions were slightly more accurate for lower point densities. Linear equations yielded RMSEs of 3.28-5.28 m$^2$/ha and 29.41-36.04 m$^3$/ha for basal area and volume respectively. There were therefore small differences in terms of accuracy of predicted parameters for different point densities, which indicates that once a good DTM is created, future LiDAR surveys can be accomplished over the same area at lower sampling densities, and thus reducing the costs but without disregarding estimation accuracy.

**Keywords:** LiDAR, airborne laser scanner, remote sensing for forestry, forest stands, accuracy

## 1 Introduction

Forests cover more than 11% of the UK surface (Forestry Commission 2006b). A well-defined database to accurately store information about the location, extent, and composition of forests is needed to understand the potential of their resources. Remote sensing can provide information on large forested areas, allowing the retrieval of variables from all forest stands. This contrasts with time-consuming and expensive field methods, which have to target sampling using small plots. These two approaches are however complementary, in that field measurements are still essential for calibration and validation. Light detection and ranging (LiDAR), a relatively new remote sensing technique, can contribute to better forest management due to its ability to measure the vertical spatial organisation of forest stands (Behera and Roy 2002).

Although remote sensing, and LiDAR in particular, can help to reduce the costs with regard to conventional field measurements, costs of LiDAR data and field methods for validation are still expensive. Optimization of the validation plot size together with a determination of the lower density of LiDAR points required to adequately estimate forest parameters is needed to reduce the cost per unit area of LiDAR data collection and therefore contribute to a more cost-effective forestry.

LiDAR consists of an active remote sensor which emits laser pulses towards a target surface and measures the time required for each laser pulse to travel from the sensor to the Earth's surface and back to the sensor. Since the laser pulses travel at the speed of light, by multiplying the return time of each pulse by this speed and halving the result, the distance between instrument and target can be determined (Bachman 1979 in Lefsky et al. (2002)). A differential GPS locates every pulse return within a known coordinate system whereas an on board Inertial Measurement Unit (IMU) accounts for the vertical and horizontal distribution of ground features (Lim et al. 2003). The laser system used in this study operates in a scanning mode where pulses are directed from side to side as the aircraft moves forward, creating a characteristic zigzag arrangement of points, as shown in Figure 1. The raw data file obtained consists of large sets of points in an ASCII XYZ format where the X, Y and Z position for each point are recoded as well as the intensity of the return (Suárez et al. 2004).

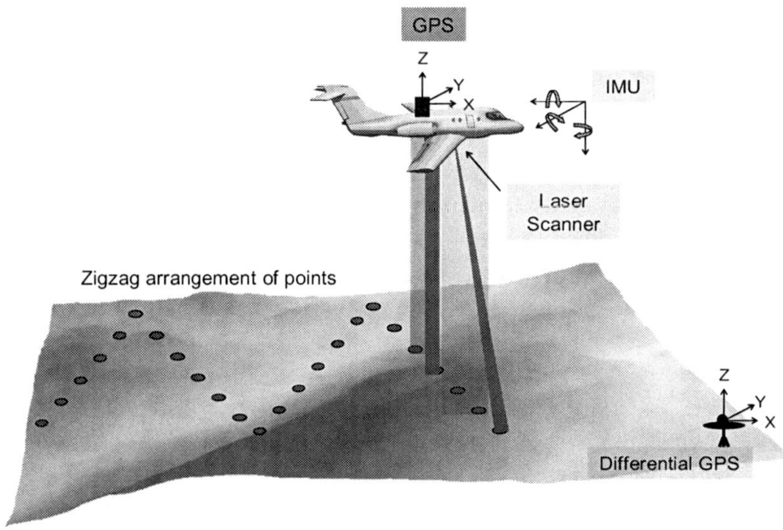


**Fig. 1:** LiDAR components.

One of the key parameters of the forest inventory is stand height. It allows for the assessment of future production potential, wood volume and treatment scheduling (Spurr 1952). While defining height for individual trees is quite straightforward, a suitable measure of the height at stand level may cause more problems, since a simple mean of tree heights within a stand could be affected by tree mortality or thinning operations, thus influenced by stand density. To deal with this issue, several height methodologies have been developed where only the "biggest" trees are selected. Such methodologies differ from each other in the number of trees selected and in the way the big trees are defined. Top height is a stand height measurement that falls into this group and it is the one that was used in this study and is defined as either the average of the 100 trees with largest diameter at breast height (DBH) or the average of the 100 tallest trees per hectare (Philip 1998).

Top height is a function of the plot size (Matern 1976; Rennolls 1978 in Lovell *et al.* (2005)), which is an integral part of the top height definition. However, as not all the plots are a hectare, it is common practice to use a smaller plot with a proportional number of trees. Several studies have investigated the effect of plot size upon forest parameters. García (1998) states that if the area of the plot used as a sample differs from that on which the definition is based, then estimated top heights are biased. Magnussen (1999) estimated the percentage of reduction in top height when this parameter was determined in plots with areas smaller than 1 ha. Substantial differences in DBH in relation to the plot size were also found in forests stands of central Canada (García 2006).

According to Bortolot and Wynne (2005) there are two approaches to estimate forest variables with small-footprint LiDAR; individual tree-based and distribution-based. The former relies on segmentation techniques to locate and measure individual trees using a canopy height model (CHM), whereas the latter relies on the extraction of height distributional parameters, such as percentiles, mean or mode, from raw returns or a CHM. Previous studies have shown the effectiveness of distribution-based approaches on the retrieval of forest stand parameters. Stand height (Magnussen and Boudewyn 1998; Means et al. 2000), basal area (Naesset 2002) and volume (Holmgren et al. 2003b; Maltamo et al. 2006) can be estimated with similar accuracies to those achieved with conventional methods (Naesset 2004b).

As previously mentioned, LiDAR can help to reduce costs in forest data collection with regard to field methods. Evans et al.(2001) pointed out that trees do not need a maximum sampling density to be detected and also that research should be developed to improve LiDAR operational parameters. Several studies have subsequently addressed this knowledge gap by working mainly with changes in platform altitude (Goodwin et al. 2006; Naesset 2004a; Yu et al. 2004) or variations in scanning angle (Holmgren et al. 2003a). Using a simulated dataset, Lovell et al.(2005) found that height retrieval was less accurate towards the ends of the swath due to uneven spacing of points. Holmgren (2004) tested the effects of different point densities on the estimation errors concluding that even low densities (0.1 returns/$m^2$) are enough to accurately predict forest parameters. Similar results have been achieved by Maltamo et al.(2006) using a synthetic reduction of points over mixed species of coniferous.

The objectives of this study were therefore two fold: (i) estimation of the spatial variability of forest stand using the top height derived from airborne laser scanner data and (ii) assessment of the relationship between prediction accuracy and sampling point density for the estimation of top height, basal area and volume at plot level. A brief description of the materials used in this study will be given, followed by a detailed explanation of the methodology. The results will be displayed and subsequently analysed, closing with a summary of the conclusions.

# 2 Materials

## 2.1 Study Area

The study area is located at 56°10' N, 40°22' W within the Trossachs-Ben Lomond Forest District, around the village of Aberfoyle (see Figure 2). The dominant tree species in the study area is Sitka spruce (*Picea sitchensis* Bong.Carr) which represents the 68.3% of the surveyed area covered by forest. Stands of mixed broadleaves, 4.5%, oak (*Querqus robur*), 2.7%, and European larch (*Larix decidua*), 2.6%, are also found. The highest part of the surveyed area rises to 323 m above sea level whereas the lowest point is found at 16 m (over an area of 14.4 km$^2$). In January 2005, the surveyed area suffered extensive wind damage, affecting several plots used for the validation of the LiDAR data, which created an excellent opportunity to study the variability of forest stands in both affected and non-affected areas.

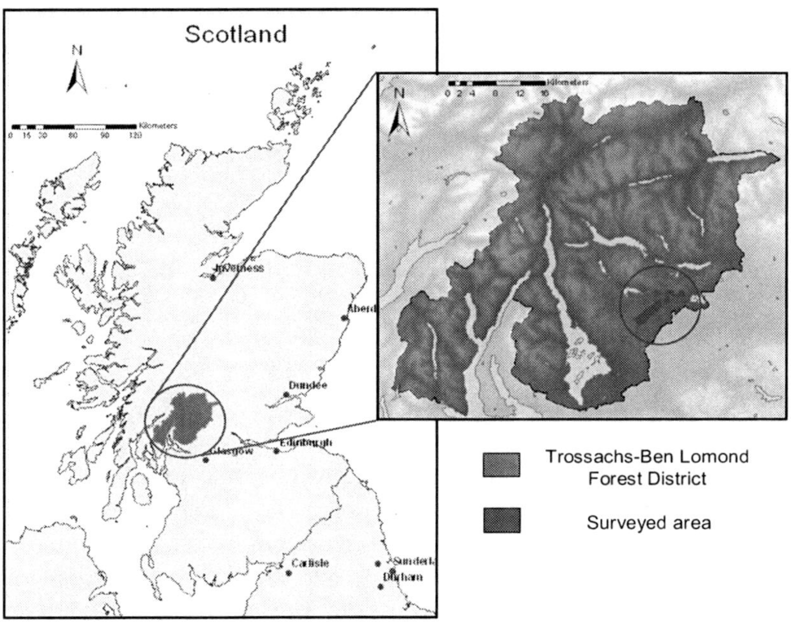


**Fig. 2:** Location of the study area

## 2.2 Stand and Plot Inventory

Field data acquisition for validation of the LiDAR analysis was performed by the Forestry Commission in twelve $50 \times 50$ m$^2$ plots, covering thinned and unthinned Sitka spruce stands, however, one of the plots, number 1, was clear-felled and thus it could not be used as a reference for validation. The stands, planted between 1969 and 1971, are located on a relatively flat terrain with mean slope gradient varying from 0% to 5%. Tree heights, tree diameters, tree position and dominance were measured for each plot, whereas top height, mean DBH, basal area and volume of each stand were subsequently derived from field data (see Table 1).

**Table 1** Summary of reference data: 11 $50 \times 50$ m$^2$ plots

| Characteristic | Range | Mean |
|---|---|---|
| Sitka Spruce (n = 11) | | |
| $h_{top}$ (m) | 24.25 - 31.2 | 27.04 |
| *d* (cm) | 25.53 - 37.46 | 30.78 |
| N (ha$^{-1}$) | 236 - 652 | 480 |
| G (m$^2$/ha) | 24.73 - 48.47 | 37.34 |
| V (m$^3$/ha) | 304.61 - 657.15 | 439.74 |
| Age (years) | 36 -38 | 37 |

$h_{top}$ = top height, $d$ = mean DBH, $N$ = steam number, $G$ = basal area
$V$ = volume

## 2.3 Laser Scanner Data

LiDAR data were acquired by the Environmental Agency on the 31$^{st}$ of May 2006 using a small footprint Airborne Laser Terrain Mapper (Optech ALTM3100). The instrument, which was flown on a plane, was fully calibrated after installation in the aircraft and the subsequent data obtained passed the control steps developed by the Environmental Agency to allow a high quality output.

Flight altitude above the ground level was approximately 800 m which created a footprint diameter of 1 m at nadir, with a scan angle of $\pm$ 10° and a laser pulse frequency of 100,000 Hz. The former pulse frequency of the sensor together with the overlap of flight passes yielded a high density of laser hits (10-12 per m$^2$). The instrument allows 4 return measurements for each pulse, however only the first and the last were used in this study, since first return data is more likely to be reflected from the canopy, that is, leaves and branches from trees, whereas last return data has more chances to hit the ground. LiDAR data was delivered as OSD files, separated in different folders for each of the 4 returns, that contained a list of X, Y and Z coordinates as well as return intensity.

**Table 2** Summary of LiDAR survey characteristics

| Parameter | Value |
|---|---|
| Sensor | Optech ALTM3100 |
| Laser pulse frequency | 100,000 Hz |
| Flying altitude | 800 m |
| Footprint diameter at nadir | 1 m |
| Scanning angle | 10 degrees |
| Sampling density | 10-12 returns per m2 |
| Elevation accuracy* | Z < 10 cm |
| Horizontal accuracy* | X, Y < 15 cm |

* Accuracies estimated according to the sensor provider and for the specified flight characteristics. Accuracies do not include GPS errors.

# 3 Methodology

## 3.1 Normalisation of Laser Data

First pulse returns are usually related to canopy height and last pulse returns to the ground. Therefore just by interpolating the last returns to a surface and subtracting the height of such surface from the first return should be enough to normalise canopy heights. The problem is that the last returns do not always penetrate the canopy: only a small proportion of them will reach the ground, and therefore a filtering of last return hits intercepted by the vegetation has to be accomplished in order to get a good estimation of the ground surface.

Filtering of raw last return LiDAR points was achieved using TreesVis. The algorithm used iteratively selects the lowest points within kernels of increasing variable size until non-ground points are completely removed. Kriging interpolator without anisotropy was subsequently applied to the filtered returns to create a digital terrain model (DTM) of $1 \times 1$ m resolution (Suárez et al. 2005). Once the DTM was obtained, it was used to normalise the height of the raw LiDAR data.

## 3.2 Stand's Spatial Variability

The spatial variability of the stand was assessed by predicting a forest parameter (top height) derived from LiDAR data over several area sizes. Accuracy and precision of the estimations were eventually used to evaluate the efficiency of different area sizes, giving an idea of how forest parameter estimations vary within a stand according to the area chosen for the estimation.

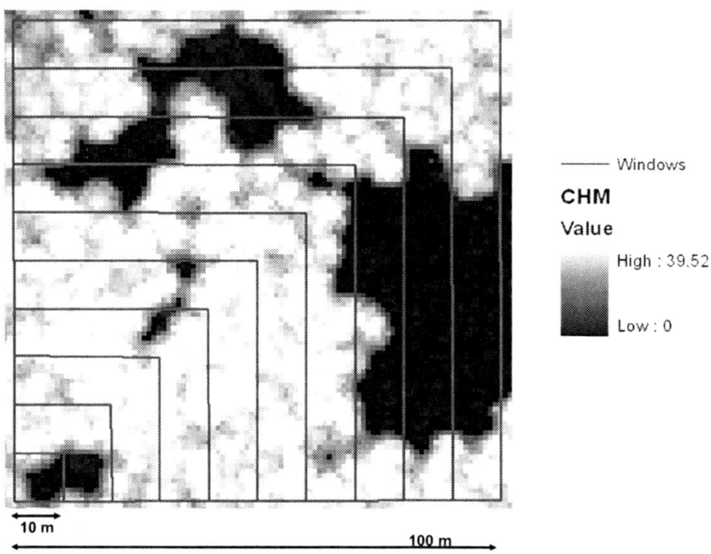


**Fig. 3:** Set of windows superimposed on the CHM, in metres, of stand 7

Squared windows, in increasing steps of 10 m and sharing the bottom-left corner (see Figure 3), were located on the centre of the stands. Window sizes started with $10 \times 10$ m$^2$, with $100 \times 100$ m$^2$ for the largest window. Where the shape of stands (long and narrow) did not allow for fitting the largest windows within them, the following approach was used: (i) squared windows were set as before until a particular size, say $70 \times 70$ m$^2$, where the window did not fit within the stand, (ii) from such size onwards, rectangular windows, which still share area with smaller windows, were located to best fit the stand shape. Since in this part of the study 11 stands are being evaluated, a total of 110 windows, 10 for each stand, were created. Note that windows do not necessarily cover the area of the stands where the field data was collected. The first pulse dataset was spatially registered to match the windows. Points falling outside the window's boundaries were excluded from analysis in this section.

Eventually the top height derived from LiDAR data, using the 99$^{th}$ percentile of a normalised set of points (Forestry Commission 2006a), was estimated for every single window. Due to the iterative nature of the task, a Java program was created to automatically subtract the height of the DTM from the set of first return points enclosed within the windows, to afterwards calculate the 99$^{th}$ percentile of the already normalised set of points.

Field plot measurements are meant to be representative of the whole stand since each stand presents relatively homogeneous characteristics. Validation of this section was carried out by comparing field top height against LiDAR-estimated top height (99$^{th}$ percentile) over the several windows of each stand.

## 3.3 Density Reduction

In order to investigate the effects of laser sampling density on forest variables accuracy (top height, basal area and volume), first and last return laser sampling density were reduced. This part of the study was accomplished over 10 windows of about $50 \times 50$ m$^2$ which matched the plots used in field measurements. Returns for the first and last pulse dataset which fell out of the boundaries of the sample plots were excluded from further analysis.

Again, a Java program was developed to randomly reduce the original density of points. A random reduction was applied because according to Holmgren (2004) if the reduction of points is done just by a minimum spatial distance, then such reduction might depend on scanning geometry and forest structure. Starting with an original average density of 11.57 returns per m$^2$, the sampling density was reduced to 10, 5, 4, 3, 2, 1, 0.50, 0.33, 0.25 and 0.20 returns per m$^2$. Data reductions were always carried out from the original data and not from any previously reduced sampling density. It should be said that this synthetic reduction of data is theoretical since the DTM used to normalise the first return points for all densities was always the same and was created with the original density of points.

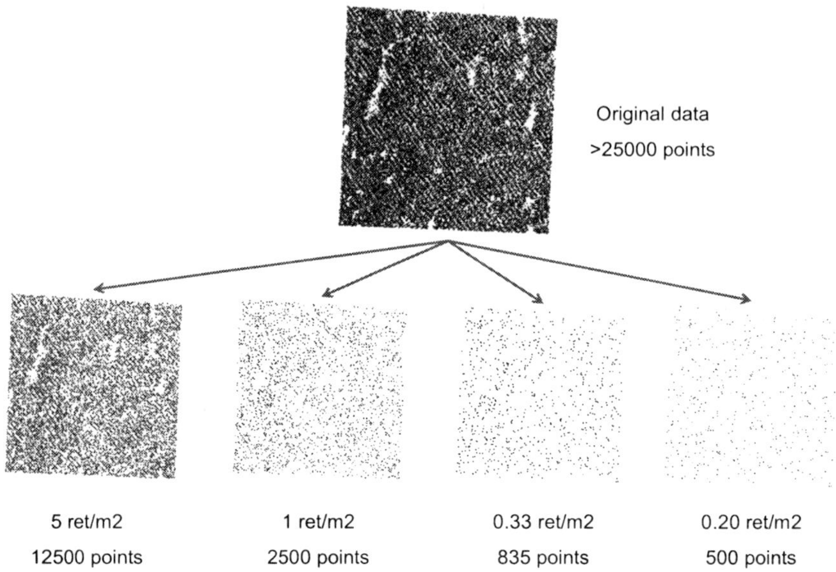


**Fig. 4:** Examples of data reduction on plot 8.

In the end, 11 different densities were obtained for each of the areas covering the 10 plots available and for each return. Top height, basal area and volume were subsequently estimated for each of the densities and for each of the plots.

Top height was estimated simply by using the 99[th] percentile of a normalised canopy height distribution. This seemed a robust method to predict top height in Sitka spruce stands, presenting negligible variations even after thinning the stand (Juan C. Suárez, personal communication). The robustness of this method is further tested in the course of this study.

Basal area and volume were obtained by linear regression analysis to establish relationships between laser and field measurements. Laser height distributions of the first and last return dataset were used to derive a large number of LiDAR metrics over each of the sample plots. Since they were found to be suitable to build regression models by others (Holmgren 2004; Maltamo et al. 2006; Naesset 2002, 2004b), the following metrics were obtained: (i) 10, 20, …, 90, 95,99 percentiles ($h_{10}$, …, $h_{99}$), (ii) the mean values ($h_{mean}$), (iii) the maximum values ($h_{max}$) and (iv) the coefficients of variation ($h_{cv}$) of the distributions. Furthermore, the percentage of points above 5, 10, 15 and 20 m of the canopy distribution ($p_5$, …, $p_{20}$), regarding the total number of points, were also derived. The last percentages were obtained under the assumption that basal area and volume could be defined not only by height metrics, but also by some parameters related to canopy area.

Linear regression models were created to estimate basal area and volume for every single sample density. Out of 10 plots available, 5 of them were chosen randomly to build the model and the rest were used to validate it. Only the independent variable which showed the smallest p-value, always inferior to 0.05, was placed in the model. Eventually, differences between observed and predicted values were assessed. Paired two-tailed t-tests were used to evaluate whether there is a statistically significant difference between the means of the two former groups.

In order to test the reliability of the estimations RMSE, bias and accuracy were applied to the results. Absolute RMSE and bias were calculated as in Equations 1 and 2 respectively, and accuracy as in Equation 3. The last term was necessary to compare predictions of the same parameter from different plots or window sizes.

$$RMSE = \sqrt{\frac{\sum_{i=1}^{n}(X_i - \hat{X}_i)^2}{n}} \tag{1}$$

$$bias = \frac{\sum_{i=1}^{n}(\hat{X}_i - X_i)}{n} \tag{2}$$

$$accuracy = \frac{\hat{X} \times 100}{X} \tag{3}$$

Where: $X_i$ = parameter observed in the field

$\hat{X}_i$ = parameter estimated from laser scanner data

n = number of samples

Note that accuracy over 100% just means that the predicted parameter is overestimated. The relative RMSE (rRMSE) and bias (rBias) are the result of converting the absolute RMSE and bias into percentages by dividing them by the mean of the field-measured parameters under evaluation (Maltamo et al. 2006), obtaining therefore a more explanatory exposition of errors since they are related to averaged field measurements.

# 4 Results

## 4.1 Top Height Estimations Using 99th Percentile

The 99th percentile plotted against field measurements yielded a biased although consistent regression line with an $R^2$ of 0.94 (Figure 5). The RMSE of laser predictions was 1.72 m which represents a 6.38% of the average field top height. While all predicted top heights were underestimated, the mean differences between observed and predicted values were nevertheless found to be significant with a p-value inferior to 0.01, (i.e. it is likely that such differences are not due to chance). Accuracy on the estimations was not affected by different levels of wind damage in field plots.

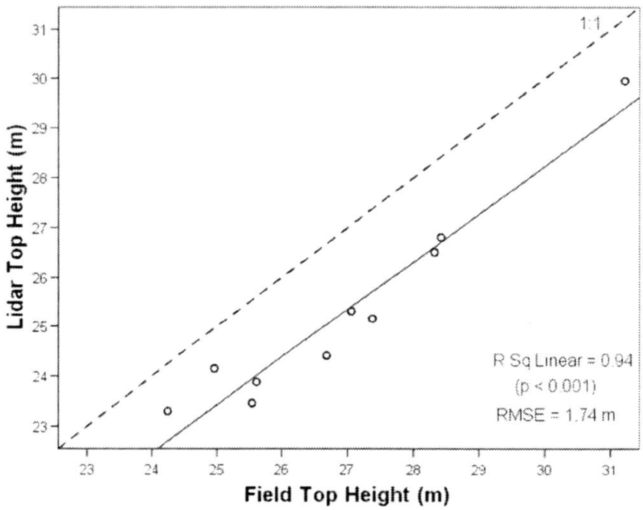


**Fig. 5:** Comparison between observed and estimated top height.

## 4.2 Stand's Spatial Variability

Figure 6(A) shows variations in predicted top height accuracy on different window sizes for individual stands. These variations are small within the stands, with the exception of $10\times10$ $m^2$ window for stands 7 and 9, and $10\times10$ and $20\times20$ $m^2$ windows for stand 5. The last 2 windows, from stand 5, are not represented on Figure 6(A) for scaling purposes, and hold accuracy values of just 0.50% and 0.83% respectively. Standard deviations of their accuracies range from 39.41% in stand 5 to 0.96% in stand 12, however almost all the areas present a standard deviation under 4%.

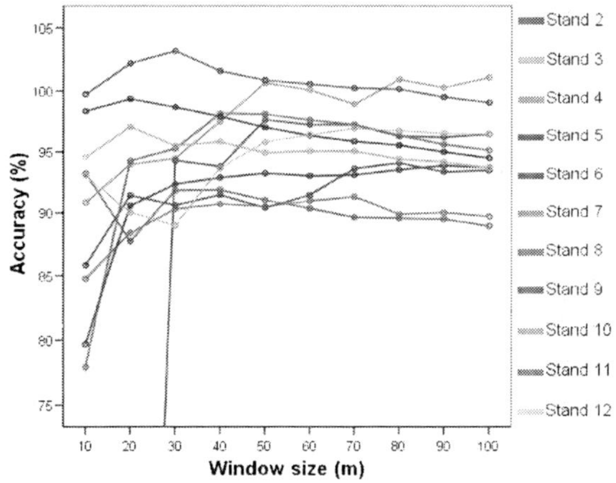


(A)

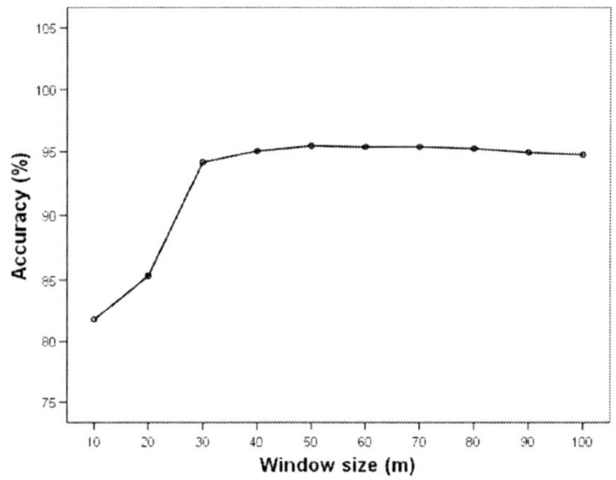


(B)

**Fig. 6:** Effect of window size on estimation of top height for each stand (A) and on average (B). Accuracy values for 10 and 20 m windows in plot 5 are 0.50% and 0.83% respectively.

As for the average top height accuracy regarding different window sizes, Figure 6(B) depicts how squared windows of 10 and 20 m show a notably lower average accuracy with values of 81.83% and 85.32% respectively. Average accuracy and standard deviation of the remaining windows are close to constant, although the interpolated line of Figure 6(B) slightly decreases towards higher window sizes after the maximum accuracy is achieved. Windows of $50 \times 50$ m$^2$ present the higher average accuracy with 95.48% whereas the lowest standard deviation of the average accuracy is obtained for $70 \times 70$ m$^2$ windows with 3.21%.

## 4.3 Different Point Density

### *4.3.1 Top Height*

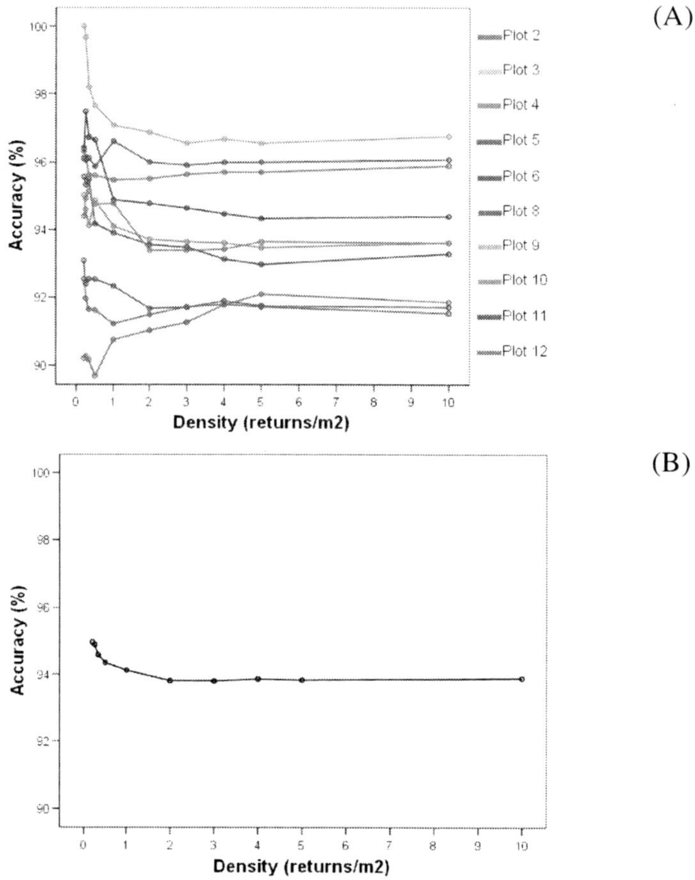


**Fig. 7:** Effect of point density on the estimation of top height for each plot (A) and on average (B)

Average accuracy of top height regarding point sample density was almost constant (ranging from 93.80% to 93.87%) for densities between 10 and 2 returns/m$^2$ (see Figure 7(B)). These values fit perfectly with the average accuracy for the original density (93.87%). Inferior point densities tend to slightly increase their average accuracy with reduction in point density, such that the highest average accuracy, 94.97%, was achieved with the lowest point density (0.20 returns/m$^2$). Plot number 2, in Figure 7(A), is the only plot in which higher densities are related to higher accuracies. The rest of the plots present a similar although more abrupt pattern than the average accuracy.

Standard deviation of average accuracy follows a similar trend which can be observed in Figure 7(A) by comparing the spread of the results regarding the point density. Differences on standard deviation for densities of 3 returns/m$^2$ or higher varied from 1.78% to 1.87%, whilst it constantly increased towards lower sample point densities. The 0.25 returns/m$^2$ density produced the higher standard deviation with a value of 2.76%.

#### 4.3.2 Basal Area

**Table 3** Coefficients and reliability figures of the linear regression models for the estimation of basal area.

| | Point density (ret/m$^2$) | C | Independent variables | | | $R^2$ | RMSE (m$^2$/ha) | Acc. (%) | MD |
|---|---|---|---|---|---|---|---|---|---|
| | | | $h_{70}$l | $h_{mean}$f | $h_{mean}$l | | | | |
| Orig. | 11.57 | 0.615 | - | - | 0.832 | 0.99 | 1.01 | 92.85 | $0.65^{ns}$ |
| Data Reduction | 10.00 | 0.650 | - | - | 0.830 | 0.99 | 1.00 | 92.95 | $0.64^{ns}$ |
| | 5.00 | -2.445 | 0.657 | - | - | 0.99 | 0.82 | 97.50 | $0.22^{ns}$ |
| | 4.00 | -2.726 | 0.673 | - | - | 0.99 | 0.85 | 97.16 | $0.25^{ns}$ |
| | 3.00 | -0.366 | - | 0.601 | - | 0.98 | 0.98 | 93.10 | $0.62^{ns}$ |
| | 2.00 | 0.772 | - | - | 0.816 | 0.99 | 1.09 | 91.72 | $0.76^{ns}$ |
| | 1.00 | 0.381 | - | - | 0.836 | 0.99 | 1.32 | 89.76 | $0.94^{ns}$ |
| | 0.50 | 0.687 | - | - | 0.807 | 0.99 | 1.25 | 89.87 | $1.10^{ns}$ |
| | 0.33 | 0.783 | - | - | 0.801 | 0.99 | 1.16 | 91.01 | $0.82^{ns}$ |
| | 0.25 | -0.503 | - | 0.608 | - | 0.98 | 1.13 | 92.68 | $0.45^{ns}$ |
| | 0.20 | -0.385 | - | 0.603 | - | 0.99 | 1.08 | 92.78 | $0.67^{ns}$ |

MD = mean difference between observed and predicted values; Level of significance: ns = not significant (>0.05), * < 0.05, *<0.01; $h_{70}$l = 70$^{th}$ percentile of the distribution of the last return; $h_{mean}$f/$h_{mean}$l = mean of the distribution of the first/last return; C = constant; Acc.= accuracy; Orig. = original data.

Using the linear regression model created at plot level, a high correlation between field and laser-predicted basal area values was found for all sampling densities. $R^2$ for linear regression models ranged from 0.98 to 0.99, where $h_{70}$l, $h_{mean}$f and $h_{mean}$l were the predictor variables which showed the lowest p-value (see Table 3). The mean of the normalised height distribution for the last return is the independent variable which appears more often in the models; however the 70$^{th}$ percentile of the last return is the Li-

DAR metric that yields lower bias and better accuracies. Figure 8 shows how relative RMSEs varied from 9.15% (4 returns/m$^2$) to 13.43% (0.50 returns/m$^2$), without any remarkable trend regarding point density. Basal area predictions were biased in all point densities. There was no evidence of a statistically significant difference between the means of observed and predicted basal area values, that is, the probability of obtaining such difference simply by chance is high.

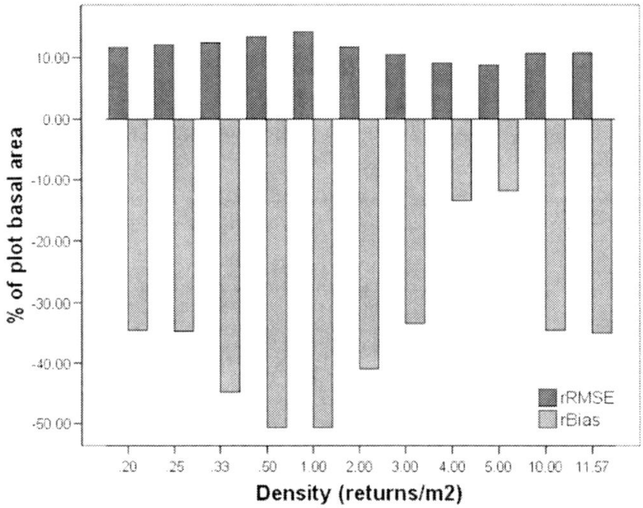


**Fig. 8:** Effect of data reduction on the prediction of basal area from linear regression models.

### *4.3.3 Volume*

High correlation between field and laser-predicted volume was also found by means of linear regression at plot level. This time, only $h_{70}$l and $h_{mean}$f were suitable to build regression models for different laser densities. The former seems to be more related to higher density samplings, whereas the latter to the lower densities, as shown in Table 4. Again $h_{70}$l was the LiDAR metric which yielded lower bias and better accuracies. Models presented negligible variations in $R^2$ (from 0.98 to 0.99). Figure 9 depicts similar RMSEs, ranging from 6.74% (2 returns/m$^2$) to 8.26% (0.50 returns/m$^2$). Volume predictions were biased in all densities and there was no evidence of a statistically significant difference between observed and predicted volume values, meaning that the difference between the two paired groups of data is likely to be due to chance.

**Table 4** Coefficients and reliability figures of the linear regression models for the estimation of volume.

| | Point density (ret/m2) | C | Independent variables | | $R^2$ | RMSE ($m^3$/ha) | Acc. (%) | MD |
|---|---|---|---|---|---|---|---|---|
| | | | $h_{70}l$ | $h_{mean}f$ | | | | |
| Orig. | 11.57 | -44.230 | 8.614 | - | 0.99 | 7.46 | 100.13 | $0.96^{ns}$ |
| Data Reduction | 10.00 | -40.866 | 8.445 | - | 0.99 | 7.73 | 100.51 | $0.65^{ns}$ |
| | 5.00 | -35.657 | 8.180 | - | 0.98 | 8.19 | 101.16 | $0.07^{ns}$ |
| | 4.00 | -39.767 | 8.407 | - | 0.99 | 8.12 | 100.79 | $0.41^{ns}$ |
| | 3.00 | -10.244 | - | 7.503 | 0.99 | 8.06 | 96.22 | $5.06^{ns}$ |
| | 2.00 | -44.735 | 8.656 | - | 0.99 | 7.35 | 99.54 | $1.56^{ns}$ |
| | 1.00 | -8.381 | - | 7.391 | 0.99 | 8.86 | 95.94 | $5.49^{ns}$ |
| | 0.50 | -10.431 | - | 7.497 | 0.99 | 9.01 | 96.08 | $5.02^{ns}$ |
| | 0.33 | -12.803 | - | 7.654 | 0.98 | 8.54 | 95.74 | $0.72^{ns}$ |
| | 0.25 | -11.855 | - | 7.592 | 0.98 | 8.33 | 95.63 | $5.36^{ns}$ |
| | 0.20 | -10.005 | - | 7.496 | 0.99 | 8.77 | 95.82 | $5.35^{ns}$ |

MD mean difference between observed and predicted values; Level of significance: ns = not significant (>0.05), * < 0.05, *<0.01; $h_{70}l = 70^{th}$ percentile of the distribution of the last return; $h_{mean}f/h_{mean}l$ = mean of the distribution of the first/last return; C = constant; Acc.= accuracy; Orig. = original data.

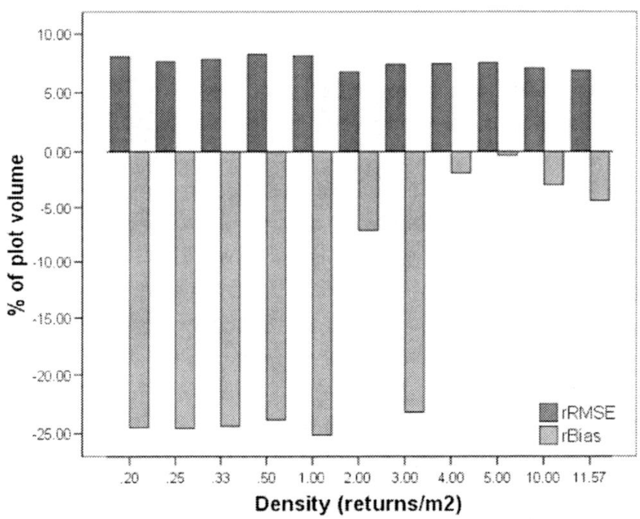


**Fig. 9:** Effect of data reduction on the prediction of volume from linear regression models

# 5 Discussion

The use of highest percentiles to estimate measurements of canopy height, such as mean, Lorey (basal area weighted mean tree height), predominant, dominant or top height, has been found effective in other studies as predictor variable in stepwise regression analysis. Means et al.(2000) used the 90$^{th}$ percentile on the estimation of mean tree height, Naesset (2004b) predicted Lorey and dominant height using 80$^{th}$ and 90$^{th}$ percentiles, whereas Holmgren (2004) used the 95$^{th}$ percentile in mean tree height prediction. However, it was Lovell et al.(2003) who moved height estimations apart from regression models stating that the average height of the 100 highest hits trapped in a hectare are a reasonable equivalent to predominant height in Australian forests. In this study, the 99$^{th}$ percentile of a normalised dataset was further tested as a simple and efficient top height predictor. It is hardly affected by large gaps in the canopy, which means that wind throw, clear felling tasks or different planting density have a negligible effect on it.

The bias of this method can be attributed to the fact that the percentile whose value is closer to the field top height might vary with the age of the Sitka spruce stand, that is, stands of notably different ages will yield different LiDAR point distributions which might vary the percentile that is closer to the observed top height. However the 99$^{th}$ percentile is a good standard equivalent. In the context of this study, it was found that the 99.9th percentile best predicted top height with a RMSE of 0.60 m, which represents only 2.24% of the average field top height, and with a notably lower bias. Nonetheless, the 99.9$^{th}$ percentile may not be effective for stands of different ages. Similarly, the effectiveness of the 99$^{th}$ percentile should be tested on other species, since crown shape affects laser metrics (Nelson 1997).

The assessment of stand's spatial variability using top height shows that, on average, standard deviation of the accuracy is reduced in windows of $30\times30$ m$^2$ and constantly low for larger areas (see Figure 6(B)). Conversely, small squared windows (10 and 20 m) have shown to be unrepresentative of the stands since they present a remarkable variability on top height prediction accuracy. In this study it is suggested that high variability in small windows is due to either (i) the stand sample found within the window not being representative of stand top height or (ii) the area within the window simply having no trees because of wind throw. Magnussen (1999) also showed the effects of the area chosen on the estimation of top height by observing that derived top height for plots of 0.01, 0.03 and 0.05

ha were 5.6%, 3.0%, and 2.5% lower regarding predictions based on 1 ha plots.

The spatial variability of the stand gives an idea of how top height varies as a function of area, providing an approximation of the minimum size to which a field plot could be surveyed in order to validate LiDAR predictions. Therefore, according to this study, a plot size of $30 \times 30$ m$^2$ is already quite reliable in terms of accuracy and standard deviation, with $40 \times 40$ m$^2$ being the optimum size with regard to cost-efficiency, since larger plot sizes present almost the same average accuracy and standard deviation.

Detection of individual trees requires a high density of points (Hyyppä and Inkinen 1999), which is currently expensive. Nevertheless, distribution-based approaches for estimations at plot and stand level have shown promising results with sampling intensities of 1 return/m$^2$ (Naesset 2002), which in this study, was found to be true even for lower densities.

Top height with 99$^{th}$ percentile yielded better accuracies for lower sampling densities (see Figure 7(B)), which is exactly the opposite of the expected result, however, the difference in average accuracy between the higher and the lower laser intensities was just about 1%, which represents 0.27 m on the average field top height. It must be noted that a percentile is just a position in a ranked dataset. In addition, the accuracies obtained with different densities are merely theoretical, since their normalisation was carried out using the best DTM available, i.e., the one created with the original density. The random extraction of points may have caused the higher 99$^{th}$ percentiles values for lower sampling densities by producing a larger percentage of high hits in lower sampling densities for most of the plots.

Despite the fact that most of the authors use multiple linear regression models for the estimation of basal area and volume (Holmgren 2004; Maltamo et al. 2006; Naesset 2002, 2004b), stepwise regressions were always stopped after the first step (that is, using just one independent variable) because it was observed that the predictions over plots used to validate the models performed better. This fact was attributed to the scarce number of plots available for both building and validating the model. Since the 5 plots used to build the model were randomly chosen, they might not collect the variability of the forest parameters within the whole population of stands. Therefore if more independent variables are added to the model (in case they are significant), it slightly improves the $R^2$ of the models created at the expense of poorer predictions of the parameters of the plots used for validation, which potentially might have fallen out of the range of variability of the plots used for building the model. If a larger number of plots were available, the set of plots used to create the models would be more likely to be more representative of the whole population of plots and thus, a multiple linear regression model would be recommended in the event that more than one predictor variable is found significant.

In the same manner the significance of the mean differences between the observed and predicted values of both basal area and volume might have been affected by the reduced number of plots, or samples, used in the paired two-tailed t-test, since the sample size influences the calculation of the p-value. With a large sample size very small differences will be detected as significant, whereas with a small sample size differences will have to be bigger in order to be detected as significant.

Naesset (2002) achieved relative RMSEs ranging between 9% and 10% for basal area and between 11% and 14% for volume, whereas relative RMSEs for basal area and volume in Holmgren (2004) where 10% and 11% respectively. In this study, for the original density, results were similar for basal area (rRMSE=10.84%) and notably better for volume, with a relative RMSE of 6.83%. This improvement in volume estimation might be due to the fact that the range of ages of the stands used in this study differs in just 2 years, as oppose to the previous studies which work with different species and a broader range of ages. Also, despite the remarkable difference in the number of plots used to build and validate the regression models, 144 in Naesset (2002) and 464 in Holmgren (2004), the relative RMSEs obtained with just 10 plots fall within a similar range of values.

This study indicates that errors for prediction of basal area and volume do not differ too much for densities between 0.2 and 11.57 returns/m$^2$. Such errors do not follow any trend regarding point density; they just present small variations depending on the predictive variable included in the model. For the same independent variable, predictions are quite close, regardless sampling density. At this point it should be remembered that all sampling densities have been normalised using the DTM created with the original density. What is more, variations in platform altitude, which affect footprint diameter, are normally used to reduce point density and it could not be simulated. However, Yu et al.(2004) found that footprint size does not have much influence on height estimations. Holmgren (2004) comprised a density reduction (4.29, 0.55, 0.17 and 0.10 returns/m$^2$) using different platform altitudes, and also concluded that there were small variations on the estimation of basal area and volume for the range of densities under study. Similarly, Maltamo et al.(2006) carried out a synthetic density reduction (12.7, 6.3, 1.3, 0.6 and 0.13 returns/m$^2$) on data retrieved from Finnish forests, stating that the effect of density reduction on volume is negligible and that small differences found are more likely random.

The fact that estimation accuracy of forest parameters hardly differs for different laser sampling densities is an important point in commercial forestry. Once a good DTM is created, future LiDAR surveys over the same forested area, for growth monitoring or wind thrown assessment, can be comprised at lower sampling densities, thus reducing costs without degrading accuracy on the estimations.

## 6 Conclusions

The overall LiDAR analysis cost depends mainly on 3 parameters, namely acquisition of LiDAR data (e.g. extent of surveyed area and pulse densities), number and size of the field plots used for validation, and amount of processing accomplished on the data. Therefore reduction of the cost of these individual parameters will contribute to a more effective use of LiDAR in forestry.

Results in this study have shown: (i) The $99^{th}$ percentile of a normalised canopy distribution is a robust method to predict top height on Sitka spruce, but it still has to be tested on different tree species. Likewise, research on finding other LiDAR metrics directly related to forest parameters should be developed, since it contributes to simpler and faster data processing. (ii) Plots used to validate LiDAR predictions should not have an area smaller than $30 \times 30$ m$^2$, since spatial variability for smaller areas is high in terms of average accuracy and standard deviation. The best results, regarding both accuracy achieved and size of the area, were for areas of $40 \times 40$ m$^2$. Larger sizes would merely increase the cost of field data collection without improving the accuracy. (iii) Simulated reduction of sampling density had a negligible effect on estimation accuracy of top height, basal area and volume from a distribution-based approach, where the small differences found were rather related to the independent variable used in each linear regression model. From the economic point of view, this means that LiDAR acquisitions over the same area can be accomplished more often as lower point densities present lower costs. The next step will be testing different densities over other forest parameters such as biomass.

Although LiDAR in forestry is still an expensive technology, particularly to cover large areas, current research is allowing more cost-effective sampling densities through the optimisation of LiDAR survey characteristics. Systems accuracy will increase, which means a better estimation of forest parameters as well as higher flying altitudes and thus larger areas surveyed. Ultimately, the amount and accuracy of information retrieved with LiDAR sensors relative to the costs, will lead to the adoption of LiDAR as a fully operational tool in forestry.

## References

Behera MD, Roy PS (2002) Lidar remote sensing for forestry applications: The Indian Context. Current Science 83: 1320-1328.

Bortolot ZJ, Wynne RH (2005) Estimating forest biomass using small footprint LiDAR data: An individual tree-based approach that incorporates training data. ISPRS Journal of Photogrammetry and Remote Sensing 59: 342-360.

Evans DL, Roberts D, McCombs JW, Harrintong RL (2001) Detection of Regularly Spaced Targets in Small-Footprint LIDAR Data: Research Issues for Consideration. Photogrammetric Engineering and Remote Sensing 67: 1113-1136.

Forestry Commission (2006a) Annual Report and Accounts 2005-2006.

Forestry Commission (2006b) Forestry Facts & Figures 2006.

García O (1998) Estimating top height with variable plot sizes. Canadian Journal of Forest Research 28: 1509-1517.

García O (2006) Scale and spatial structure effects on tree size distributions: impllications for growth and yield modelling. Canadian Journal of Forest Research 36: 2983-2993.

Goodwin NR, Coops NC, Culvenor DS (2006) Assessment of forest structure with airborne LiDAR and the effects of platform altitude. Remote Sensing of Environment 103: 140-152.

Holmgren J (2004) Prediction of Tree Height, Basal Area and Stem Volume in Forest Stands Using Airborne Laser Scanning. Scandinavian Journal of Forest Research, vol 19, pp 543-553.

Holmgren J, Nilsson I, Olsson H (2003a) Simulating the effects of lidar scanning angle for estimation of mean tree height and canopy closure. Canadian Journal of Remote Sensing 29: 623-632.

Holmgren J, Nilsson I, Olsson H (2003b) Estimation of Tree Height and Stem Volume on Plots Using Airborne Laser Scanning. Forest Science 49: 419-428.

Hyyppa J, Inkinen M (1999) Detecting and estimating attributes for single trees using laser scanner. Photogrammetric Journal of Finland 16: 27-42.

Lefsky MA, Cohen WB, Parker GG, Harding DJ (2002) Lidar Remote Sensing for Ecosystem Studies. BioScience 52: 19-30.

Lim K, Treitz P, Wulder M, St-Onge B, Flood M (2003) LiDAR remote sensing of forest structure. Progress in Physical Geography 27: 88-106.

Lovell JL, Jupp DLB, Culvenor DS, Coops NC (2003) Using airborne and ground-based ranging lidar to measure canopy structure in Australian forests. Canadian Journal of Remote Sensing 29: 607-622.

Lovel JL, Jupp DLB, Newnham GJ, Coops NC, Culvenor DS (2005) Simulation study for finding optimal lidar acquisition parameters for forest hight retrieval. Forest Ecology and Management 214: 398-412.

Magnussen S (1999) Effect of Plot Size on Estimates of Top Height in Douglas-Fir. Western Journal of Applied Forestry 14: 17-27.

Magnussen S, Boudewyn P (1998) Derevations of stand heights from airborne laser scanner data with canopy-based quantile estimators. Canadian Journal of Forest Research 28: 1016-1031.

Maltamo M, Eerikainen K, Packalén P, Hyyppa J (2006) Estimation of stem volume using laser scanning-based canopy height metrics. Forestry 79: 217-229.

Means JE, Acker SA, Brandon JF, Renslow M, Emerson L, Hendrix CJ (2000) Predicting forest stand characteristics with airborne scanning lidar. Photogrammetric Engineering and Remote Sensing 66: 1367-1371.

Naesset E (2002) Predicting forest stand characteristics with airborne scanning laser using a practical two-stage procedure and field data. Remote Sensing of environment 80: 88-99.

Naesset E. (2004a) Effects of different flying altitudes on biophysical stand properties estimated from canopy heitht and density measured with a small-footprint airborne scanning laser. Remote Sensing of Environment 91: 243-255.

Naesset E (2004b) Practical Large-scale Forest Stand Inventory Using a Small-footprint Airborne Scanning Laser. Scandinavian Journal of Forest Research 19: 164-179.

Nelson R (1997) Model Forest Canopy Heights: The Effects of Canopy Shape. Remote Sensing of Environment 60: 327-334.

Philip MS (1998) Measuring Trees and Forests, Oxon, CABI Publishing.

Spurr SH (1952) Forest inventory, New York, Ronald Press.

Suárez JC, Ontiveiros C, Smith S, Snape S (2004) The Use of Airborne LiDAR and Aerial Photography in the Estimation of Individual Tree Heights in Forestry. AGILE Conference on Geographic Information Science. Heraklion, Greece.

Suárez JC, Snape S, Ontiveiros C, Smith S (2005) Use of airborne LiDAR and aerial photography in the estimation of individual tree heights in forestry. Computers and Geosciences 31: 253-262.

Yu X., Hyyppa J, Hyyppa H, Maltamo M (2004) Effects of flight altitude on tree height estimation using airborne laser scanning. In laser scanners for forest and landscape assessment. IN THIES, M., KOCH, B., SPIECKER, H. & WEINACKER, H. (Eds.) Proceedings of the ISPRS working group VIII/2. Freiburg, Germany, October, 3 – 6, 2004. Int. Arch. Photogramm. Remote Sens. Spatial Inf. Sci. 36 (8/W2), pp 96 – 101.

# Model Based Optimization of Mobile Geosensor Networks

Alexander C. Walkowski

Insitute for Geoinformatics, University of Münster, Robert-Koch-Str. 26-28, 48149 Münster, Germany, walkowski@uni-muenster.de

**Abstract.** An approach for monitoring network optimization is presented which estimates the characteristics of a spatial phenomenon based on the measured values in order to perform the optimization. Based on a phenomenon model it is determined where additional information is needed. During this optimization process also the limitations and constraints of the geosensor network are taken into account. After deriving the model based approach from theoretical considerations, the presented approach is evaluated by means of a table top experiment.

**Keywords:** mobile geosensor networks, network optimization, geostatistics

## 1 Introduction

In order to observe natural and also man-made environmental phenomena the use of sensors is essential. Only the sensor based capturing enables humans to observe and comprehend these phenomena.

The temporal dimension of a phenomenon describes how it behaves at a fixed location in the course of time. For analyzing this dimension the time series of single sensors are examined. Usually environmental phenomena (e.g. air temperature or concentration of certain air pollutants) do not occur pointwise. Instead they can be best described using continuous value surfaces. This makes it necessary to take another dimension into account: space.

Regarding the monitoring of the two exemplary mentioned phenomena, at the moment only in-situ sensors are used which deliver measurements that are valid for the location where they are installed. In order to capture the spatial dimension, it is necessary to deploy sensors as a spatially distributed monitoring network. An example for such a monitoring network is the Air Quality Monitoring System (LUQS) which is operated by the Environmental Agency of North Rhine-Westphalia, Germany (LUA-NRW 2001). This network, consisting primarily of in-situ sensors, aims at capturing several air quality parameters (particulate matter emission, nitrogen dioxide, ozone concentration, and others).

Traditional data capturing techniques for monitoring spatial phenomena (e.g. the previously mentioned LUQS monitoring network) are based on a small number of fixed sensors (Szewczyk et al. 2004). These sensors are installed in a controlled environment using a previously defined deployment strategy. This strategy allows for example to find the locations where the sensors need to be deployed or to determine calibration parameters.

Because of the advancing development of sensor technologies the way of capturing data about spatiotemporal phenomena is revolutionized. This development and especially the evolution of micro-electro-mechanical-systems (MEMS) causes a change in the way spatiotemporal phenomena are captured. Currently a change from a centralized approach based on isolated sensors to an approach using distributed sensors is performed (Nittel and Stefanidis 2005). This makes the aspect of monitoring network optimization in order to enhance efficiency more and more important.

Additionally to the hardware development current research focuses on constructing communication infrastructures that are adapted to the special needs of sensor networks (Nittel et al. 2004) and that allow the integration of large numbers of sensor nodes into sensor networks. Exemplarily the development of special networking protocols (Goldin et al. 2005; Wang and Ramanathan 2004), concepts for self organization (Brooks 2004), and mobility strategies (Sliwinski and Simonis 2005) shall be mentioned.

In this paper we present an approach for monitoring network optimization that takes into account both, the characteristics of the phenomenon and those of the geosensor network. A phenomenon model describing the characteristics of the phenomenon is built using geostatistical methods. The phenomenon model allows determining where additional information is needed. The limitations and constraints of the geosensor network are considered in the optimization process.

The rest of the paper is structured as follows: First the boundary conditions of this work are presented. Subsequently a scenario is presented which will be used as an example in order to illustrate this purpose. After this, existing approaches for optimizing monitoring networks are shown and subsequently the new combined solution is presented. The model based approach is evaluated by means of a table top experiment. Finally, a conclusion and an outlook on potential future developments are given.

## 2 Boundary Conditions

In this section we define the fundamental terms used throughout the paper. The first subsection is devoted to the underlying phenomenon definition. Subsequently the term mobile geosensor network is delineated and an illustrating scenario is presented at the end of this section.

### 2.1 Phenomenon

Within the paper in hand the term phenomenon is defined as a spatiotemporal stochastic process exhibiting a spatial autocorrelation (persistence). According to (Cressie 1993) such a spatiotemporal stochastic process is defined as the set Z of random variables Z(s,t)

$$Z = \{Z(s,t) | s \in D(t), t \in T\} \quad (1)$$

with $s$ indicating the localization of the random variable in the spatial domain $D$ and $t$ signifying the random variables localization in time. Without loss of generality we restrict the spatial domain $D$ to $\Re^2$. Alternatively the set of random variables can also be denoted as a random function that assigns to every point in space a corresponding random variable. Basically every random variable has its own distribution function. Only under the assumption of stationarity the distribution functions of all random variables are identical. The realization of a random function is a value surface that is termed regionalized variable (Matheron 1971).

The set of measured values

$$z = \{z_i | z_i \in \Re; i = 1, \ldots, n; i \in N\} \quad (2)$$

at the locations $s_1, \ldots, s_n \in D$ with $n \in N$ and $s_i = (x_i, y_i)$, $x, y \in \Re$ is a sample of the regionalized variable. On two counts the measured values are a finite sample of the stochastic process: measured values are available only for a limited number of locations within the investigation and for all of these locations only one measured value exists. These samples are gathered by the geosensor network which observes the phenomenon.

### 2.2 Mobile Geosensor Networks

A distributed sensor network (DSN) consists of a great number of heterogeneous intelligent spatially distributed sensors that are interconnected by a communication network and which report occurring events to a user

(Iyengar et al. 2004). The term geosensor network (GSN) describes distributed sensor networks that are specialized in order to capture spatiotemporal phenomena (Nittel and Stefanidis 2005).

If the nodes (sensors) of a GSN are actively or passively moving through the space they form a mobile geosensor network. Based on different stimuli (e.g. observations, messages from users or network optimization components) these sensors are able to adapt their behavior. It has to be noted that the sensor behavior is influenced by several limitations like the amount of energy available and mobility constraints in the vein of speed. The application of a GSN is an important tool which provides an insight into spatiotemporal phenomena that was previously not available (Szewczyk, et al. 2004; Shepherd and Kumar 2004).

## 2.3 Scenario

The following scenario is assumed: In a given area a geosensor network for monitoring the air quality is installed. This GSN consists of two subnetworks (see Fig.1):

- traditional stationary monitoring network ($GSN_s$)
- a mobile GSN ($GSN_m$)

The users of $GSN_m$ are able to define spatiotemporal way points. This means that a sensor can be tasked to execute one or more measurements at a certain point in time or time interval at a certain location. In Fig. 1 there are two sensors at time $t_1$ that measure a certain position, but they are already scheduled for $t_2$. Sensor 1 is planned to observe another location at $t_2$. The schedule for sensor 2 prescribes to observe the phenomenon at the new location for the time interval from $t_2$ until $t_6$. Due to the spatiotemporal waypoint sensor 1 could not be used in the optimization process, at least for time $t_2$. Sensor 2 is bound to his position for the entire time period $t_2$ to $t_6$.

The scenario consists of a chemical accident that results in a toxic plume. The GSN shall be used to capture the pollutant concentration in an optimal way. As the toxic cloud disperses in the course of time it is necessary to optimize $GSN_m$ in order to gather as much information about the spatiotemporal phenomenon as possible. This is illustrated in Fig. 1 by sensor 3, which moves into the plume.

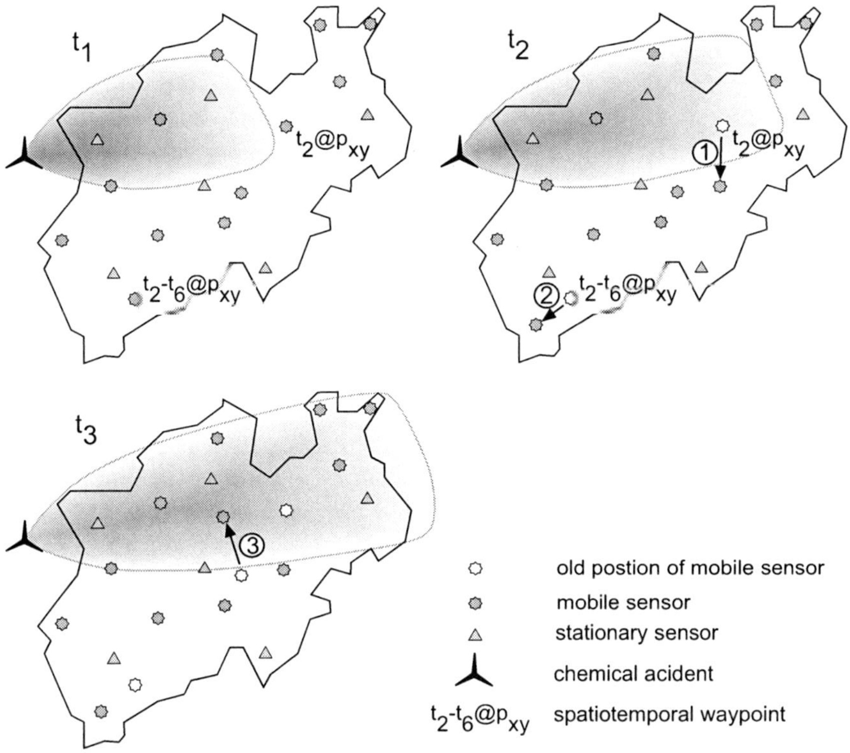


**Fig. 1:** Scenario illustrating the geosensor network observing a toxic plume

## 3 Monitoring Network Optimization

The goal of monitoring network optimization is to adapt the constellation of the monitoring network in such a way that the sensors are used as efficiently as possible. The GSN produces samples of the regionalized variable and a model of the phenomenon is calculated in order to represent the basic population as good as possible. It is desired that the sensors perform their measurements at those locations, where they are able to gather information that is not known within the model. Consequently, sensors are used efficiently, if they reduce the lack of information within the phenomenon model. The lack of information relates to the basic population.

In the following sections different approaches for optimizing monitoring networks are presented. At first, we describe some conventional approaches before the model based approach developed by the author is presented.

## 3.1 Conventional Approaches

Based on a literature review existing approaches can be classified into coverage-oriented approaches relying on geometric considerations and phenomenon-oriented approaches considering the observed values.

### *Coverage-Oriented Approach*

The term coverage of a geosensor network describes a quality parameter that indicates how good an area of interest is covered by sensors (Meguerdichian et al. 2001; Howard et al. 2002). The existence of areas with weak coverage means that events occurring in the area of interest may remain undetected by the GSN.

In (Zou and Chakrabarty 2003) for example a so called Virtual Force Algorithm (VFA) is presented. Starting with a number of randomly distributed sensors this algorithm allows reaching an optimal coverage after a single repositioning phase. In order to avoid clustering of sensors a force is defined that leads to repulsion between too closely located sensors. Similarly obstacles create a repulsive force which helps the mobile sensors to avoid collisions. Furthermore an attractive force is defined that drags sensors into areas with a lack of coverage. In the end these forces create an even distribution of sensors across the area of interest.

Coverage oriented approaches mainly rely on geometric considerations. The deficit of this type of strategy is that the measured values are not taken into account for the sensor network optimisation. Thus only a solution is produced which is optimally adapted to the geometry of the area of interest but not to the phenomenon.

### *Phenomenon-Oriented Approach*

Unlike coverage oriented approaches which do not take into account the characteristics of the phenomenon, phenomenon based methods perform the sensor network optimization based on the values that are measured.

In (Sliwinski and Simonis 2005) an autonomous sensor strategy is presented which makes a sensor stay as long at a certain location as it is able to register the phenomenon there. If the phenomenon is not detected anymore the sensor moves randomly through the space. This approach explicitly considers the phenomenon. An optimization of the sensor network in an integrated way is not performed, though.

Another approach is presented by (Pardo-Igúzquiza and Dowd 2005). They present a method which is based on the interpolated value surface and the associated estimation errors. The approach makes use of the kriging variance for iteratively positioning a single sensor in a way that the spatial extend of a cloud is optimally captured. The drawback of this solution is that it is usually not realistic to assume a single sensor within a GSN. Furthermore it is not clearly defined which sensor needs to move to which area with a strong lack of information (high kriging variance).

## 3.2 Model Based Approach

The previously presented approaches for optimizing monitoring networks are only to a limited extend suited for optimizing GSN. Coverage oriented methods do not take into account the realization of the dynamic phenomenon and thus do not adapt the monitoring network to processes like the dispersion of toxic clouds. As a result, the capturing of these phenomena can not be executed in an effective manner. In the following section a model based approach is presented which combines a phenomenon model with a GSN model in order to optimize the monitoring network.

### *Phenomenon Model*

The monitoring of the phenomenon by means of a geosensor network yields in a sample of the regionalized variable. In the context of this paper it is assumed that all observations (samples) are propagated to a base station which computes a model of the monitored phenomenon.

The centralized phenomenon model has to fulfill the following *requirements* in order to facilitate the optimization of the geosensor network. The model

- must provide a measure for the information deficit and
- is obliged to be autoprojective.

According to (Nipper and Streit 1982) autoprojective models use only the specific information about the phenomenon in question. In other words, the model representing the regionalized variable is build out of the samples of the regionalized variable without any further information. Aside a continuous representation of the phenomenon a measure is needed, which indicates areas where further information is needed.

A literature review yields that simple interpolation methods – like inverse distance weighting (Shepard 1968), nearest neighbor (Thiessen 1911), and triangulation (Isaaks and Srivastava 1989) – do not fulfill the requirements. Neither allow the simple interpolation methods the definition of a measure for the information deficit, nor are they autoprojective. Thus a geostatictical approach is chosen for the computation of the phenomenon model. The geostatistical approach was developed during the 1960ies parallel by two different scientists. In France Georges Matheron (Matheron 1963) developed it for utilization in connection with mining application whereas Lev Gandin (Gandin 1963) in the Soviet Union performed his development against the background of meteorology (Cressie 1990). The semivariogram $\gamma$ is defined as half of the variance of the difference between values at locations separated by the lag vector h (Cressie 1993):

$$2\gamma(h) = Var[Z(s+h) - Z(s)] \tag{3}$$

The semivariogram represents a statistical measure that describes the persistence of a regionalized variable. The model of the spatial stochastic process and the semivariogram as a measure for the spatial persistence allow estimating the values for regionalized variables at those points without measured values.

The term Kriging identifies a set of linear estimation algorithms that use the calculation of a weighted spatial average. These weights are optimized so that the variance of the estimated values is minimized and that the estimated values are unbiased (Heinrich 1992), yielding in so called best linear unbiased estimator (BLUE).

The ordinary Kriging estimator is defined as the weighted average of the neighboring values (Matheron 1971):

$$\hat{z}(s_0) = \sum_{i=1}^{n} \lambda_i \, z(s_i) \tag{4}$$

The number of neighbors n is determined according to the semivariograms range (Webster and Oliver 2001). The weights are calculated using the semivariogram so that

- the Kriging estimator is unbiased
- the variance of the estimation error is minimized

Due to the assumption of an unbiased estimation, the variance of the estimation error (kriging variance) must be minimized under the condition that the sum of weights is 1 (Heinrich 1992; Matheron 1971). The kriging equation system is obtained by minimizing the kriging variance using the Lagrange multiplicator method. For further details concerning the kriging equation system and the calculation of the weights the author refers to (Webster and Oliver 2001; Huijbregts 1975).

The Kriging variance is according to (Journel and Huijbregts 1997, Webster and Oliver 2001) defined as follows:

$$\sigma_{kr}^2(s_0) = \sum_{i=1}^{n} \lambda_i \gamma_{i0} + \mu \tag{5}$$

As this definition shows the Kriging variance is influenced only by the values of the semivariogram which rely in a further step on the configuration of the monitoring network. Because the Kriging variance is independent of locally measured values it can not be used as a measure for the local estimation precision (Journel 1986). Instead it is an indicator for the lack of information which can be used for optimizing monitoring networks.

The *phenomenon model* used for the geosensor network optimization comprises of the following components:

- continuous representation of the phenomenon
- measure for the information deficit
- forecast of the information deficit

A model is computed from the values $z(t_j)$ observed by the GSN at the point in time $t_j$ in order to represent the continuous value surface. For the generation of the model kriging methods are used, which yield not only in the value surface, but also in the kriging variance that determines the lack of information at each point in the area of interest for the time $t_j$. If the optimization of the GSN would be performed only on the basis of this information, the dynamic of the phenomenon would not be taken into account. The optimization would be based solely on the lack of information at $t_j$. Instead the change of the spatiotemporal phenomenon between measurements at time $t_j$ and $t_{j+1}$ must be considered. At a first glance, methods of time series analysis would be applied in order to calculate the temporal variability which then could be used for the geosensor network optimization. But in mobile GSN there are no long lasting time series available. As the sensors move around in most cases only single measurement values are available for one location.

The following approach is proposed within this paper to solve the abovementioned issue. Based on the method of trend surface analysis (TSA) a two dimensional polynomial is fitted to each of the last n kriging variance fields. This results in a vector (time series) for each parameter of the polynomial. In the next step we apply time series analysis tools to the parameter vectors in order to forecast the parameter values. The choice of method depends upon the length of the time series. For short time series simple trend models will be applied; for longer time series Box-Jenkins (Box et al. 1994) models might be used. Based on the predicted parameters of the polynomial for time $t_{j+1}$ a surface of the future information deficit is computed.

The surface of the future information deficit allows for the identification of sensors which are able to reach the centers of regions with a high lack of information. These regions are defined as areas in which the information deficit exceeds a predefined threshold. Thus the network constellation will be adapted in order to minimize the (future) information deficit.

### *GSN Model*

As mentioned in the previous section, the phenomenon model allows for the identification of sensors which are able to reach the regions with high lack of information. A simple GSN optimization approach would send the sensor which is the closest one to the area with a high lack of information. But this approach is not suited for heterogeneous GSN which may consist of mobile and stationary sensors. Additionally this approach does not take into account spatiotemporal waypoints. This means that sensors may have to execute measurements at defined locations at certain points of time. Thus spatiotemporal waypoints limit the radius where sensors are able to move to. As a result the determination of the sensor that is best suited for moving into a region with a lack of information is more complicated. For these reasons a GSN model is proposed which fulfills the following *requirements*:

- Description of the GSN's current state.
- Characterization of the potential activity area of each sensor.

The concept of Hägerstrand's time geography (Hägerstrand 1970) will be used as GSN model. The *space-time path* depicts the movement of individuals – in the context of the GSN model sensors – in space over time (see Fig. 2); it allows for the recording of the history of each sensor.

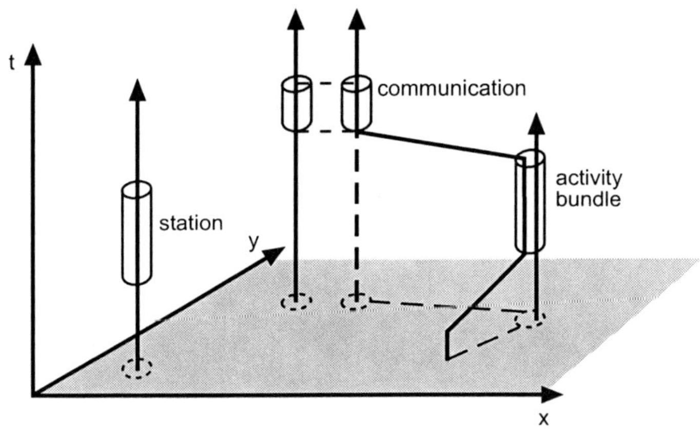


**Fig. 2:** Schematic illustration of space-time path, which group at stations to so called activity bundles (Hägerstrand 1970)

Hägerstrand formulates three major types of constraints that limit an individual's activities in space and time. Transferred to the GSN model these constraints could be formulated as follows:

- *Capability constraints* are those which limit the activity of a sensor because of its physical construction, e.g. communication hardware and mobility potential (speed, range, etc.).
- *Coupling constraints* define where, when and for how long the sensor has to join other assets or has to measure at certain locations.
- *Authority constraints* subsume those general rules and laws which determine if a sensor is allowed to enter a certain area or not (e.g. nature protection areas).

Due to coupling constraints, sensors will group together for certain time intervals at certain locations. This could be a sensor which has to move in to the communication range to the next sensor or to the base station in order to recharge the battery.

These constraints yield in the construction of the *space-time prism* for each sensor. It describes the latitude of a sensor located at position $s(t_i)$ at time $t_i$. In the context of time geography this is called the potential path space (PPS). The projection of the potential path space onto the geographic space results in the potential path area (PPA), which represents all locations that are in reach of the sensor during the time interval $\Delta t = t_j - t_i$ (see Fig 3.).

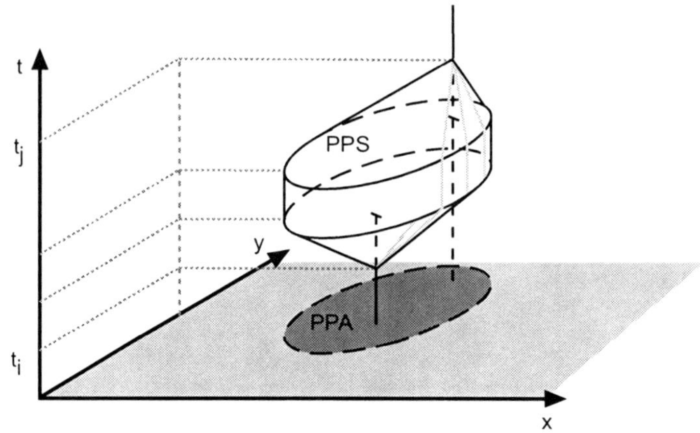


**Fig. 3:** Space-time prism after (Lenntrop 1976)

Using the concepts of time geography, it is possible to determine those sensors that are able to reach the centre of a region with a high lack of information ($R_{IM}$). This is the case if the centre of $R_{IM}$ lies within the time-space-prism of a sensor. If the centre of $R_{IM}$ is contained in the space-time-prisms of more than one sensor it is necessary to select the sensor that has to cover the smallest distance. Thus it is ensured that the GSN is optimally organized using a minimal amount of energy.

## 4 Evaluation

In order to test the presented approach a spatiotemporal phenomenon was simulated by means of a table top experiment. The dispersion of a toxic plume was simulated by trickling red ink on a 50 by 50cm plane covered with a water film. The current of the water film imitates the drift of the toxic plume in the direction of the wind. We took every three seconds a picture of the plane, representing the population that is sampled by 100 sensors. At the beginning of the test, the sensors were placed randomly. They observe the concentration of the red ink by evaluating the RGB color space.

The observations were fed into the phenomenon model, resulting in a surface representing the phenomenon and a surface of the information deficit (kriging variance). After an initialization phase of n time steps the optimization was executed for the first time. That means the future information deficit was predicted and one sensor was relocated based on the GSN model. In this first evaluation the GSN model was not implemented as a software module, but was executed manually. Fig. 4 depicts some exemplary results.

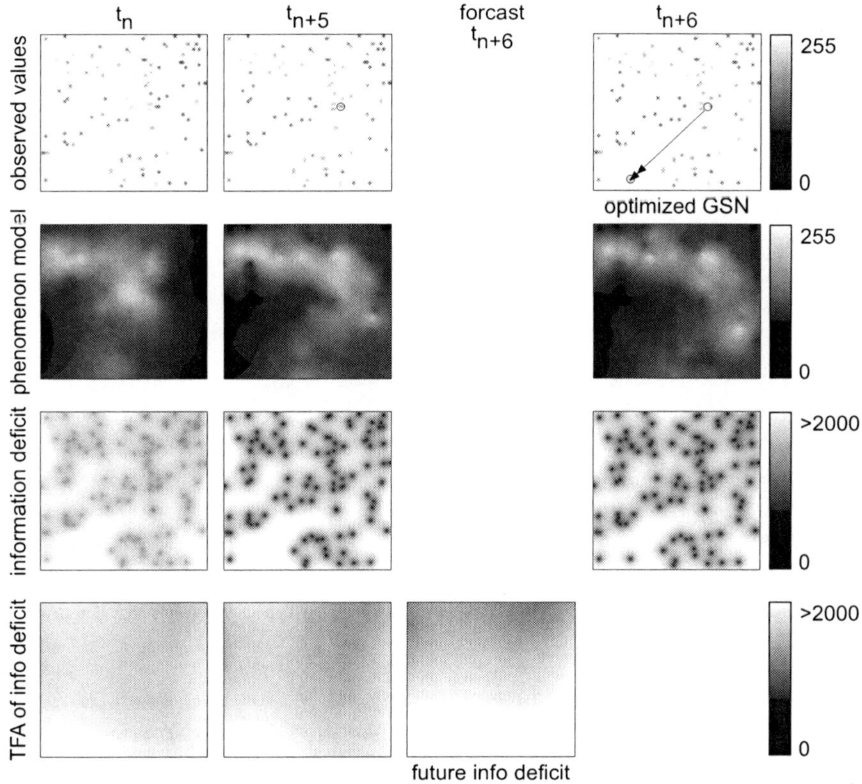


**Fig. 4:** Exemplary results of the first test. The legend of the first two rows refers to the observed ink concentration and in the last two rows to the calculated kriging variance

The first row shows values observed by 100 sensors for three points in time. For $t_n$ the sensors were located randomly, but they were not moved from $t_n$ to $t_{n+5.}$ The phenomenon model – shown in the second row – was calculated by means of an ordinary kriging, which allows for the computation of the information deficit (kriging variance), depicted in the third row. The last row in Fig. 4 lists the results of a trend surface analysis for the information deficit. Based on the future information deficit – shown in the third cell of the last row in Fig. 4 – the geosensor network is optimized. In this evaluation procedure it was assumed that only the sensor marked by a circle could be relocated (see second cell of the first row in Fig. 4). All others are either fixed, or have spatiotemporal waypoints, which prohibit a

relocation of the sensor into the region with high information deficit. In the last column the optimized GSN, the resulting phenomenon model, and the information deficit are presented.
The example shows, that the model based GSN optimization yields in a more efficient observation scheme. The sensors are relocated in order to observe the phenomenon at locations where information is needed. The reduced information deficit in Fig. 4 (fourth cell of the third row) shows that the sensors are used more efficiently after the optimization of the network constellation.

## 5 Conclusion

The optimization method for GSN presented in this paper allows the explicit integration of phenomenon characteristics and geosensor network properties in order to reach an optimal solution. It relies on the minimization of the kriging variance which is used as a measure for the lack of information. By modeling the properties of sensor nodes using time geography it is made possible to consider the limitations of sensors during the optimization process.

The work presented is at an early stage. Future research will have to deal with the following steps:

- Implementation of the GSN model. For the evaluation presented in the previous section only the phenomenon model was implemented into a software module. The GSN model was realized manually.
- Practical evaluation in a real scenario. At the moment the moment the phenomenon was simulated by means of a table top experiment. As a next step an evaluation based on real measurements of a mobile GSN is planned.
- Comparison of the phenomenon model with the real phenomenon. In order to evaluate the usefulness of the kriging variance as single indicator for information deficit, the real phenomenon and the interpolated phenomenon model should be compared.
- Comparison of the model based approach with conventional approaches. Such a comparison will show, if the effort of the presented approach can be justified.

Nevertheless, the presented work shows that the monitoring process of spatiotemporal phenomena could be optimized by combining a phenomenon model with a GSN model.

## Acknowledgements

This material is based upon work partially supported by the German Federal Ministry of Education and Research under grant number 01|S07009A and the SAP AG. The author owes special thanks to colleagues from the Institute for Geoinformatics, who have provided valuable comments on the author's ideas. In addition the author thanks the anonymous reviewers who helped to improve the paper.

## References

Box G. E. P., Jenkins G. M. and Reinsel G. (1994) Time Series Analysis - forecasting and control. Upper Saddle River

Brooks R. R. (2004) Need for Self-Configuration. In: Iyengar S. S. and Brooks R. R. (eds) Distributed Sensor Networks. Boca Raton, pp 847-854

Cressie N. A. C. (1990) The Origins of Kriging. Mathematical Geology 22:239-252

Cressie N. A. C. (1993) Statistics for spatial data. New York

Gandin L. S. (1963) Objective Analysis of Meteorological Fields. Leningrad (Translated by Israel Program for Scientific Translations, Jerusalem, 1965)

Goldin D., Song M., Kutlu A., Gao H. and Dave H. (2005) Georouting and Delta-Gathering: Efficent Data Propagation Techniques for GeoSensor Networks. In: Stefanidis A. and Nittle S. (eds) GeoSensor Networks. Boca Raton, pp 73-95

Hägerstrand T. (1970) What About People in Regional Science? Papers of the regional science association 24:6-21

Heinrich U. (1992) Zur Methodik der räumlichen Interpolation mit geostatistischen Verfahren - Untersuchungen zur Validität flächenhafter Schätzungen diskreter Messungen kontinuierlicher raumzeitlicher Prozesse. Wiesbaden

Howard A., Mataric M. J. and Sukhatme G. S. (2002) Mobile Sensor Network Deployment using Potential Fields: a Distributed, Scalable Solution to the Area Coverage Problem. Proceedings of: 6th International Symposium on Distributed Autonomous Robotics Systems, June 25-27, 2002. Fukuoka.

Huijbregts C. J. (1975) Regionalized Variables and Quantitative Analysis of Spatial Data. In: Davis J. C. and McCullagh M. J. (eds) Display and Analysis of Spatial Data. London, pp 28-53

Isaaks E. H. and Srivastava R. M. (1989) An Introduction to Applied Geostatistics. Oxford

Iyengar S. S., Tandon A. and Brooks R. R. (2004) An Overview. In: Iyengar S. S. and Brooks R. R. (eds) Computer and Information Science Series. Boca Raton, pp 3-10

Journel A. G. (1986) Geostatistic: Models and Tools for the Earth Sciences. Mathematical Geology 18:119-140

Journel A. G. and Huijbregts C. J. (1997) Mining Geostatistics. London

Lenntrop B. (1976) Paths in space-time environments - a time-geographic study of movement possibilities of individuals. Lund

LUA-NRW (2001) Luftqualität in Nordrhein-Westfalen. LUQS-Jahresbericht 1999. City

Matheron G. (1963) Principles of Geostatistics. Economic Geology 58:1246-1266

Matheron G. (1971) The Theory of regionalized variables and its applications. Fontainebleau

Meguerdichian S., Koushanfar F., Potkonjak M. and Srivastava M. B. (2001) Coverage Problems in Wireless Ad-hoc Sensor Networks. Proceedings of: InfoCom 2001. 20th Annual Joint Conference of the IEEE Computer and Communications Societies, April 22-26, 2001. Anchorage.

Nipper J. and Streit U. (1982) A comparative study of some stochastic methods and autoprojective models for spatial processes. Environment and Planning A 14:1211-1231

Nittel S., Duckham M. and Kulik L. (2004) Information Dissemination in Mobile Ad-Hoc Geosensor Networks. In: Egenhofer J. M., Freksa C. and Miller H. J. (eds) Geographic Information Science - Third International Conference, GIScience 2004 Adelphi, MD, USA, October 2004 Proceedings. Berlin and Heidelberg, pp 206-222

Nittel S. and Stefanidis A. (2005) GeoSensor Networks and Virtual GeoReality. In: Stefanidis A. and Nittle S. (eds) GeoSensor Networks. Boca Raton, pp 1-9

Pardo-Igúzquiza E. and Dowd P. A. (2005) Multiple indicator cokriging with application to optimal sampling for environmental monitoring. Computers & Geosciences 31:1-13

Shepard D. (1968) A two-dimensional interpolation function for irregularly-spaced data In: Blue R. B. and Rosenberg A. M. (eds) Proceedings of the 23rd ACM national conference New York, pp 517-524

Shepherd D. and Kumar S. (2004) Microssensor Applications. In: Iyengar S. S. and Brooks R. R. (eds) Distributed Sensor Networks. Boca Raton, pp 11-27

Sliwinski A. and Simonis I. (2005) An Experiment on Geosensor Mobility Strategies in the Planar Space. Proceedings of: 8th AGILE Conference on GIScience, May 26-28 2005. Estoril, Portugal.

Szewczyk R., Osterweil E., Polastre J., Hamilton M., Mainwaring A. and Estrin D. (2004) Habitat Monitoring with Sensor Networks. Communications of the ACM 47:34-40

Thiessen A. H. (1911) Precipitation average for large area. Monthly weather review 39:1082-1084

Wang K.-C. and Ramanathan P. (2004) Location-Centric Networking in Distributed Sensor Networks. In: Iyengar S. S. and Brooks R. R. (eds) Distributed Sensor Networks. Boca Raton, pp 555-571

Webster R. and Oliver M. A. (2001) Geostatistics for Environmental Scientists. Chichester

Zou Y. and Chakrabarty K. (2003) Sensor Deployment and Target Localization Based on Virtual Forces. Proceedings of: InfoCom 2003, Twenty-Second Annual Joint Conference of the IEEE Computer and Communications Societies. March 30-April 3, 2003. San Francisco.

# Evaluation of the Geometric Accuracy of Automatically Recorded 3D – City Models Compared to GIS-Data

Gerald Gruber, Christian Menard, Bernhard Schachinger

Department of Geoinformation, Carinthian University of Applied Sciences, Europastrasse 4, A-9524 Villach/St. Magdalen, g.gruber@fh-kaernten.at

**Abstract.** In this paper we propose methods for evaluating the geometric accuracy of three-dimensional city models. The approach based on the concept of an error matrix, statistical analysis of height differences and a buffer-overlay-statistic leads to accuracy parameters for an automated city modelling workflow from aerial images[1]. We study the theoretical properties of our approach and we show that in practice the concept of the paper behaves very well on real test data from the 3D-model of the inner city of Graz. The work concludes with a summary and an outlook to future work.

**Keywords:** 3D city models, accuracy, buildings, geostatistics

## 1 Introduction

3D-city modelling is a strong growing and dynamic research area in the field of geographic information science (Hu et al. 2003). 3D-city models are useful in various industry lines, e.g. in telecommunication, utility management, municipalities, security, tourism and marketing (Batty et al. 2000). The data acquisition techniques for these models are using photo-

[1] The modelling workflow comes from *Microsoft Photogrammetry* which is a research and development unit of Microsoft. It emerged from the merger of Microsoft and Vexcel Imaging GmbH.

grammetry, active sensors or hybrid sensors. Moreover hybrid systems can combine for example aerial images with CAD-data (Hu et al. 2003).

The diverse modelling techniques vary in the extent of the human interaction required in the modelling process. Kocaman et al. (2006) present a manual workflow, while Suveg and Vosselman (2000) elaborate on a semi-automatic modelling workflow. Karner et al. (2006) and Zebedin et al. (2006) present an automatic workflow without any human interaction throughout the entire modelling process. The models proposed by Karner et al. (2006) are incorporated into Microsoft® Virtual Earth™.

In an automated modelling workflow the accuracy of the 3D models are of utmost importance. In this paper we propose a standardized method for accuracy evaluation of 3D-city models generated by Karner et al. (2006). The test dataset used in this work is the city model of Graz provided by *Microsoft Photogrammetry* (see Fig. 1).

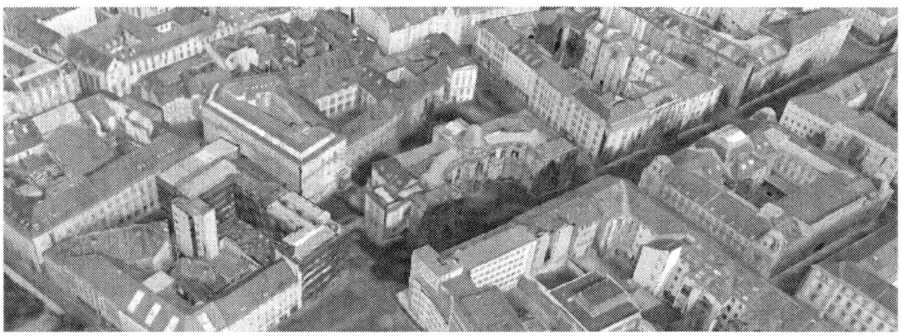

**Fig. 1:** Detail from the 3D – city model of Graz.

## 2 Overview of 3D – City Modelling and Microsoft Photogrammetry's Modelling Workflow

A 3D-city model is a digital representation of spatial georeferenced data. A city model usually is composed of the terrain, buildings, vegetation and transportation. The main purpose of it is presenting, analyzing and managing the data of a city. According to this, the model can either show a high degree of photorealism or can be abstracted to a thematic representation (Döllner et al. 2006). Overviews about 3D-city models are given e.g. in Baltsavias (2004), Hu et al. (2003) and Batty et al. (2000). They discuss topics like the usage of 3D-city models, the source data, techniques for producing such models and difficulties when generating and using these models. In this work we use data from intermediate steps of the automatic modelling workflow of *Microsoft Photogrammetry,* see Zebedin et al. (2006) and Karner et al. (2006), which is described in the following paragraph.

The input data for the modelling workflow of *Microsoft Photogrammetry* are panchromatic, RGB and Near Infrared images provided by the airborne camera UltraCamX of Vexcel Imaging GmbH. These aerial images are recorded at a flight height of about 1000 meters above ground with a ground sampling distance of 8 cm and an overlapping area of 80% in flight direction and 60% sideways. The test data set concerning the inner city of Graz/Austria was acquired in summer 2005. An example is shown in Fig. 2.

**Fig. 2:** Image data delivered by UltraCamX (Zebedin et al. (2006)).

The initial step in the modelling workflow involves manual input for training a classification algorithm. This algorithm classifies the test area into the categories solid objects, roofs, soil, water, vegetation, dark shadows and swimming pools. In the next step an aerial triangulation is established automatically by extracting points occurring in several images, which leads to a *Digital Surface Model* (DSM) calculation. This DSM is used for generating true ortho images. True ortho images are aerial images having no distortions caused by height differences of the underlying terrain and do not show opaque walls of buildings. These true ortho images are used for refining the initial classification from the first step. The DSM together with the refined classification is then used for fitting planes and extracting building blocks in vector form. Finally a *Digital Terrain Model* (DTM) is calculated by subtracting the objects found in the previous step from the DSM. The final model can be textured with the data from the input images.

Fig. 3 shows an example of a final building block which was modelled using the proposed technique.

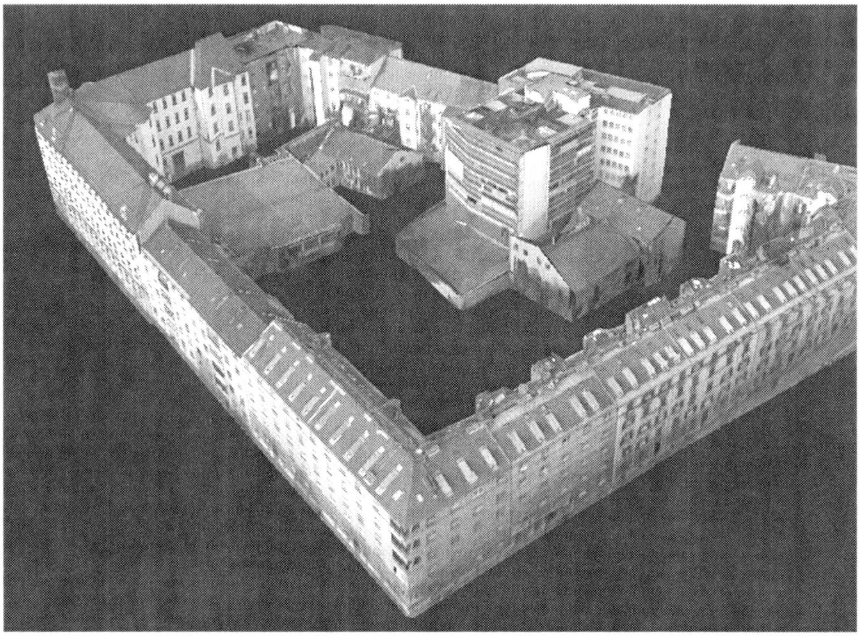

**Fig. 3:** Building block modelled from aerial images.

The following data sets are used for the accuracy assessments:

- results of the initial classification,
- the height model (DSM) and the
- modelled building footprints.

These datasets are chosen for the evaluation due to the effect that inaccuracies in these datasets causes incorrectly modelled objects in the 3D-model output. If there are erroneous classifications in the images the objects like buildings or vegetation would be extracted at wrong positions. Inaccurate height values in the DSM would cause wrong fitting of roof and façade planes of the buildings, which would result in footprints that do not represent the real buildings. A study by the *European Organization for Experimental Photogrammetric Research* (OEEPE) showed that 95% of the users of a city model are most interested in representations of buildings (Förstner 1999). For that reason our accuracy assessment methods concentrate mainly on the feature class buildings.

The classification results are compared to a reference raster by analyzing the consistency of the data sets in a confusion matrix. This approach is described in Longley et al. (2001) and Schneider (2004). The height model is analyzed by applying the methods proposed in Czegka et al. (2005) and Gspurnig and Sulzer (2004). The main idea of this method is to statistically analyse the height difference of sample points between the height model and reference data. Finally the positional accuracy of the modelled building footprints is assessed by performing techniques introduced by Goodchild and Hunter (1997) and Tveite and Langaas (1999). These methods are calculating the percentage of lines to be tested lying inside of buffer zones, which have varying width around the reference data.

The underlying concept of these accuracy assessment methods, the process of comparing test data and reference data can also be found in Kaartinen et al. (2005), where similar assessment methods are used to determine the suitability of various kinds of input data.

## 3 Methods for Accuracy Assessment

According to Schneider (2004) accuracy is a measure for the consistency of test data and reference data, which is assumed to represent the reality correctly. Thus the presented assessment methods require appropriate reference data, which are provided by the municipal surveying office of Graz and acquired through conventional surveying methods.

The results of the classification step are analyzed using an error matrix which is a common tool for comparing two raster datasets and analyzing the consistency (Longley et al. 2001). It shows the number of the correctly classified and the misclassified pixels and results in the *Percentage of Correctly Classified* pixels (PCC). Additionally the Kappa-Index is calculated, which gives information about the correctly classified pixels excluding the percentage of pixels classified correctly by random (Schneider 2004). Thus the accuracy of the classification is assessed by calculating the PCC and the Kappa-Index.

The accuracy of the height model is determined by calculating the height difference between reference points and height values on manually selected positions. The manual selection is essential to avoid an erroneous comparison of height values on positions with disturbing objects like cars. According to Gspurning and Sulzer (2004) and Czegka et al. (2005) the statistical measures of mean, standard deviation and *Root Mean Square Error* (RMSE) are sufficient for an analysis of the height differences. A detailed explanation of these measures can be found in Longley et al. (2001).

In general the positional accuracy of an object is determined by the difference between the represented position and the true position (Goodchild and Hunter 1997). The analysis of the positional accuracy of the building footprint lines requires a more sophisticated approach since there are no comparable elements available. The difficulty arises from the fact that the numbers of vertices in the test data set and in the reference data set must not be equal and thus the number of lines can vary as well. According to Tveite and Langaas (1999) the positional accuracy of lines is defined as

- the correspondence of well-defined points which can be identified in both datasets and the
- consistency of the shape of the lines.

The first component can be determined by manually selecting well-defined corresponding points in both datasets (e.g. building corners) and calculating the radius of a mean error circle according to Cialek et al. (1999):

$$r_{error} = \sqrt{\frac{\sum_{i=1}^{n}((x_{reference_i} - x_{test_i})^2 + (y_{reference_i} - y_{test_i})^2)}{n}} \tag{1}$$

where $x_{reference}$ and $y_{reference}$ are the XY-coordinates of the *i-th* reference point, $x_{test}$ and $y_{test}$ are the XY-coordinates of the *i-th* tested point, $n$ denotes the number of points to be tested and $r_{error}$ is the radius of the mean error circle. The shape of the lines is determined using a method based on Buffer-Overlay-Statistics (BOS) by Tveite and Laangas (1999), which is illustrated in Fig. 4.

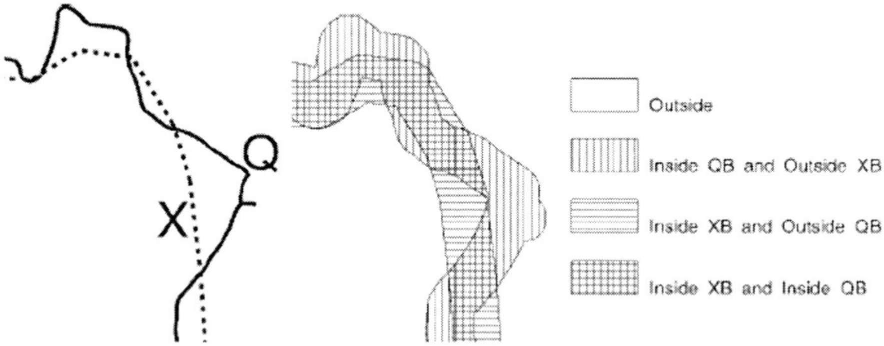


**Fig. 4:** Generation of buffer areas around the line features and calculation of the overlaying percentage (Tveite and Langaas, 1999)

The positional accuracy of the lines can be calculated by using the following equation for varying buffer width:

$$AverageDisplacement_i = \pi \cdot b_i \cdot \frac{Area(\overline{XB_i} \cap QB_i)}{Area(XB_i)} \tag{2}$$

where $b_i$ is the *i-th* buffer width, $Area(\overline{XB_i} \cap QB_i)$ is the area outside the buffer around the line to be tested (*X*) but inside the buffer around the reference data (*Q*), $Area(XB_i)$ is the buffer area around the data to be tested and *AverageDisplacement* is the positional accuracy. Additionally these areas are used for calculating values for *Completeness*, *Miscodings* and *Oscillation* for each buffer width. The value for average displacement gives the positional accuracy when either all of the lines to be tested lie within the buffer areas or a previously defined percentage of lines is covered by the buffer areas. A visual inspection of the test data shows that many differences arise between these data sets due to different acquisition situations and missing corresponding representation rules (e.g. the test data contains balconies whereas those are not in the reference data). This inspection shows that a distance of more than 1.20m in between the two datasets indicates a difference in the data which is not necessarily an error. These distant line segments should be excluded before the algorithm of Tveite and Langaas (1999) is applied.

## 4 Results of the Accuracy Assessment

The methods for accuracy assessment presented in the previous section are applied on a 3D-city model of Graz, which is produced by *Microsoft Photogrammetry* in an automated workflow using only aerial images. The evaluation of the classification results is applied on the categories *Buildings* and *Roads*. Table 1 shows the results for *Buildings* and gives a PCC-value of 95% and a Kappa-Index of 89%. The results for the category *Roads* are presented in Table 2 and show a PCC-value of 89% and a Kappa-Index of 71%.

Table 3 gives a statistic of the height differences of 100 manually selected positions in the test area. The mean error is 18 cm. The spatial distribution of these points is depicted in Fig. 5. White points indicate positions with minimum error. Due to less overlapping input images the data inaccuracy increases towards the borders of the test area (see Fig. 5). The positional accuracy of well-defined points of the building footprints results in a mean error radius of 72cm. According to the method of Tveite and Langaas (1999) the positional accuracy of the line shapes is 75cm as shown in Figure 6.

**Table 1** Results of accuracy assessment for the classification raster of category *Buildings*.

| | Others | Buildings | Total | User's A |
|---|---|---|---|---|
| Others | 58% | 2% | 60% | 97% |
| Buildings | 3% | 37% | 40% | 91% |
| Total | 61% | 39% | 100% | |
| Producer's A | 94% | 95% | | 95% (89%) |

**Table 2** Results of accuracy assessment for the classification raster of category *Streets*.

| | Others | Streets | Total | User's A |
|---|---|---|---|---|
| Others | 70% | 5% | 69% | 92% |
| Streets | 5% | 20% | 31% | 79% |
| Total | 75% | 25% | 100% | |
| Producer's A | 93% | 77% | | 89% (71%) |

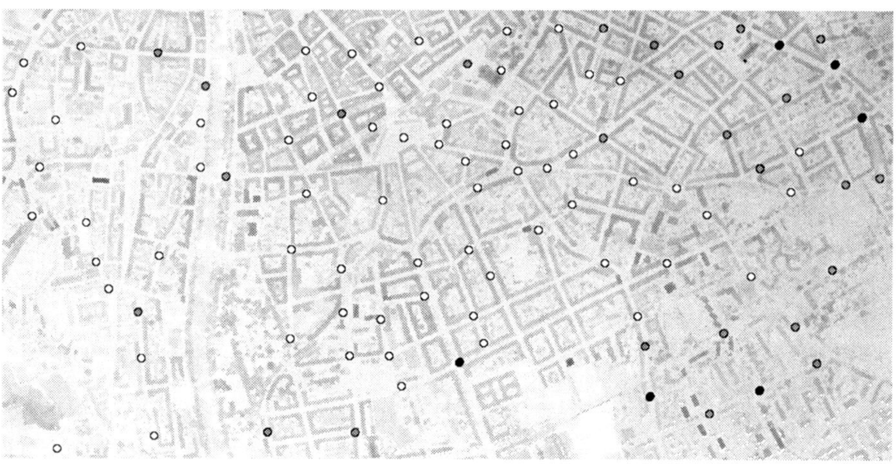

**Fig. 5:** Manually selected points which are used for calculating the height difference. White points show the minimum error.

**Table 3** Statistics of the height differences of the selected points.

| Manual Selection | |
|---|---|
| # Points | 100 |
| Minimum Error | 0.0 m |
| Maximum Error | 0.6 m |
| Mean Error | 0.18 m |
| Standard Deviation | 0.15 m |
| RMSE | 0.24 m |

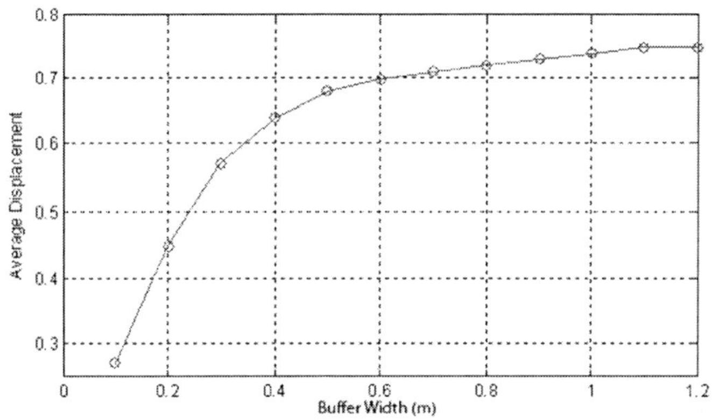


**Fig. 6:** Graph of the average displacement per buffer width.

## 5 Interpretation of the Results

The results of the accuracy assessments must be handled with care. On the one hand the reference data provided has only a positional accuracy of +/-10cm and a height accuracy of +/-20cm. On the other hand there are differences in between the test data and the reference data that influence the accuracy results but are not really errors. These differences come from different acquisition situations and different acquisition criteria like e.g. considering a carport as a building or not. An overview about these error sources in the context of classification analyis is given in Foody (2002).

## 6 Summary and Outlook

Digital 3D-city models are used in various fields and business areas and thus can be built from many different input data sources. So far there exists no standardized method for evaluating the accuracy of a 3D-city model. This work proposes methods for calculating accuracy parameters for a city model which is generated in an automatic modelling workflow by *Microsoft Photogrammetry* only from aerial images.

The methods presented here are chosen, because an error in any of the examined data sets causes a dramatic change in the resulting 3D-model. The correspondence of the classification of the input images in buildings and roads is compared to reference data with a confusion matrix and the Kappa-Index. The PCC-value of the category buildings is 95% while the Kappa-Index is 89%. The category roads has a PCC of 89% compared to a Kappa-Index of 71% The height model is analyzed by comparing it to reference points on selected positions and calculating statistics from the height differences. The mean error of the height model is 18cm. The positional accuracy of building footprints is calculated on the one hand by the mean error radius of well-defined points and on the other hand by a Buffer-Overlay-Statistics Method proposed by Tveite and Langaas (1999). This results in a mean error radius of 72cm and an average displacement of the lines of at most 75cm. Nevertheless the most important quality criterion for the tested 3D-city model for use in Microsoft® Virtual Earth™ a proper visual appearance, is fulfilled. The accuracy assessment shows that data, which is acquired by automatic methods from aerial images corresponds well to conventionally acquired data. The current assessment workflow was performed using various GIS software tools.

One potential in the future is that these tools should be replaced by an incorporation of the accuracy assessment workflow into the modelling software of *Microsoft Photogrammetry*. Further development is directed towards the adaption of the proposed evaluation methods to the data of 3D-city models provided by competitors of *Microsoft Photogrammetry*.

### Acknowledgements

We thank the Austrian research company VRVis GmbH for funding this research and *Microsoft Photogrammetry* and the municipal surveying office of Graz for providing sample data.

## References


Baltsavias, E. P. (2004) Object Extraction and Revision by Image Analysis using existing Geodata and Knowledge: Current Status and Steps towards operational Systems. International Journal of Photogrammetry and Remote Sensing, Vol. 58:pp. 129 – 151.

Batty, M., Chapman, D., Evans, S., Haklay, M., Kueppers, S., Shiode, N., Smith, A. and Torrens P. M. (2000) Visualizing the City: Communicating Urban Design to Planners and Decision-Makers. CASA Working Papers, No. 26, Centre for Advanced Spatial Analysis, London, United Kingdom.

Cialek, C., Elwood, D., Johnson, K., Kotz, M., Krafthefer, J., Maxwell, J., Radde, G., Schadauer, M. and Wencl, R. (1999) Positional Accuracy Handbook. Online: http://server.admin.state.mn.us/pdf/1999/lmic/nssda_o.pdf. Minnesota Planning (Land Management Information Center), Last date accessed: 30.04.2007.

Czegka, W., Braune, S. and Behrends, K. (2005) Validierung der freien C-Band-SRTM-Höhendaten in Hinblick auf Anwendungsmöglichkeiten in den Geo- und Umweltwissenschaften. In Beiträge zum AGIT-Symposium Salzburg 2005, pp. 106 – 111. Wichmann Verlag, Heidelberg.

Döllner, J., Baumann, K. and Buchholz, H. (2006) Virtual 3D City Models As Foundation of Complex Urban Information Spaces. In Proceedings of CORP 2006 & Geomultimedia06, pp. 107 – 112.

Förstner, W. (1999) 3D-City Models: Automatic and Semiautomatic Acquisition Methods. In Proc. Photogrammetric Week '99, pp. 291 – 303. University of Stuttgart, Institute for Photogrammetry.

Goodchild, M. F. and Hunter, G. J. (1997) A Simple Positional Accuracy Measure for Linear Features. International Journal of Geographical Information Science, Vol. 11, No. 3:pp. 299 – 306.

Gspurning, J. and Sulzer, W. (2004) DEM-Generierung aus ASTER-Daten und Evaluierung. In Beiträge zum AGIT-Symposium 2004 Salzburg, pp. 190 – 195. Wichmann Verlag, Heidelberg.

Hu, J., You, S. and Neumann, U. (2003) Approaches to Large-Scale Urban Modeling. IEEE Computer Graphics & Applications, Vol. 23, No. 6:pp. 62 – 69.

Kaartinen, H., Hyyppä, J., Gülch, E., Vosselman, G., Hyyppä, H., Matikainen, L., Hofmann, A. D., Mäder, U., Persson, Å., Söderman, U., Elmqvist, M., Ruiz, A., Dragoja, M., Flamanc, D., Maillet, G., Kersten, T., Carl, J., Hau, R., Wild, E., Frederiksen, L., Holmgaard, J., and Vester, K. (2005) Accuracy of 3D City Models: EuroSDR Comparison. ISPRS Workshop Laser Scanning 2005, pp. 227 - 232.

Karner, K., Hesina, G., Maierhofer, S. and Tobler, R. F. (2006) Improved Reconstruction and Rendering of Cities and Terrains based on Multispectral Digital Aerial Images. In Proceedings of CORP 2006 & Geomultimedia06, pp. 299 – 304.

Kocaman, S., Zhang, L., Grün, A. and Poli, D. (2006) 3D City Modeling from High-Resolution Satellite Images. In Proceedings of ISPRS Workshop on Topographic Mapping from Space 2006, Proceedings in CD-ROM.

Longley, P. A., Goodchild, M. F., Maguire, D. J. and Rhind, D. W. (2001) Geographic Information Systems and Science. John Wiley & Sons, Ltd.

Schneider, S. (2004) Evaluation of new Classification Methods for X- and P-Band SAR Images. Diploma Thesis, Carinthia University of Applied Sciences.

Suveg, I. and Vosselman, G. (2000) 3D Reconstruction of Building Models. In Proceedings of the XIXth Congress of ISPRS, volume XXXIII-Part BL, pp. 538 – 545.

Tveite, H. and Langaas, S. (1999) An Accuracy Assessment Method for Geographical Line Data Sets Based On Buffering. International Journal of Geographical Information Science, Vol. 13, No. 1: pp. 27 – 47.

Zebedin, L., Klaus, A., Gruber-Geymayer, B. and Karner, K. (2006) Towards 3D Map Generation from Digital Aerial Images. International Journal of Photogrammetry and Remote Sensing, Vol. 60, No. 6:pp. 413 – 427.

# Lifting Imprecise Values

Gerhard Navratil, Farid Karimipour, Andrew U. Frank

Institute for Geoinformation and Cartography, Vienna University of Technology, Gusshausstr. 27-29, A-1040 Vienna, Austria,
[navratil,karimipour,frank]@geoinfo.tuwien.ac.at

**Abstract.** The article presents a conceptual framework for computations with imprecise values. Typically, the treatment of imprecise values differs from the treatment of precise values. While precise computations use a single number to characterize a value, computations with imprecise values must deal with several numbers for each value. This results in significant changes in the program code because values are represented, e.g., by expectation and standard deviation and both values must be considered within the computations. It would be desirable to have a solution where only limited changes in very specific places of the code are necessary. The mathematical concept of lifting may lead to such a solution.

**Keywords:** imprecise values, error propagation, lifting

## 1 Introduction

The integration of quality descriptions is one of the most important practical problems for the GIS research and development community. All data in a GIS have limited precision leading to a corresponding level of uncertainty about the true values. These uncertainties spread if the data are used for other computations. The determination of the result's uncertainty is crucial for the user. Results are only meaningful if the possible deviations of the result do not change the result significantly. Assessment of the uncertainty of the result requires knowledge on the uncertainty of the original data. This knowledge must then be carried along with each processing step.

We propose a conceptual framework that deals with the problem of error propagation in a mathematically clean and simple way. Frank demonstrated the use of functors to lift map algebra from a pure spatial context to a spatio-temporal context (Erwig and Schneider 1999; Frank 2005). The same concept has been used to lift the basic algebraic operations for numbers to normally distributed values (Navratil 2006). Since normally distributed values are only one kind of description for imprecision we extend the concept to other kinds of descriptions.

The article is structured as follows: In section we discuss different approaches to describe imprecise values. Section 3 shows how errors propagate and how the result can be computed. Sections 4 and 5 introduce the concept of lifting and show an implementation of lifting for error propagation. The paper concludes with a discussion of aspects that need to be addressed in the future.

## 2 Imprecise Values

The results of measurements are not precise numbers (Viertl 2002). The limitation of precision propagates when using these values in mathematical models. Several methods have been developed to cope with that problem. We arbitrarily selected the three different models normally distributed values, intervals, and fuzzy values for the discussion in the remainder of the paper. Other models, e.g., for skewed models were left for future research.

### 2.1 Normally Distributed Values

Observations of geometric qualities like distances or angles are usually assumed to be normally distributed. Maybe even other physical qualities like density can be assumed to be normally distributed. The reason for this assumption is the central limit theorem. It states that, if the sum of many independent and identically distributed variables has a finite variance, then the sum will be approximately normally distributed. Two variables are independent if the probability for the occurrence of one event is independent from the result of the other event. An example for independent variables is the numbers resulting from rolling dices. Weather conditions in a specified location on successive days are dependent events.

Normal distribution is defined by two parameters, the expected value $\mu$ and the statistical dispersion $\sigma$. Measures for the statistical dispersion are variance and standard distribution as the positive square root of the variance. In the following the variance will be used.

Normally distributed values follow the density function

$$f(x)=\frac{1}{\sigma\sqrt{2\pi}}e^{-\frac{(x-\mu)^2}{2\sigma^2}} \tag{1}$$

The area below the density function is a probability measure. The area for an arbitrary interval is the probability that a randomly picked value belonging to this distribution lies within the interval. The density function is defined for the interval ]-∞,+∞[ and is symmetric (compare Fig. 1).

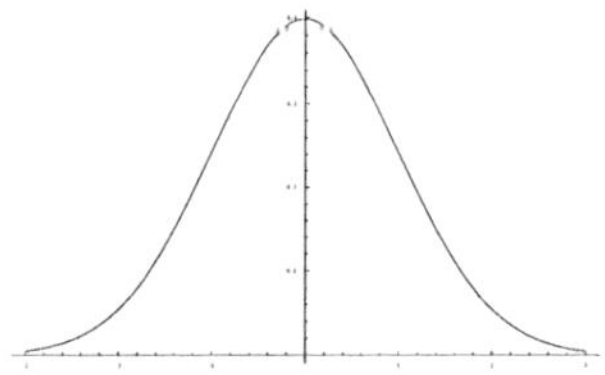

**Fig. 1:** Density function for a normalized (μ = 0, σ = 1) normal distribution. The horizontal axis is the outcome *x* and the vertical axis specifies the density *f(x)*.

Measures for dependency are covariance and correlation. Covariance describes the common variation of two observations. The units of measurement depend on the units of measurement of the observations. The correlation on the other hand is dimensionless and describes the linear dependence between the two observations. The Pearson product-moment correlation coefficient $r_{xy}$ is defined as:

$$r_{xy}=\frac{\sigma_{xy}}{\sqrt{\sigma_x^2}\sqrt{\sigma_y^2}} \tag{2}$$

Independent parameters have a correlation $r_{xy}=0$ and a covariance $\sigma_{xy}=0$.

## 2.2 Fuzzy Values

Fuzzy values or fuzzy numbers are fuzzy sets whose members are real numbers in which uncertainty is represented through a non-probabilistic way (Siler and Buckley 2005). For example, suppose you are driving and trying to keep the speed at exactly 90 km/h. In practice the speed will vary and it will be a fuzzy value around 90.

Mathematically, a fuzzy value is a function (called membership function) $\mu(x): R \to [0,1]$, which relates each number to its grade of membership. Generally, this function may have any shape. The complexity of the operations on fuzzy values depend on the shape: The more irregular the membership function the more complicated the calculations (Fodor and Bede 2006).

The so-called *L-R fuzzy values* are one of the most important and practical types of fuzzy numbers. For an L-R fuzzy value $a$, membership function is defined as follows (Fodor and Bede, 2006):

$$\mu(x) = \begin{cases} L(\frac{\hat{a}-x}{\underline{a}}) & x \leq \hat{a} \\ R(\frac{x-\hat{a}}{\overline{a}}) & x > \hat{a} \end{cases} \tag{3}$$

where $L, R : [0,+\infty[ \rightarrow [0,1]$ are two continuous, decreasing functions fulfilling $L(0) = R(0) = 1$ and $L(1) = R(1) = 0$, $\hat{a}$ is a real number with $\mu(\hat{a}) = 1$ and $\overline{a}$, $\underline{a}$ are two positive real numbers for which $\mu(\overline{a}) = \mu(\underline{a}) = 0$.

If functions $L$ and $R$ are linear, a *trapezoidal fuzzy value* will be obtained (Figure 2) represented by a quadruple $(a,b,c,d), a \leq b \leq c \leq d$.

$$\mu(x) = \begin{cases} 0 & x < a \\ \frac{x-a}{b-a} & a \leq x < b \\ 1 & b \leq x \leq c \\ \frac{d-x}{d-c} & c < x \leq d \\ 0 & x > d \end{cases} \tag{4}$$

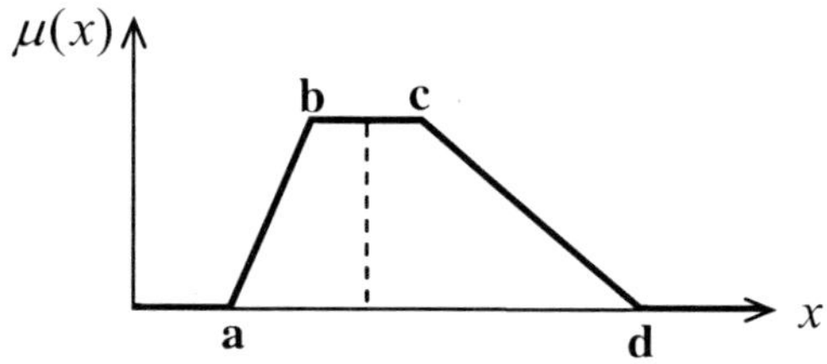


**Fig. 2:** A trapezoidal fuzzy number represented by (a, b, c, d)

As a particular case, a triangular fuzzy value (Figure 3) is a trapezoidal fuzzy value in which the values $b$ and $c$ are equal. A triangular fuzzy value can be represented either by a triple $(a,b,c)$ or a quadruple $(a,b,b,c), a \leq b \leq c$. However, quadruple representation allows us to have a unified algebra.

$$\mu(x)=\begin{cases}0 & x<a\\ \dfrac{x-a}{b-a} & a\le x<b\\ \dfrac{c-x}{c-b} & b\le x<c\\ 0 & x>c\end{cases} \tag{5}$$

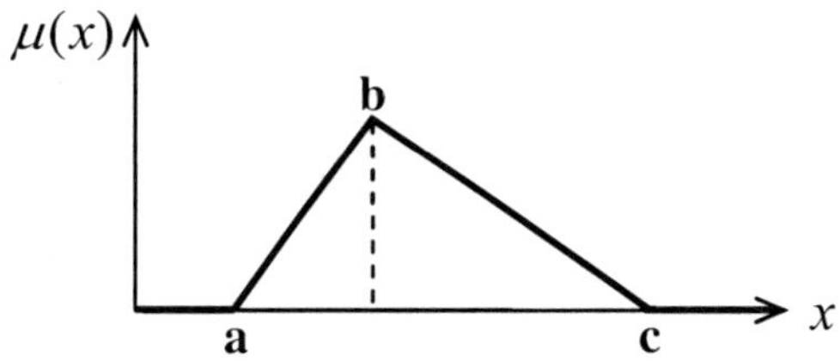


**Fig. 3:** A triangular fuzzy number represented by (a, b, c) or (a, b, b, c)

## 2.3 Interval Values

If we can state with equal confidence that a value lies somewhere between 'a' and 'b', it can be referred to as an interval [a, b] in which the occurrence probability for all elements are the same (Hayes 2003). For example, if you are measuring a distance, you can never be certain about the result, but you may be able to say that it is certainly between 76 and 78 meters.

As Figure 4 shows, an interval value can be considered as a very simple fuzzy value with a binary membership function $\mu(x): R \to \{0,1\}$ :

$$\mu(x)=\begin{cases}1 & a\le x\le b\\ 0 & elsewhere\end{cases} \tag{6}$$

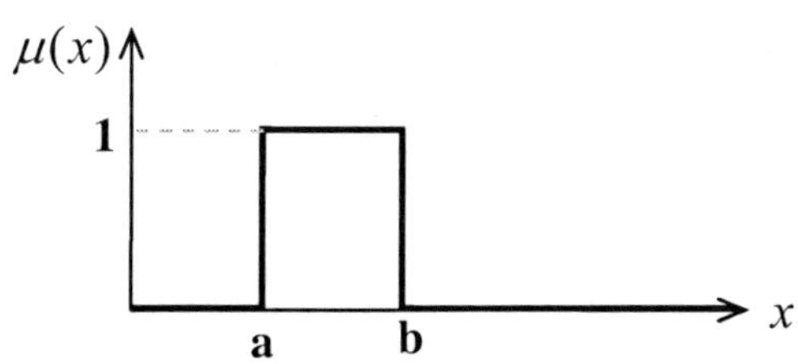


**Fig. 4:** An Interval value represented by [a, b]

Sometimes an interval is shown by the notation $[\underline{x}, \overline{x}]$ to emphasize the lower and upper limitations (Hayes 2003).

## 3 Propagation of Imprecision

Observations are used for analysis and the results are often used in decision-making processes. Imprecision in the observations will propagate through the analysis and result in imprecise results. Knowledge on the imprecision is necessary for consideration in the decision-making. Much work has been done in the field of imprecision and error propagation (Heuvelink 1998; Bachmann and Allgöwer 2002; Heuvelink and Burrough 2002; Karssenberg and de Jong 2005). In the following we show the basic operations used for the three types of imprecise values.

### 3.1 Operations on Normally Distributed Values

Let us assume we have normally distributed values $x_1, \ldots, x_n$ and apply a function to these values. The function result is determined by applying the function to the expected values, but what will be the variation? There are several methods to assess the result of error propagation (Heuvelink 1998, pp. 36-42). The mathematically correct solution is the strict, algebraic computation of the distribution, which results from applying the function. However, the practicality of the solution is hampered by its complexity.

A different approach is the first order Taylor series method, which assumes that the functions can be differentiated. The first order Taylor series method replaces the function by its tangent and produces useful results if the values contain only small deviations. The computation of the first order Taylor series method for a function *f* is defined as

$$\sigma_f^2 = \mathbf{F}^T \mathbf{\Sigma}_{xx} \mathbf{F} \tag{7}$$

where **F** is the Jacobi matrix for the function *f* and $\mathbf{\Sigma_{xx}}$ is the variance-covariance matrix for the parameters.

The assumption of independent parameters simplifies the computation. The variance-covariance matrix becomes a matrix in diagonal form and (7) can be simplified to

$$\sigma_f^2 = \sum_i \sigma_i^2 \left( \frac{\partial f}{\partial x_i} \right)^2 \tag{8}$$

Monte Carlo simulation is a numerical solution and leads to the same result as the mathematically correct solution if the number of computations is large enough. However, computational costs are high since computations must be repeated frequently to assess the distribution of the result.

## 3.2 Operations on Fuzzy Values

To operate on fuzzy values with general membership functions, we need the concept of ▯-cut interval, which is defined for a fuzzy value $u$ and $0 < \alpha \leq 1$ as shown in Figure 5. The definition is (Fodor and Bede 2006):

$$[u]^\alpha = [\underline{u}^\alpha, \overline{u}^\alpha] = \{x \in R \mid \mu(x) \geq \alpha\} \tag{9}$$

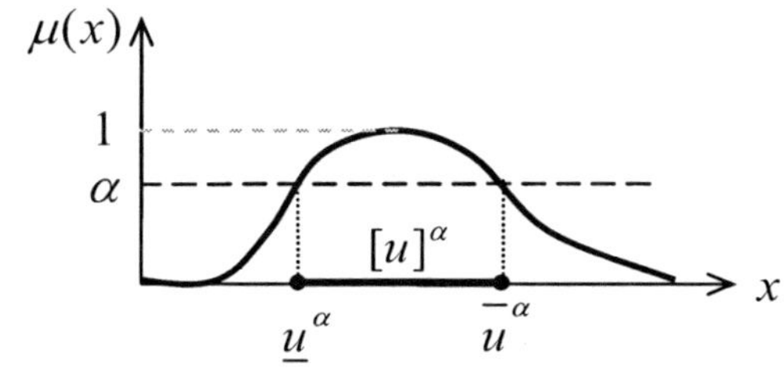


**Fig. 5:** The concept of an ▯-cut interval

Then if $u$ and $v$ are two fuzzy values and $f$ and $\circ$ are two operators with one and two operands, respectively, the following rules are used to find the result for an ▯-cut (Fodor and Bede 2006):

$$\begin{aligned} \underline{f(u)}^\alpha &= \min\left\{ f(\underline{u}^\alpha), f(\overline{u}^\alpha) \right\}, \\ \overline{f(u)}^\alpha &= \max\left\{ f(\underline{u}^\alpha), f(\overline{u}^\alpha) \right\}, \end{aligned} \tag{10}$$

$$\begin{aligned} \underline{(u \circ v)}^\alpha &= \min\left\{ \underline{u}^\alpha \circ \underline{v}^\alpha, \underline{u}^\alpha \circ \overline{v}^\alpha, \overline{u}^\alpha \circ \underline{v}^\alpha, \overline{u}^\alpha \circ \overline{v}^\alpha \right\}, \\ \overline{(u \circ v)}^\alpha &= \max\left\{ \underline{u}^\alpha \circ \underline{v}^\alpha, \underline{u}^\alpha \circ \overline{v}^\alpha, \overline{u}^\alpha \circ \underline{v}^\alpha, \overline{u}^\alpha \circ \overline{v}^\alpha \right\}. \end{aligned} \tag{11}$$

The lists in (11) result from a Cartesian product. The result of a Cartesian product $u \times v$ using an operation $\circ$ is a list of all possible combinations $(u_i \circ v_i)$ of elements $u_i$ from $u$ with elements $v_i$ from $v$. The elements of $u$ in this case are

$$u = \left\{ \underline{u}^\alpha, \overline{u}^\alpha \right\} \tag{12}$$

and the elements of $v$ are

$$v = \min\left\{\underline{v}^{\alpha}, \overline{v}^{\alpha}\right\}. \tag{13}$$

The results of applying an operation on fuzzy values are as follows (Fodor and Bede 2006):

$$f(u) = sort\left\{\underline{f(u)}^{0}, \overline{f(u)}^{0}, \underline{f(u)}^{1}, \overline{f(u)}^{1}\right\}, \tag{14}$$

$$(u \circ v) = sort\left\{\underline{(u \circ v)}^{0}, \overline{(u \circ v)}^{0}, \underline{(u \circ v)}^{1}, \overline{(u \circ v)}^{1}\right\}. \tag{15}$$

In the case of trapezoidal fuzzy values, $[u]^0 = [a,d]$ and $[u]^1 = [b,c]$. So the above formulas can be simplified (Fodor and Bede 2006):

$$f(u) = sort\{f(a), f(b), f(c), f(d)\}, \tag{16}$$

$$\begin{aligned}
\underline{(u \circ v)}^{0} &= \min\{a_u \circ a_v, a_u \circ d_v, d_u \circ a_v, d_u \circ d_v\},\\
\overline{(u \circ v)}^{0} &= \max\{a_u \circ a_v, a_u \circ d_v, d_u \circ a_v, d_u \circ d_v\},\\
\underline{(u \circ v)}^{1} &= \min\{b_u \circ b_v, b_u \circ c_v, c_u \circ b_v, c_u \circ c_v\},\\
\overline{(u \circ v)}^{1} &= \max\{b_u \circ b_v, b_u \circ c_v, c_u \circ b_v, c_u \circ c_v\},
\end{aligned} \tag{17}$$

$$(u \circ v) = sort\left\{\underline{(u \circ v)}^{0}, \overline{(u \circ v)}^{0}, \underline{(u \circ v)}^{1}, \overline{(u \circ v)}^{1}\right\}. \tag{18}$$

The result of some operations (e.g., + and -) on trapezoidal fuzzy values are also trapezoidal. However, for other operations (e.g., * and ÷) it is non-trapezoidal. In such cases, the non-trapezoidal shape of the result can be usually opted with a trapezoid although repeated operations may increase the uncertainty (Fodor and Bede 2006).

### 3.3 Operations on Interval Values

Operations on interval values are defined as follows (Hayes 2003):

$$f(u) = f[\underline{u},\overline{u}] = [\min\{f(\underline{u}), f(\overline{u})\}, \max\{f(\underline{u}), f(\overline{u})\}], \tag{19}$$

$$(u \circ v) = [\underline{u},\overline{u}] \circ [\underline{v},\overline{v}] =$$
$$[\min\{\underline{u} \circ \underline{v}, \underline{u} \circ \overline{v}, \overline{u} \circ \underline{v}, \overline{u} \circ \overline{v}\}, \max\{\underline{u} \circ \underline{v}, \underline{u} \circ \overline{v}, \overline{u} \circ \underline{v}, \overline{u} \circ \overline{v}\}]. \tag{20}$$

## 4 Lifting Error Propagation for Imprecise Values

### 4.1 Mathematical Concept Lifting

Lifting is a mathematical concept emerging from category theory (MacLane and Birkhoff 1999). It uses functors to map objects and functions between these objects from one class to another class (Moggi 1989).

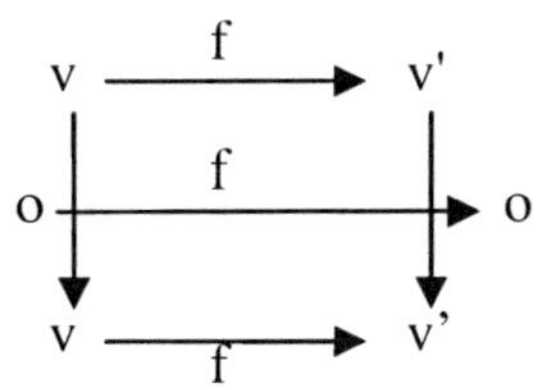


**Fig. 6:** Mapping of precise values to imprecise values and functions on precise values to functions on imprecise values.

Figure 6 shows the basic idea. A function *o* has an object *v* as the argument and returns another object. Assume there is another object *v'* with a mapping function *f* from *v* to *v'*. Lifting then is the concept of using the mapping function *f* not only for the objects *v* and *v'* but also for the function *o*. Thus we can write (compare Marquis 2006)

$$f(o(v)) = o'(f(v)) = o'(v') \tag{21}$$

or in case of two objects *a* and *b*

$$f(o(a,b)) = o'(f(a), f(b)) = o'(a', b'). \tag{22}$$

The mapping function *f* is called a functor. The application of a functor on a given object or function is called lifting.

The functors we deal with are pointed functors. A pointed functor has a specified natural equivalence. In our case the pointed functor maps from precise to imprecise values.

## 4.2 Programming Paradigm Lifting

Lifting is a concept that can simplify programming. Assume we specify a function, which takes two points and computes the distance between these points. The points are defined as pairs of floating point numbers. The function will work correctly if it is applied to the correct data type but how can we apply it to different data types, e.g., data types for moving points? In this case the coordinates of the points and the distance between the points depend on an additional parameter 'time'. In an imperative programming language that does not support polymorphism and generic programming (such as C) we may use one of the following solutions:

- we copy the function and edit the copy such that it works with the new data type or
- we take care of the differences whenever we call the function, by first clarifying the temporal aspect and then computing the distance.

In a programming environment the first method will inevitably result in a long list of slightly different versions of the same function. Each version is applicable to only one data type. The second method may result in strange errors if one of the programmers does not know about the limitations or makes a mistake in the calling routine.

Lifting provides an elegant solution to that problem. The language must be capable of two concepts:

- The language must be able to overload a function. Overloading a function allows using the same function name for different data types. Typical examples known from standard imperative languages are the basic mathematical operations, which can be applied to different numerical data types. A language used for lifting must provide this capability for all types of functions including user-defined functions.
- The language must be a second-order language, i.e., the function must support the use of functions as parameters for other functions.

Lifting only requires the definition of a functor. This functor can then be used to automatically adapt functions to the data type.

## 4.3 Application of Lifting for Imprecise Values

How can we apply the concept of lifting to precise and imprecise values? Starting point is a data type a for precise values. Parts of the definition for basic mathematic operations in Haskell (e.g., Bird 1998) are shown here:

```
class Num a where
   (+)    :: a -> a -> a
   (*)    :: a -> a -> a
   abs    :: a -> a
   negate :: a -> a

class (Num a) => Fractional a where
   (/) x y :: a -> a -> a
   recip   :: a -> a

class (Fractional a) => Floating a where
   sqrt :: a -> a

class (Floating a) => Numbers a where
   sqr :: a -> a
```

Applying these functions to a new data type requires building the corresponding instances. These instances can be constructed in a traditional way by rewriting the code. The result for a new data type could look like the following:

```
data MyNumbers a = MyNum a

instance (Floating (MyNumbers Float)) =>
         Numbers (MyNumbers Float) where
   sqr (MyNum x) = MyNum (x * x)
```

The disadvantage of this method has been shown in section 4.2. A functor eliminates the necessity to rewrite the code. The definition for a class of functors mapping from a data type a to a data type b a is

```
class Lifts b a where
   lift0 :: a -> b a
   lift1 :: (a -> a) -> b a -> b a
   lift2 :: (a -> a -> a) -> b a -> b a -> b a
```

The function `lift0` maps a value from one class to another. The functions `lift1` and `lift2` provide the same for functions with one or two parameters. Functions with more than two parameters must be mapped recursively.

We can now define the instance of above class `Numbers` for the data type `MyNumber a` as follows

```
instance Lifts MyNumber a where
   lift0 x          = MyNum x
   lift1 op x       = MyNum (op x)
   lift2 op x1 x2 = MyNum (op x1 x2)

instance (Floating MyNumber) => Numbers MyNumber
where
   sqr x = lift1 sqr x
```

The lifting function maps the functionality of the existing data type to the new data type. This avoids rewriting the code. The benefit becomes evident if an error must be fixed in the implementation. Instead of fixing all instances, only the original instance must be fixed and this change automatically affects all other instances.

# 5 Sample Implementation

In this section we describe how to implement lifting for the imprecise concepts defined in section 2. The mapping of functions uses the error propagation concepts shown in section 3. We now want to use these definitions to define lifting functions as introduced in section 4.

The first step in each section is the definition of a data type. The operations necessary to propagate imprecision are concept dependent. The concepts also use different parameters to describe the imprecision. These parameters are collected in corresponding data types.

## 5.1 Lifting Normally Distributed Values

Normally distributed values are defined by the expected value and the variance, which are stored in this order in the data set. ND is a constructor function, which creates the data set using the provided values.

```
data NormDist v = ND v v
```

Error propagation for normally distributed values requires the computation of partial derivatives. Numerical differentiation using linearization is a simple solution that provides an approximation. For a function $f$ with parameters $x_1$, $x_2$, ... the derivative in $x_1$ is

$$\frac{\partial f}{\partial x_1} = \frac{f(x_1 + \varepsilon, x_2, \ldots) - f(x_1, x_2, \ldots)}{\varepsilon} \tag{23}$$

where $\varepsilon$ is a small value specifying the length of the interval used for the linearization. More sophisticated strategies can be found in the literature

(Press et al. 1988). The local functions `diff` respectively `diff1` and `diff2` provide lifting function (23).

```
instance Lifts NormDist Double where
   lift0 v = ND v 0
   lift1 op (ND v s) = ND (op v) (diff * s) where
      diff = ((op (v+epsilon)) - (op v)) / epsilon
   lift2 op (ND v1 s1) (ND v2 s2) =
     ND (op v1 v2) (diff1^2*s1 + diff2^2*s2)) where
      diff1 = ((op (v1+epsilon) v2) - (op v1 v2)) /
                                                 epsilon
      diff2 = ((op v1 (v2+epsilon)) - (op v1 v2)) /
                                                 epsilon
```

Lifting a precise value requires an assumption for a variance. Since the value shall be precise, the variance is set to zero. Lifting a mathematical operation `op` with one or two parameters uses numerical differentiation and formula (8) to compute the variance of the function result.

This instance then provides the functionality to easily lift the basic mathematical operations. This is done as shown in section 4.3.

## 5.2 Lifting Fuzzy Values

The data type for fuzzy values must store the four points necessary to define the trapezoidal representation shown in Figure 2. Other realizations like triangular distribution functions may use different data types. `F` is again the constructor function.

```
data Fuzzy    v = F  v v v v
```

The implementation for fuzzy values is equal to the examples above:

```
instance Lifts Fuzzy Double where
   lift0 v = F v v v v
   lift1 op (F a b c d) =
     listToFuzzyNum (sort cartProduct) where
       cartProduct = [(op a), (op b), (op c), (op d)]
   lift2 op (F a1 b1 c1 d1) (F a2 b2 c2 d2) =
     listToFuzzyNum (sort cartProduct) where
        cartProduct = [minimum cartProduct1,
                       maximum cartProduct1,
                       minimum cartProduct2,
                       maximum cartProduct2]
        cartProduct1 = [(op a1 a2), (op a1 d2),
                        (op d1 a2), (op d1 d2)]
        cartProduct2 = [(op b1 b2), (op b1 c2),
                        (op c1 b2), (op c1 c2)]
```

The functions `cartProduct`, `cartProduct1`, and `cartProduct2` define the Cartesian product as introduced in section 3.2.

### 5.3 Lifting Interval Values

The data type for the interval stores begin and end of the interval. `I` is again a constructor function.

```
data Interval v = I  v v
```

As seen in section 3, interval arithmetic is rather simple. Again we use the Cartesian product to determine the boundaries of the resulting interval.

```
instance Lifts Interval Double where
   lift0 v = I v v
   lift1 op (I a b) = I (minimum cartProduct)
                        (maximum cartProduct) where
      cartProduct = [(op a), (op b)]
   lift2 op (I a1 b1) (I a2 b2) =
    I (minimum cartProduct) (maximum cartProduct)
      where cartProduct = [(op a1 a2), (op a1 b2),
                           (op b1 a2), (op b1 b2)]
```

## 6 Examples for Using the Lifted Operations

How can we now use the lifted function? The simplest example is using a basic mathematic operation, e.g., by adding two numbers. This requires typing `a+b` on the command line. Depending on the definition of the parameters `a` and `b`, we get different results. Table 1 shows the results.

**Table 1** Results of addition for different types of values.

| Type | Values | Result of a+b |
|---|---|---|
| precise values | a = 5.0<br>b = 2.0 | 7.0 |
| normally distributed | a = ND 5.0 0.05<br>b = ND 2.0 0.04 | ND 7.0 0.0899 |
| interval | a = I 4.85 5.15<br>b = I 1.88 2.12 | I 6.73 7.27 |
| fuzzy | a = F 4.85 4.95 5.05 5.15<br>b = F 1.88 1.96 2.04 2.12 | F 6.73 6.91 7.09 7.27 |

A more complex example is the computation of the distance between two points. The points are defined by a pair of coordinates in a plane coordinate system. The distance is then defined as

$$s_{12} = \sqrt{(x_2 - x_1)^2 + (y_2 - y_1)^2} . \tag{24}$$

In Haskell the definition is

```
data Point f  = Pt f f deriving Show

dist :: Numbers a => (Point a) -> (Point a) -> a
dist (Pt x1 y1) (Pt x2 y2) =
                        sqrt (sqr(x1-x2) + sqr(y1-y2))
```

We can now again use different implementations of imprecise values to test the implementation. Table 2 shows the results.

**Table 2** Results of distance computation for different types of values

| Type | Values | Result of `dist` |
|---|---|---|
| precise values | ptD1, ptD2 :: Point Double<br>ptD1 = Pt 20 20<br>ptD2 = Pt 12 17 | 8.544 |
| normally dis-<br>tributed | ptN1, ptN2 :: Point (NormDist Double)<br>ptN1 = Pt (ND 20 0.02) (ND 20 0.04)<br>ptN2 = Pt (ND 12 0.06) (ND 17 0.01) | ND 8.544 0.076 |
| interval | ptI1, ptI2 :: Point (Interval Double)<br>ptI1 = Pt (I 19.94 20.06) (I 19.88 20.12)<br>ptI2 = Pt (I 11.82 12.18) (I 16.97 17.03) | I 8.267 8.822 |
| fuzzy | ptF1, ptF2 :: Point (Fuzzy Double)<br>ptF1 = Pt (F 19.94 19.98 20.02 20.06)<br>(F 19.88 19.96 20.04 20.12)<br>ptF2 = Pt (F 11.82 12.94 12.06 12.18)<br>(F 16.97 16.99 17.01 17.03) | F 7.595 8.302 8.524<br>8.822 |

## 7 Conclusions and Future Work

In this paper we use the domains of precise and imprecise values as an example. We showed different models of imprecise values in section 2 and how to propagation imprecision in each model in section 3. In section 4 we discussed the concept of mapping between different domains. We have shown that it is possible to construct functors, which map precise values and functions on these values to imprecise values. This allows the specification of functions, which work with different kinds of values. As shown in section 5.5 the defined functors allow a simple mapping of functions from a simple domain to an extended domain. It can be generalized from the results how we can use functors to map data and functions from one domain to another domain.

## References


Bachmann, A. and Allgöwer, B. (2002). Uncertainty Propagation in Wildland Fire and Behaviour Modelling. International Journal of Geographic Information Science 16(2): 115-127.

Bird, R. (1998). Introduction to Functional Programming Using Haskell. Hemel Hempstead, UK, Prentice Hall Europe.

Erwig, M. and Schneider, M. (1999). Developments in Spatio-Temporal Query Languages. Proceedings of 10th International Workshop on Database and Expert Systems Applications, Florence, Italy.

Fodor, J. and Bede, B. (2006). Arithmetics with Fuzzy Numbers: A Comparative Overview. Proceeding of 4th Slovakian-Hungarian Joint Symposium on Applied Machine Intelligence, Herl'any, Slovakia.

Frank, A.U. (2005). Map Algebra Extended with Functors for Temporal Data. Proceedings of the ER Workshop 2005 (CoMoGIS'05). J. Akoka et al. Klagenfurt, Austria, Springer-Verlag Berlin Heidelberg. LNCS 3770: 193-206.

Hayes, B. (2003). A Lucid Interval. American Scientist 91(6): 484-488.

Heuvelink, G.B.M. (1998). Error Propagation in Environmental Modelling with GIS. London, Taylor & Francis.

Heuvelink, G.B.M. and Burrough, P.A. (2002). Developments in Statistical Approaches to Uncertainty and its Propagation. International Journal of Geographic Information Science 16(2): 111-113.

Karssenberg, D. and de Jong, K. (2005). Dynamic Environmental Modelling in GIS: 2. Modelling Error Propagation. International Journal of Geographic Information Science 19(6): 623-637.

MacLane, S. and Birkhoff, G. (1999). Algebra (3rd Edition), AMS Chelsea Publishing.

Marquis, J.-P. (2006). Category Theory, Stanford Encyclopedia of Philosophy.

Moggi, E. (1989). A Category-Theoretic Account of Program Modules. Category Theory and Computer Science, Springer-Verlag, Berlin.

Navratil, G. (2006). Error Propagation for Free? GIScience, Münster, Germany, Institute for Geoinfomatics, University Münster.

Press, W.H., Flannery, B.P., Teukolsky, S.A. and Vetterling, W.T. (1988). Numerical Recipes in C - The Art of Scientific Computing. Cambrigde, Cambridge University Press.

Siler, W. and Buckley, J.J. (2005). Fuzzy Expert Systems and Fuzzy Reasoning, Wiley Press.

Viertl, R. (2002). On the Description and Analysis of Measurements of Continuous Quantities. Kybernetika 38(3): 353-362.

# GeoSR: Geographically Explore Semantic Relations in World Knowledge

Brent Hecht, Martin Raubal

Department of Geography
University of California, Santa Barbara
1832 Ellison Hall, Santa Barbara, CA 93106-4060
{bhecht, raubal}@geog.ucsb.edu

**Abstract.** Methods to determine the semantic relatedness (SR) value between two lexically expressed entities abound in the field of natural language processing (NLP). The goal of such efforts is to identify a single measure that summarizes the number and strength of the relationships between the two entities. In this paper, we present *GeoSR*, the first adaptation of SR methods to the context of geographic data exploration. By combining the first use of a knowledge repository structure that is replete with non-classical relations, a new means of explaining those relations to users, and the novel application of SR measures to a geographic reference system, GeoSR allows users to geographically navigate and investigate the world knowledge encoded in Wikipedia. There are numerous visualization and interaction paradigms possible with GeoSR; we present one implementation as a proof-of-concept and discuss others. Although, Wikipedia is used as the knowledge repository for our implementation, GeoSR will also work with any knowledge repository having a similar set of properties.

**Keywords:** semantic relatedness, natural language processing, geographic reference system, GeoSR, Wikipedia

## 1 Introduction and Related Work

In today's information-overloaded world, researchers in both the academic and professional community, students, policy analysts and people in many other fields frequently find themselves in the position of trying to locate a useful needle of information in a haystack of data. This search is often aided by the use of a spatial lens, as up to 80 percent of human decisions affect space or are affected by spatial situations (Albaredes 1992). For example, a student doing a project on Judaism, love, George W. Bush, Berlin or any other concept or named entity will definitely want to know the places that are most related to these concepts and named entities and why. *GeoSR* provides users with a novel method of easily accomplishing this task.

## 1.1 GeoSR and Wikipedia

GeoSR uses Wikipedia as its knowledge repository. The introduction of every paper produced by the burgeoning Wikipedia research community has its own way of describing the phenomenon that is Wikipedia. However, they all seem to agree on several vital properties. First, Wikipedia is a free encyclopedia that is produced via a collaborative effort by its contributors. Second, Wikipedia is highly multilingual, with hundreds of available languages. Third, Wikipedia is enormous and is, by far, the largest encyclopedia the world has ever seen. Indeed, as of October 2007, Wikipedias in 14 languages had over 100,000 articles and the largest Wikipedia, English, had over 2.05 million. Finally, many researchers argue that Wikipedia "has probably become the largest collection of freely available knowledge" (Zesch et al. 2007a, p. 1).

The above facts are all relatively well known among people who use Wikipedia, which in the U.S. amount to 36 percent of the Internet-using population (Rainie and Tancer 2007). However, what is less understood in the general and scientific communities are the opportunities presented by the massive knowledge repository of ubiquitously available information that Wikipedia represents. The research here is part of the first work (Hecht 2007) that explores the *spatio-temporal possibilities* of this knowledge repository, as well as others in the future that could offer similar content, structure, and size (for example, *Citizendium*[2]). Several authors have conducted other research projects in this area including Minotour (Hecht et al. 2007a), WikEye (Hecht et al. 2007b), and WikEar (Schöning et al. 2007a).

It is important to note that because this research uses Wikipedia as a data source, it is vulnerable to the risks of Wikipedia information as identified by Denning et al. (2005). However, we believe these risks apply only minimally to GeoSR for the following reasons: (1) GeoSR is not tied to the

2 http://www.citizendium.org

editorial policies of Wikipedia, only its structure and size and, as such, the research is much more general than the data set it relies on, (2) GeoSR provides a novel and useful method for visualizing and exploring data people are already accessing in massive numbers despite the risks, and (3) Giles (2005) has shown that the accuracy of Wikipedia, at least in the scientific context, is comparable to that of more conventional encyclopedias.

## 1.2 GeoSR and Semantic Relatedness

Semantic relatedness (SR), which is at the heart of GeoSR, is a well-known topic in the field of natural language processing (NLP). There are many applications of SR in NLP, including word sense disambiguation, text summarization, information extraction and retrieval, and correction of word errors (Budanitsky and Hirst 2006). There are two general methodological families of SR measures; SR measures based on graph- or network-based lexical resources, from which this research derives inspiration, and SR measures based on distributional similarity, which implement bag-of-word techniques. However, it has been argued that the distributional similarity family "is not an adequate model for lexical semantic relatedness" (Budanitsky and Hirst 2006, p. 30).

SR is often confused with semantic similarity. While many fields use the concept of semantic similarity differently, in the world of NLP, similarity measures are identical to SR measures if and only if the only relationships being examined are hypernymy and hyponymy (the *isA* relationship viewed from both sides). Similarity is thus a special case of SR (Budanitsky and Hirst 2006).

While members of the NLP community have presented myriad SR measures, most of these are designed for WordNet (Miller 1995), GermaNet (Kunze 2004), or older knowledge repositories. Very recently, some researchers have been investigating the modification of these methods for Wikipedia. Wikipedia has three structures that can be used to measure semantic relatedness: the Wikipedia Category Graph (WCG), the Wikipedia Article Graph (WAG), and the text of the Wikipedia entries (WT) (see (Zesch et al. 2007a) and (Hecht 2007)). Strube and Ponzetto (2006) presented the first effort to estimate SR using Wikipedia, *WikiRelate!*. It uses the WCG and reported slightly better correlation with human judgments – the so-called "gold standard" of SR measures, even though many researchers have taken issue with available datasets – than similar WordNet-based measures for some test sets.

Very recently, Gabrilovitch and Markovitch (2007) developed *Explicit Semantic Analysis* (*ESA*), which used the WT structure with much improved results over WikiRelate! (as well as methods developed using other knowledge repositories) in terms of correlation with the gold standard. However, ESA relies exclusively on distributional similarity mechanisms.

Both ESA and WikiRelate! use the English Wikipedia as its knowledge repository. Zesch et al. (2007b) compared GermaNet and the German WCG for use in semantic relatedness applications. They concluded that Wikipedia excels at SR, while GermaNet is better for similarity applications (as defined by the NLP community).

All of the aforementioned SR measures were designed for traditional NLP applications. Because of the data exploration needs of the GeoSR project and especially because of the importance of spatial-entity-to-spatial-entity and spatial-entity-to-non-spatial entity relationships, it was necessary to develop a novel SR measure and corresponding algorithm for this research. We have called this measure, which is the first to use the Wikipedia Article Graph (WAG), *ExploSR* (pn: "explosure").

## 1.3 Overview of Paper and System Framework

The framework of GeoSR is as follows: Wikipedia provides the world knowledge and the ExploSR semantic relatedness measure is responsible for assigning relative weights to the myriad relationships found in the Wikipedia repository. Based on some input named entity or concept (such as Judaism, love, George W. Bush, or Berlin), these values are then visualized *geographically* in one of several ways using *spatial articles* as anchors in a geographic reference system. Users can employ these visualizations as a context from which to engage in data exploration. Figure 1 demonstrates one possible visualization and interaction schema, which is discussed in more detail in section five. Only the top 100 locations are shown. For the location "Tuxplan (Veracruz)" (in Mexico), the explanation information is found in the "Identify" window in the "Explanatio" field, and can be seen in greater detail in figure 2. This data has been generated using the German Wikipedia, with the "Explanatio" field manually populated with English information. Missing links have not been included in this iteration of GeoSR due to implementation issues that are discussed in section 4.2.

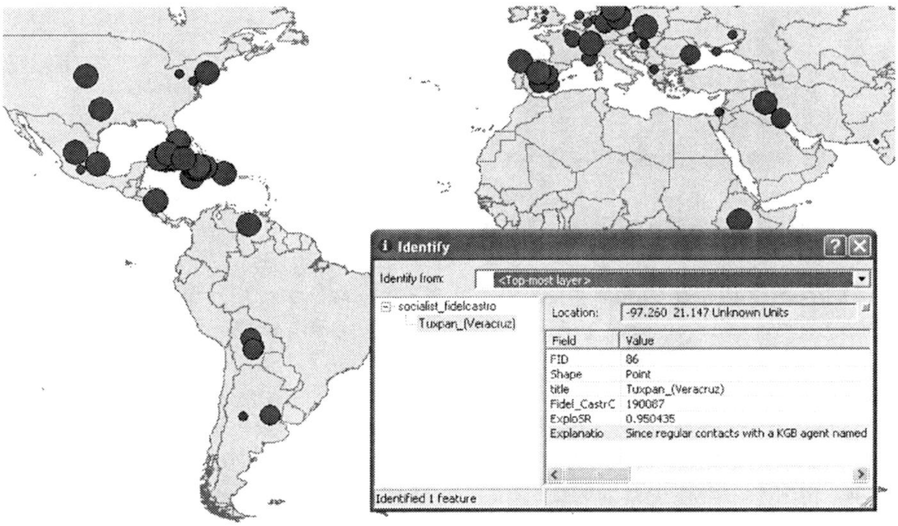


**Fig. 1:** A visualization of GeoSR data in which *Fidel Castro* was the input entity. Large dots represent the most related locations to *Fidel Castro* and smaller dots represent less important locations (within the top 100 locations).

**Fidel Castro -> Tuxpan (Veracruz)**

*Fidel Castro -> Cuban Revoluion -> 26th of July Movement*

Since regular contacts with a KGB agent named Nikolai Sergeevich Leonov in Mexico City had not resulted in the hoped for weapon supply, they decided to go to the United States to gather personnel and funds from Cubans living there, including Carlos Prío Socarrás, the elected Cuban president deposed by Batista in 1952. Back in Mexico, the group trained under a Spanish Civil War Veteran, Cuban-born Alberto Bayo[35] who had fled to Mexico after Francisco Franco's victory in Spain. On November 26, 1956, Castro and his group of 81 followers, mostly Cuban exiles, set out from **Tuxpan, Mexico** aboard the yacht Granma for the purpose of starting a rebellion in Cuba.

**Fig. 2:** An expansion of the content of the explanation field seen in figure 1.

Section two of this paper describes the pre-processing of Wikipedia required before its use in GeoSR and lays out a spatio-temporal framework with which to view Wikipedia. The advantages of using the Wikipedia Article Graph (WAG) over other structures in the encyclopedia in this context are discussed in section three. Section four covers ExploSR in detail, highlighting its strengths and weaknesses. In section five, several applications for GeoSR are presented and one is fully demonstrated as a proof-of-concept. Finally, we wrap things up with a conclusion and describe directions for future research in section six.

## 2 Wikipedia Knowledge Repository

### 2.1 Preprocessing and API Access

While the Wikipedia knowledge repository has made implementing this research possible, a large number of pre-processing steps are necessary before Wikipedia can be used efficiently by GeoSR. Wikipedia data is received in the form of the XML "database backup dumps" provided by the Wikimedia Foundation[3], which runs Wikipedia. Dumps are made available every three to four weeks in every language in which there is Wikipedia. These dumps represent an enormous amount of text; the English Wikipedia dump from October 23, 2007 weighs in at 12.3 GB and the October 10, 2007 dump from second largest Wikipedia, that of German, is a sizeable 3.63 GB.

Once these dumps are downloaded, they must be processed by our Wikipedia parser and API, WikAPIdia, which we are considering releasing in the near future. During the parsing stage, structured information about each article including data about links, text, titles, title aliases (redirects), and much more is stored in a series of MySQL tables. Due to its size, the parsing step for the English Wikipedia can take a moderately-powered computer up to two to three days.

The MySQL database forms the data model from which the API portion of WikAPIdia operates. This API is the back end of all the Wikipedia-related projects in which our research group has participated, including GeoSR. It is important to note that while the only Wikipedias currently supported by our software are that of English, German and Spanish, we have constructed the software such that support for other Wikipedias is quite simple to add for a native speaker of that language.

### 2.2 Spatio-Temporal Wikipedia data

In addition to processing lexical structures, WikAPIdia has special facilities for mining the spatial and temporal data in Wikipedia. Spatial data mainly comes in the form of explicitly "geotagged" articles, or articles

3 http://www.wikimedia.org

with spatial reference information that describes the location of their subjects. We have labeled articles with spatial references as *spatial articles* and those without *non-spatial articles*. The distinction between spatial and non-spatial articles plays a critical role in this research. Spatial articles are the intersection between "geographic space" and "Wikipedia space". As such, as will be explicated further in section five, spatial articles can essentially represent SR value samples in the real world.

The corollary to spatial articles in the temporal domain are what we call *pure temporal articles*, which, through their titles, contain references to a temporal reference system. While some of these references, such as the article titled "October 29" are ambiguous, others are not (such as "1983" or "April 29, 1983"). The pure temporal article construct plays an important role in our ExploSR algorithm (shown in section four), although its reference system utility is not emphasized in this research. Hecht (2007) provides a more general description of our Wikipedia spatio-temporal framework.

## 3 Advantages of the Wikipedia Article Graph

As noted in the introduction, ExploSR is the first SR methodology designed explicitly for data exploration use. However, it is also unique in that it is the first Wikipedia-focused measure to use the Wikipedia Article Graph (WAG). The WAG is the graph that is composed of the set of articles in a Wikipedia (set $A$), and the standard links between them (set $L$), which are defined using brackets in the Wiki markup language. Formally, graphs are usually defined as an ordered double, where a graph $G = (V, E)$. $V$ is the set of vertices in the graph, and $E$ is the set of edges (Piff 1991). In this case, $A = V$ and $L = E$.

The WAG has two essential properties. First and foremost, the WAG is the ideal Wikipedia structure to use for data exploration SR measures because it is a simple matter to explicitly explain to users the relationships that resulted in the measure value between any two concepts. Secondly, the WAG contains much broader and deeper relation information than the knowledge repositories commonly used in SR research as well as other structures embedded in Wikipedia. This fact proves vital to examining relations between two spatial features and those between a spatial feature and non-spatial entity. The rest of this section is dedicated to explaining these two advantages in detail.

### 3.1 The Wikipedia Snippet – Paragraph Independence Facilitates Data Exploration

Nearly all articles in Wikipedia have uniquely independent paragraphs, which we term *snippets*. The Wikipedia snippet is a distinctive natural text phenomenon in that we have found qualitatively that nearly all Wikipedia snippets are entirely independent of other snippets within the same article. In other words, snippets rarely contain ambiguous text that the reader is expected to disambiguate using knowledge acquired from other snippets on the same Wikipedia page. This is important because it signifies that the meaning of a link is almost always contained within the snippet that hosts the link (see figure 2). Additionally, this property ensures that snippets can be safely rearranged or presented independently without severely reducing their understandable information content. We have found that the only context necessary for fully comprehending nearly all snippets is the title of the Wikipedia article in which they appear. Most of the remaining snippets can be completely framed by providing the hierarchy of titles, headings, and subheadings under which a snippet appears (i.e., for the *United States* article, *United States* -> History of the United States -> Revolutionary War).

Thus far, two possible causes of the unique snippet substructure in Wikipedia have been identified. The first is the collaborative nature of Wikipedia. Buriol et al. (2006) found that the average Wikipedia article has at least seven authors. This means that, in many cases, different parts of an article are written by different contributors, surely adding to the disjointedness of the text. This disjointedness, however, is desired in the Wikipedia community because of the encyclopedic nature of the writing style in Wikipedia. This writing style, termed *WikiLanguage* by Elia (2006), is the second identified cause of the independence of snippets. Wikipedians do not seek to create prose that flows from paragraph to paragraph; they seek to inform about facts in an organized fashion.

In summary, the independence of snippets provides an easy way to identify and present to the user the subset of text on any Wikipedia page that can explain the meaning of a link between two pages: the snippet in which the link resides. Explaining the meaning of links in the WCG in a similar manner would be impossible, as the meaning of WCG relationships is never explicitly explained. ESA, which is a distributional similarity measure, identifies relationships essentially by measuring the similarity between the unique words of Wikipedia articles. As such, using ESA to provide the full meaning of relationships between these articles in human-readable form would require a process entirely exogenous to the relatedness measure.

## 3.2 Depth and Breadth of Encoded Relations in the WAG

The second advantage of a WAG-based measure in the context of this research relates to the unique spatial needs of GeoSR. It has been qualitatively found that both ESA and WCG-based methods alone do not work well for spatial/spatial and spatial/non-spatial relationships. While this, along with the effectiveness of ExploSR outside the data exploration context, will be investigated in detail in future research, it is believed that the failure of WCG- and WT-based methods in the spatial context results from two characteristics of those two data structures: missing *classical relations*, and the worse offender, missing *non-classical relations*.

Morris and Hirst (2004) define classical relations as relations that depend on the sharing of properties of classical categories (Lakoff 1987). Common classical relations include hypernymy/hyponymy (*isA*), meronymy/holonymy (*hasA*), synonymy (*likeA*) and antonymy (*isNotA*). WordNet, the lexical resource focus of most semantic relatedness research, offers only relations of this type. The vast majority of relations in the WCG are classical, and in fact are limited almost entirely to *isA* relations with a sprinkling of meronymy/holonymy (Zesch et al. 2007b). The WCG contains a large number of missing important *hasA* relations (not to mention displaying a complete lack of antonymy, synonymy, etc.), making the WCG weak in both breadth and depth of classical relational coverage. In sum, the WCG is essentially a semantic similarity resource, not a SR resource (as defined by the NLP community). This is a critical problem when it comes to spatial entities: a hypernymy/hyponymy-only (*isA*-only) path in a taxonomy in which one endpoint entity is a spatial entity essentially limits the path to spatial entities. For instance, a spatial entity such as *California* is no doubt closely related to *Gold Rush*, but it is difficult to imagine a short hypernymy/hyponymy path between the two entities in a graph, even though the meronymy/holonymy relation is direct. Similarly, in the case of the WT, the unique word vectors of the *Gold Rush* article and that of the *California* article are highly dissimilar; the *Gold Rush* article focuses on the details of gold rushes in general and the California article is a broad overview of the state. As such, distributional measures also fail to understand the important *California-Gold Rush* meronymy/holonymy relation, which is captured at a simple path distance of 1 in the WAG.

Spatial/spatial and spatial/non-spatial article relationships also tend to display a large number of *non-classical* relations. Non-classical relations are associative or ad-hoc in nature (Budanitsky and Hirst 2006) and are defined by Morris and Hirst (2004) as relations that “do not depend on the shared properties required of classical relations” (p. 2). Budanitsky and

Hirst (2006) list the following examples of these types of relations: *isUsedTo* (*bed-sleep*), *worksIn* (*judge-court*), *livesIn* (*camel-desert*), and *isOnTheOutsideOf* (*corn-husk*). The WAG is absolutely replete with these types of relations. For instance, all of the above relations are encoded as at least unidirectional links in the English WAG (*judge-court* is bidirectional). Despite the fact that non-classical relations have been found to be an extremely important aspect of lexical relationships (Budanitsky and Hirst, 2006; Morris and Hirst 2004), all graph-based SR research on Wikipedia thus far has focused on the WCG, which encodes almost none of these relations. The extent to which a distributional measure such as ESA understands non-classical relations is unclear.

Of course, non-classical relationships in which at least one of the entities involved is a spatial entity play a vital role in this research. For instance, the article on the University of California, Santa Barbara (UCSB) has numerous non-classical relations regarding the protests that occurred here against the Vietnam War, protests that shaped the character of the campus for decades. For instance, the link to former California Governor Ronald Reagan, *UCSB-Ronald Reagan* is best typed *imposedACurfewToReduceRiotingAt*, which is an archetypal non-classical relation. GeoSR would fail a user seeking to learn more about Ronald Reagan's influence in the South Coast area of California if it did not report this important relationship. As such, the WCG and the WT are insufficient resources for this research due to their near complete lack of or unclear understanding of non-classical relations.

# 4 ExploSR: Using the WAG for Semantic Relatedness

## 4.1 Microstructure of ExploSR

It is important to note that because of the relative unimportance of hypernymy and hyponymy in the WAG, the WAG is a novel challenge for semantic relatedness researchers. As of this writing, there are no peer-reviewed WAG-based measures available, let alone one that is optimized to allow for data exploration. As such, it was necessary to develop our own measure, ExploSR. We chose to approach the problem from the point of view of the Wikipedia editors, the people actually creating the link structure. We started by asking what it means about the relationship between a page *A* and a page *B* when a Wikipedian creates a link between the two pages. In section three, the generic semantic *type* of these links was analyzed, but to convert these into semantic relatedness values, it is necessary to assign a quantitative measure of the *strength* and *number* of these relations. Budanitsky and Hirst (2006) note that this "scaling" of a knowledge repository network used in a SR method is "a widely acknowledged problem". Indeed, this was the key challenge in designing ExploSR. Stated more simply, ExploSR must be able to assign a quantitative relatedness measure, or weight, to each edge in the WAG. To do so, it uses the following general formulas:

If $|OL_A| > C$,

$$ExploSR_A = 1 - \frac{|OL_{A \to B}|}{C + (1 + \log_2 |OL_A - C|)} \tag{1a}$$

Else,

$$ExploSR_A = 1 - \frac{|OL_{A \to B}|}{|OL_A|} \tag{1b}$$

And if $|OL_B| > C$,

$$ExploSR_B = 1 - \frac{|OL_{B \to A}|}{C + (1 + \log_2 |OL_B - C|)} \tag{2a}$$

Else,

$$ExploSR_B = 1 - \frac{|OL_{B \to A}|}{|OL_B|} \tag{2b}$$

with the final ExploSR value being,

$$ExploSR_{A \leftrightarrow B} = \frac{ExploSR_A + ExploSR_B}{2} \tag{3}$$

In these formulas, $|OL_A|$ and $|OL_B|$ represent the total number of outlinks (the *outdegree*, in graph theory terminology) of articles *A* and *B*. $|OL_{A \to B}|$ and $|OL_{B \to A}|$ signify the size of the set of outlinks from article *A* to article *B* and vice versa. *C* is a constant that is predefined and explained below. In all cases, if either the $ExploSR_A$ or $ExploSR_B$ value is less than zero, it is set to zero[4].

---

[4] This would occur if, for example, an article *B* has 500 outlinks and the number of links from article *B* to article *A* was greater than the denominator value. In other words, in equation 2a, if C = 5, $OL_{B\text{->}A}$ = 16 and $OL_B$ = 500, then equation

The motivation behind this approach to edge weighting is straightforward. Given the nature of Wikipedia, the percentage of outlinks directed from any article *A* to any article *B* and vice versa is a good measure of the importance of the relationship(s) between A and B. However, since longer articles generally have more *relationship content*, encoded as a larger number of outlinks, some additional scaling must be done. The reasoning for the logarithm-based schema is that it was determined through extensive experience with Wikipedia that, in general, long articles are split up into sections, in each of which a cluster of references to the same articles is likely to occur. In the case of an article *B* that is extremely closely related to a long article *A*, a significant sprinkling of references to *B* is expected outside of that cluster as well. For example, in the *United States* article, links to the *Democracy* article are going to be clustered in the section on politics. However, since Democracy is so vital to the United States, it is likely to be mentioned occasionally elsewhere as well. The value *C* is the expected size of a cluster of links (C = 5 in our current implementation) and the logarithmic part of the normalization methodology approximates the number of links external to the cluster ("the sprinkling"). If equations 1a and 2a were omitted in favor of 1b and 2b for all outlink values, long articles would almost always appear to contain only weak relationships.

It is important to note that ExploSR is technically a measure of semantic distance, or the lack of semantic relatedness. We have chosen to encode it in this manner for the purposes of easily incorporating it into a Dijkstra's shortest path (Dijkstra 1959) algorithm implementation, which is described in section 4.3.

While the formula above provides our general approach, there are a few minor data set-specific modifications. For instance, links that appear in the first paragraph – almost always a *gloss*, or summary of the article content – are treated as codifications of especially strong relationships. Similarly, we take measures to handle the unique relationships present in links between articles such as *Austria* and *Geography of Austria*.

## 4.2 The Missing Link Problem

While the Internet as a whole suffers greatly from link spam, the larger problem in Wikipedia is missing links (Adafre and Rijke 2005). This, of course, has a detrimental effect on ExploSR as a missing link essentially

---

2a evaluates to approximately 1 - 16/(5 + 1 + 8.951), which is less than one. The value is then set to 0.

represents a missing relation. In the context of ExploSR, there are two types of missing links, type one and type two, both of which are important issues. In the case of type one missing links, the target of the missing link is an article that is not linked elsewhere in the page. This affects whether or not a relationship between the pages in question is identified at all. Type two missing links occur when the target of the missing link is the target of another link elsewhere in the article. In other words type one missing links affect the recognition of relationships between entities, while type two missing links affect the ability of ExploSR to identify the relative importance of existing relationships. Of course, there are some type one "missing links" that represent relationships so unimportant or weak that we would prefer that these links not be "found". "Finding" these links would be essentially introducing link spam to the data set.

In an effort to avoid the link spam problem, we currently only target type two missing links with our missing link reduction approach, which has been implemented but not applied to the whole of a WAG due to computational complexity issues. That said, our missing link processor represents a rudimentary but sufficient algorithm for this proof-of-concept stage. Future work may improve this area quite a bit, possibly enhancing the system of Adafre and Rijke (2005), which presented qualitatively promising results. Simply stated, we do a text search for all forms of links that already appear on a page and code matching non-linked forms as links. A link's "forms" include the title of the target of the link, the set of "anchor texts" (Adafre and Rijke 2005) that are used to describe that link (i.e. the link appears as "GIS" to Wikipedia readers, but the target of the link is "Geographic Information Science"), as well as the set of redirects to the link target defined globally in the Wikipedia data set.

## 4.3 Macrostructure of ExploSR

So far, we have described how ExploSR scales the relationship between any two linked articles *A* and *B*. But how does ExploSR work across the entire WAG? How does this apply to the spatial context of this research? These are the topics of this subsection.

At the core of ExploSR's macrostructure is an implementation of Dijkstra's shortest path algorithm (Dijkstra 1959). The input to this algorithm (by a user or a system; see section five for more details) is a spatial or non-spatial article *A*. The algorithm then evaluates the relations between both the articles to which *A* links, as well as the articles that link to *A*, using the ExploSR measure. It continues according to Dijkstra, summing the ExploSR values along each path, either until the entire WAG has been explored or a certain stop condition has been met. While doing this, it is recording the snippets containing each of the links it encounters. In this fashion, every relationship has a *snippet path* of sorts, even for paths that are several edges long. These snippet paths are essential to data exploration because they almost always fully explain the relationship found by the algorithm, as is noted in section three.

We made several modifications to the standard Dijkstra algorithm in order to account for the Wikipedia data set and our spatially-focused application. First, a condition has been placed in the algorithm to stop processing paths when it encounters the pure temporal articles discussed in section two. This effectively prevents the recognition of all relationships through these articles. We have done this because pure temporal articles almost always have extraordinarily weak relationships encoded in both their inlinks and outlinks (Hecht et al. 2007a). Hecht et al. (2007a) describe the example that the pure temporal article *1979* is "essentially a list of events that occurred in 1979, a list that is so disparate that it includes the acquisition of home rule for Greenland and the premiere of 'Morning Edition' on the United States' National Public Radio." (p. 4) We have found that it is better to simply ignore the relatedness of Greenland and "Morning Edition" rather than use ExploSR to estimate its microscopic general value.

Second, a similar *optional* stop condition is made available for spatial articles, albeit for an entirely different reason. When the Dijkstra algorithm encounters a spatial article, the articles that link to this article and that are linked in this article will have a large degree of spatial autocorrelation. If the user wishes to mute this effect, she can enable this stop condition. Obviously, if the user inputs a spatial article to the algorithm, this condition is not applied on the input article.

While we have now answered the question regarding the application of ExploSR to the entire WAG, we have not explained how all these values are applied to a geographic reference system. The answer to this question lies in the output of the modified Dijkstra algorithm, which is the set of spatial articles encountered by the algorithm, along with the ExploSR values of these articles and their snippet paths. This can either be a size-limited set representing the top $n$-most related articles to the input article, a value-limited set containing all spatial articles with an ExploSR value of no more than $v$ from the input article, or, if computational complexity is no object, the entire set of spatial articles. For instance, a user who inputs the article *Fidel Castro* to GeoSR and sets $n$ to 100 will receive the 100 spatial articles with the lowest ExploSR scores from *Fidel Castro* (figure 1), along with the attribute data described above.

## 5 Applications

As noted in section three, spatial articles, or articles with a geographic reference system location, can act as "sample points" for the ExploSR semantic relatedness values in the real world. It is upon this ability that we envision a myriad of applications for GeoSR.

### 5.1 Simple Data Exploration

The most immediately obvious application of GeoSR is to use it for point-based data exploration of the knowledge contained in Wikipedia. This application has been implemented and can be seen in figures 1 and 3. Users input an entity (which must have a corresponding Wikipedia article) into the system, and a map indicating the $n$-most related spatial articles is presented, with the articles represented as points at their geotagged locations. Users can then click on the points to view the snippet paths for the clicked spatial article (figure 2).

We have implemented this system by exporting the output of GeoSR into a shapefile and loading this data into ArcGIS[5]. The shapefile contains three columns in its attribute table: name of the spatial entity, its ExploSR value, and its snippet path (snippet path functionality not yet fully implemented). The shapefile is visualized in ArcGIS using a reverse graduated symbol schema such that lower ExploSR values result in bigger symbols. As such, the visualization represents semantic relatedness and not semantic distance. Users can engage in data exploration by using the "Identify" tool in ArcGIS to view the snippet paths (figures 1 and 2).

5 http://www.esri.com

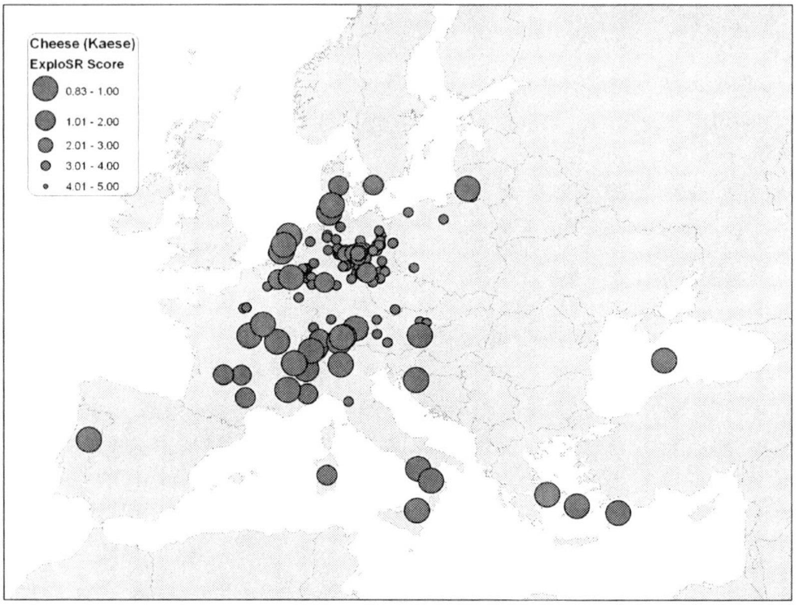


**Fig. 3:** A visualization of the output resulting from inputting the article *Kaese* (German for *Cheese*) into the GeoSR system operating on the German Wikipedia. Spatial stemming was turned on, and missing links were not included. The top 200 locations were output, but not all are located in the region depicted above.

## 5.2 Area-Based Query

If all spatial articles have been evaluated against all non-spatial articles (or a subset of non-spatial articles that are of interest), a user can query any extent and receive the most related non-spatial articles to that extent. This can be easily calculated using summary statistics of the SR values generated from the spatial articles located inside the chosen extent. It would also be a simple matter to explain the relatedness of these non-spatial articles to the extent using snippet paths.

## 5.3 Analyzing the First Law of Geography

Simply stated, the "First Law of Geography", first recognized by Tobler (1970), declares that everything is related, but nearer entities are more related than distant entities. While the nature of this "law" as actually being more of a "guideline" has been widely recognized for many years now, researchers could, by entering spatial articles as the input, have another means of exploring the degree to which this guideline holds true.

### 5.4 Regionalization

Many regionalization schemas and algorithms could be applied using the output of GeoSR as input. For instance, McKnight (2000) uses "basic features of homogeneity" as a means for regionalizing North America. Such uniform regionalizations could be completed by analyzing the variation in the most related non-spatial articles across space. Similarly, nodal regions could be made by evaluating the output of GeoSR when a spatial article is input.

### 5.5 Subsets and Algebra

While all the aforementioned applications have been explained using a single input value, there is no reason the outputs of multiple inputs cannot be combined to give new meaning to the above applications. For instance, the system described in section 5.1 could be used to examine the spatial footprint of the union of *Cheese* and *Fondue* by simply adding together the output from two iterations of GeoSR. Similarly, the applications could be used on subsets of spatial or non-spatial articles. For instance, application 5.2 could be used on the subset of non-spatial articles that are about architecture or even country music, as defined by the architecture and country music categories in the WCG.

## 6 Conclusions and Future Work

In this paper, we have presented two inter-linked innovations. First, we have demonstrated the benefits of visualizing semantic relatedness measures from the perspective of a geographic reference system. Second, we have created a semantic relatedness measure that is optimized for data exploration purposes. Integrating these innovations resulted in a novel data exploration environment that can form the basis for many useful applications. However, there is much work yet to be done.

First and foremost, there is no reason that GeoSR needs to be restricted to geographic reference systems. In theory, our reference system + data exploration methodologies could be applied to any *semantic* reference system (Kuhn 2003; Kuhn and Raubal 2003). For instance, temporal reference systems would be an easy extension as all of the above applications have simple corollaries in the temporal domain. Extending our research to semantic reference systems is the most immediate direction of future research.

Secondly, some sort of a formal evaluation is in order (we have evaluated thus far using our area knowledge of test input entities). This is a particularly difficult problem. Semantic relatedness researchers have had some difficulty evaluating their measures within their own domain, and inside the spatial domain we have the additional dilemma of the spatial dependence of opinions about relatedness between many entity pairs. Nowhere is this more evident than in the varied results of GeoSR depending on the language of the WAG. For instance, when GeoSR operates on the German WAG, no matter what its input, entities within the German-speaking world of Germany, Austria, and Switzerland always rank high, even when the input article is *Surfing* (*Wellenreiten*).

While ExploSR is currently the only WAG-based semantic relatedness measure, Zesch et al. (2007b) have expressed interest in experimenting with the WAG and surely other SR researchers will join in as well. Depending on their methodologies, it may be possible to replace ExploSR with another SR measure if that measure is proven to be higher quality and capable of producing snippet paths for data exploration. This would be another interesting area of further research.

Finally, it is our intention to analyze the extent to which relations to and from spatial entities differ from those between non-spatial entities. For instance, we would like to better investigate from a theoretical and experimental perspective why non-classical relations are so important to spatial entity relationships.

## References

Adafre, S. F. and de Rijke, M. (2005). Discovering Missing Links in Wikipedia. LinkKDD (in conjunction with SIGKDD), Chicago, IL.

Albaredes, G. (1992). A New Approach: User Oriented GIS. EGIS '92.

Budanitsky, A. and Hirst, G. (2006). Evaluating WordNet-based Measures of Lexical Semantic Relatedness. Computational Linguistics, *32*(1), 13-47.

Buriol, L. S., Castillo, C., Donato, D., Leonardi, S., and Millozii, S. (2006). Temporal Analysis of the Wikigraph. Proceedings of Web Intelligence, Hong Kong.

Denning, P., Horning, J., Parnas, D., and Weinstein, L. (2005). Inside Risks: Wikipedia Risks. Communications of the ACM, 48 (12).

Dijkstra, E. W. (1959). A note on two problems in connection with graphs. *Numerische Mathematik*, *1*(1), 269-271.

Elia, A. (2006). An analysis of Wikipedia digital writing. Conference of the European Chapter of the Association for Computational Linguistics, Trento, Italy.

Gabrilovich, E., Markovitch, Shaul (2007). Computing Semantic Relatedness using Wikipedia-based Explicit Semantic Analysis. Paper presented at the Proceedings of the Twentieth Join Conference for Artificial Intelligence, Hyberabad, India.

Giles, J. (2005). Internet encyclopaedias go head to head. Nature, http://www.nature.com/nature/journal/v438/n7070/full/438900a.html

Hecht, B. (2007). Masters Thesis. Using Wikipedia as a Spatiotemporal Knowledge Repository. University of California, Santa Barbara, California, United States.

Hecht, B., Rohs, M., Schöning, J., and Krüger, A (2007b). WikEye - Using Magic Lenses to Explore Georeferenced Wikipedia Content. PERMID 2007 (in conjunction with the Fifth International Conference on Pervasive Computing), Toronto, Ontario, Canada.

Hecht, B., Starosielski, N., and Dara-Abrams, D. (2007a). Generating Educational Tourism Narratives from Wikipedia. Association for the Advancement of Artificial Intelligence (AAAI) Fall Symposium on Intelligent Narrative Technologies, Arlington, VA.

Kuhn, W. (2003). Semantic reference systems. International Journal of Geographical Information Science, *17*(5), 405-409.

Kuhn, W. and Raubal, M. (2003). Implementing Semantic Reference Systems. AGILE 2003 - 6th AGILE Conference on Geographic Information Science, Lyon, France.

Kunze, C. Lexikalischsemantische Wortnetze. Computerlinguistik und Sprachtechnologie, *2004*, 423-431.

Lakoff, G. (1987). Women, Fire and Dangerous Things. Chicago, Illinois: University of Chicago Press.

Miller, G. A. (1995). WordNet: a lexical database for English. Communications of the ACM, *38*(11), 39-41.

McKnight, T. L. (2000). Regional Geography of the United States and Canada (3rd ed.). Prentice Hall.

Morris, J. and Hirst, G. (2004). Non-classical lexical semantic relations. Workshop on Computational Lexical Semantics, Human Language Technology Conference of the North American Chapter of the Association for Computational Linguistics (HLT-NAACL 2004).

Piff, M. (1991). Discrete Mathematics - An introduction for software engineers. Cambridge, England: Cambridge University Press.

Rainie, L. and Tancer, B. (2007). 36% of online American adults consult Wikipedia; It is particularly popular with the well-educated and college-age students. Pew Internet and American Life Project. http://www.pewinternet.org/PPF/r/212 /report_display.asp

Schöning, J., Hecht, B., Rohs, M., and Starosielski, N. (2007). WikEar – Automatically Generated Location-Based Audio Stories between Public City Maps. 9th International Conference on Ubiquitous Computing Demo Proceedings, Innsbruck, Austria.

Strube, M. and Ponzetto, S. P. (2006). WikiRelate! Computing Semantic Relatedness Using Wikipedia. AAAI 2006.

Tobler, W. (1970). A computer movie simulating urban growth in the Detroit region. Economic Geography, *46*, 234-240.

Zesch, T., Gurevych, I., and Mühlhäuser, M. (2007a). Technical Report. Analyzing and Accessing Wikipedia as a Lexical Semantic Resource. Tuebingen, Germany.

Zesch, T., Gurevych, I., and Mühläuser, M. (2007b). Comparing Wikipedia and German Wordnet by Evaluating Semantic Relatedness on Multiple Datasets. NAACL-HLT.

# A Study on the Cognitive Plausibility of SIM-DL Similarity Rankings for Geographic Feature Types

Krzysztof Janowicz, Carsten Keßler, Ilija Panov, Marc Wilkes, Martin Espeter, Mirco Schwarz

Institute for Geoinformatics, University of Münster, Germany
janowicz | carsten.kessler | i.panov | marc.wilkes | m.espeter | mirco.schwarz | @uni-muenster.de

**Abstract.** The SIM-DL theory has been developed to enable similarity measurement between concept specifications using description logics. It thus closes the gap between similarity theories from psychology and formal representation languages from the AI community, such as the Web Ontology Language (OWL). In this paper, we present the results of a human participants test which investigates the cognitive plausibility of SIM-DL, that is, how well the rankings computed by the similarity theory match human similarity judgments. For this purpose, a questionnaire on the similarity between geographic feature types from the hydrographic domain was handed out to a group of participants. We discuss the set up and the results of this test, as well as the development of the according hydrographic feature type ontology and user interface. Finally, we give an outlook on the future development of SIM-DL and further potential application areas.

**Keywords:** similarity measurement, SIM-DL, geographic feature types, Web Ontology Language

## 1 Introduction and Motivation

Human judgments of similarity have been subject to research in psychology for more than fifty years now (Goldstone and Son 2005). Different approaches to modeling similarity have been developed, including feature-based, network-based, and geometric approaches. More recently, the artificial intelligence (AI) community started investigations on computational similarity models as a new method for information retrieval (Rissland 2006). The Matching Distance Similarity Measure (MDSM) (Rodríguez and Egenhofer 2004) was the first similarity-based model that has been developed specifically for the geospatial domain. The description logic based SIM-DL similarity theory (Janowicz 2006; Janowicz et al. 2007) discussed in this paper has been introduced to overcome the gap between models from psychology and formal knowledge representation languages used in the AI community.

Numerous geospatial applications carry potential for the implementation of similarity-based information retrieval techniques. Geoportals could supplement result pages with matches that do not exactly fit the user's query, but share certain characteristics with the matches. Location-based services could derive points of interest in the user's current vicinity from similar, previously visited places. From a developer's point of view, similarity measurements bear a great potential to simplify and accelerate processes that still require tedious manual configuration, such as data integration or ontology alignment. Those are just a few examples; in general, every tool that has to deal with fuzzy or ambiguous input—either from users or from other systems—is a candidate for the application of similarity.

For this paper, an application scenario from gazetteer research has been chosen. Current activities in this research area aim at the development of a distributed gazetteer infrastructure to replace existing isolated gazetteers (Janowicz and Keßler 2008) as shown in figure 1. In this envisioned infrastructure, similarity measurement is applied to enable interoperability among different gazetteers. Moreover, similarity allows for novel user interfaces which allow for imprecise input and do not require the user to know about the internal organization of the gazetteer any longer (see Janowicz and Keßler 2008 for details).

However, to enable such new techniques, a formal representation of the geographic feature types organized in gazetteers is required. As a starting point for first tests, an ontology specifying hydrographic feature types has been created. This ontology is used for the human participants test presented in this paper. The purpose of this test is to show that the similarity rankings calculated by SIM-DL correspond to human similarity judgments. The participants were asked to rate the similarity of a number of concepts such as *Lake*, *Ocean*, or *River* to the search concept *Canal*, solely based on the given concept definitions.

The remainder of the paper is organized as follows: section 2 presents previous work on similarity measurement and gazetteers. Section 3 introduces the SIM-DL theory and presents its implementation within the SIM-DL similarity server, as well as a novel gazetteer user interface that communicates with this server. Section 4 presents the human participants test and discusses the results, followed by conclusions in section 5.

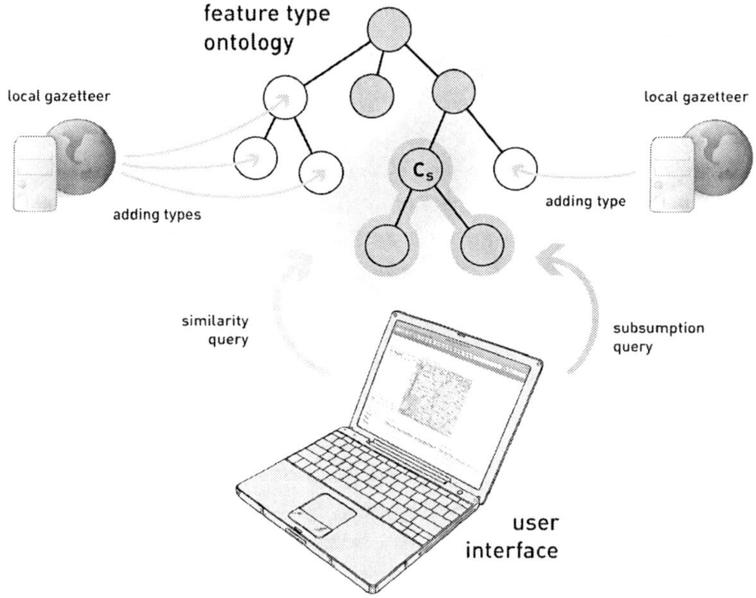


**Fig. 1:** Subsumption and similarity based information retrieval within the proposed gazetteer infrastructure (Janowicz and Keßler, 2008).

## 2 Related Work

This section points to previous work on similarity and gazetteer research with a special focus on similarity theories applied within GIScience.

### 2.1 Semantic Similarity Measurement

The theory of similarity has its origin in cognitive science and was established to determine why and how entities are grouped into categories, and why some categories are comparable to each other while others are not (Goldstone and Son 2005; Medin et al. 1993). The main challenge with respect to *semantic* similarity measurement is the comparison of meanings as opposed to purely structural comparison. A language has to be specified to express the nature of entities and a measurement theory needs to be established to determine how (conceptually) close compared entities are. While entities can be expressed in terms of attributes, the representation of entity types (concepts) is more complex. Depending on (computational) characteristics of the representation language, types are specified as sets of features, dimensions in a multidimensional space, or formal restrictions specified on sets using various kinds of description logics. While some representation languages have an underlying formal semantics (e.g., model theory), the grounding of several representation languages remains on the level of an informal description. As the compared types are representations of concepts in human minds, similarity depends on what is said (in terms of computational representation) about these types. This again is connected to the chosen language, leading to the fact that most similarity theories cannot be compared. Beside the question of representation, context is another major challenge for similarity assessments. In many cases, meaningful notions of similarity cannot be determined without defining in respect to what similarity is measured (Medin et al. 1993; Goodman 1972; Keßler 2007; Frank, 2007).

Similarity has been widely applied within GIScience over the past few years. Based on Tversky's (1977) feature model, Rodríguez and Egenhofer (2004) developed the Matching Distance Similarity Measure that supports a basic context theory, automatically determined weights, and asymmetry. Raubal and Schwering used so-called conceptual spaces to implement models based on distance measures within geometric space (Raubal 2004, Schwering and Raubal, 2005), while Sunna and Cruz (2007) applied a network based similarity measure for ontology alignment. Several measures (Janowicz 2006; d'Amato et al. 2006; Borgida et al. 2005) were developed to close the gap between (geo-) ontologies described by various kinds of description logics, and similarity theories that had not been able to handle the expressivity of such languages. Other similarity theories (Li and Fonseca 2006; Nedas and Egenhofer 2003) have been established to determine the similarity between spatial scenes. The ConceptVISTA (Gahegan et al. 2007) ontology management and visualization toolkit uses similarity for knowledge retrieval and organization.

## 2.2 Gazetteer Research

Gazetteers are knowledge organization systems for spatial information. They deliver feature types and geographic footprints for searched place names (Hill 2006). The categorization into feature types is crucial for geographic information retrieval, as it enables concept-based queries and reasoning in the first place (e.g., when looking for *villages in Catalunya*). However, current gazetteers such as the Alexandra Digital Library (ADL)

Gazetteer[6] are based on semi-formal feature type thesauri with limited support for formal reasoning methods.

The human participants test described in this paper has been carried out on a subset of a geographic feature type *ontology* (FTO). The FTO is currently being developed based on the ADL feature type thesaurus to demonstrate the benefits of subsumption and similarity based reasoning for Gazetteers (Janowicz and Keßler 2008). Such an ontology provides support for innovative user interfaces as discussed in section 3.3. Subsumption and similarity reasoning allow the user to intuitively *browse* the gazetteer, continuously being provided with all relevant information for the current query. Moreover, as outlined in section 1, they allow for new approaches towards gazetteer interoperability.

To ensure that such a user interface actually returns similar results as expected by the user, the cognitive plausibility of the similarity theory must be approved. In the following, we will outline the basic characteristics of the similarity theory SIM-DL, and discuss the human participants test and its results.

## 3 SIM-DL Theory, Implementation and Application

This section gives a brief insight into the SIM-DL similarity measurement theory, its implementation within the SIM-DL server, and the gazetteer web interface.

### 3.1 SIM-DL Theory

SIM-DL (Janowicz 2006; Janowicz et al. 2007) is an asymmetric and context aware similarity measurement theory used for information retrieval and organization. It compares a search concept $C_s$ with target concepts $\{C_t\}$ from an ontology (or several ontologies using the same shared vocabulary). The concepts themselves can be specified using various kinds of expressive description logics (Baader et al. 2003).

Within SIM-DL, similarity between concepts in canonical form (Janowicz 2006; Horrocks, 2003) is measured by comparing their definitions for overlap, where a high level of overlap indicates high similarity and vice versa. Description logics concepts are specified based on primitive concepts and roles using language constructors such as intersection, union,

---

[6] http://www.alexandria.ucsb.edu/gazetteer/

and existential quantification. Hence, similarity is defined as a polymorphic, binary, and real-valued function $C_s \times C_t \rightarrow R[0,1]$ providing implementations for all language constructs offered by the used description logics. The overall similarity between concepts is the normalized (and weighted) sum of the single similarities calculated for all parts (i.e., superconcepts) of the concept definitions. A similarity value of 1 indicates that the compared concepts cannot be differentiated, whereas 0 implies total dissimilarity. As most feature and geometric approaches, SIM-DL is an asymmetric measure, i.e., the similarity $sim(C_s, C_t)$ is not necessarily equal to $sim(C_t, C_s)$. Therefore, the comparison of two concepts does not only depend on their descriptors, but also on the direction in which both are compared.

A single similarity value (e.g., 0.67) computed between two concepts hides most of the important information. It does not answer the question whether there are more or less similar target concepts in the examined ontology. It is not sufficient to know that possible similarity values range from 0 to 1 as long as their distribution is unclear. Imagine an ontology where the least similar target concept has a value of 0.6 (compared to the source concept), while the comparison to the most similar concept yields 0.9. In this case, a similarity value of 0.67 is not high at all. Beside these interpretation problems, isolated comparison puts too much stress on the concrete similarity value. It is hard to argue that and why the result is (cognitively) plausible without other reference values (Jurisica, 1994).

Accordingly, SIM-DL focuses on similarity rankings. The search concept is compared to all target concepts derived from the measurement context (Janowicz 2006; Keßler 2007; Keßler et al. 2007; Janowicz 2008), i.e., a subset of the ontology, also referred to as the domain of application. The result is an ordered list with descending similarity values. Consequently, in the following we do not argue that single similarity values are cognitively plausible, but that the computed order correlates with human ranking judgments.

## 3.2 SIM-DL Server

The SIM-DL similarity server and a client plug-in for the Protégé Ontology Editor[7] are available as an open-source cross-platform project at SourceForge.net. The current beta version[8] supports basic reasoning ser-

[7] http://protege.stanford.edu/

[8] The current release can be downloaded at http://sim-dl.sourceforge.net/.

vices (e.g., subsumption reasoning) and similarity measurement up to *ALCHQ*[9] (Janowicz et al. 2007; Baader et al. 2003).

The reasoning component implements a tableaux algorithm to determine TBox subsumption based on ABox satisfiability, while the similarity component is based on the SIM-DL framework and theory. Each similarity request involves interaction with the reasoning component to determine all target concepts in the context. Furthermore, the reasoner is required for several similarity functions and optimization.

The SIM-DL server interprets incoming requests and starts the similarity and reasoning engines. The requests conform to the DIG 1.1 specification (Bechhofer 2003) which provides a standardized XML-based interface for reasoning services. In our previous work, the DIG interface has been extended in order to support similarity measurement between concepts (Janowicz et al. 2007). Within this paper, the server is used to compute the similarity values for the compared concepts and to interact with the new gazetteer web interface.

### 3.3 Application Scenario: Gazetteer Web Interface

The Gazetteer Web Interface connects the SIM-DL Server with the Alexandria Digital Library (ADL) gazetteer offering an enhanced search mechanism for geographic features. It is realized as a mashup combining an auto-suggest input field for feature types, an input field for feature names, and Google Maps™. The map is used to restrict a query's spatial extent and to display matching features retrieved by the gazetteer.

The intention with regard to the development of this interface is to optimize the gazetteer request procedure using similarity reasoning. In contrast to the standard ADL front-end, it does not require knowledge of the gazetteer's internal feature type thesaurus (FTT) hierarchy. The SIM-DL server uses an ontology (FTO) extending the FTT hierarchy that provides the concepts and relations being utilized within the SIM-DL similarity measurement process. The autosuggest text field used for searching feature types is based on *Asynchronous Javascript and XML* (AJAX) technology: as the user enters the name of the requested feature type, feature types matching the letters entered so far are automatically retrieved and displayed. The suggestions returned by the SIM-DL server consist of the suggested type itself, its super types, and similar types. The similar types are

[9] See http://www.cs.man.ac.uk/~ezolin/dl/ and Janowicz et al. (2007) for more details about the used description logic and its computational characteristics.

presented in different font sizes, reflecting their similarity to the suggested concept. Comparable to *tag clouds*, the bigger a concept is displayed, the more similar it is. All suggestions are hyperlinked and are shifted into the input field when clicked.

Beside the selection of a feature type and the definition of an area of interest, users can also search by place names, such as the *Dortmund-Ems Canal*. The results are displayed on the selected map extent as shown in figure 2.

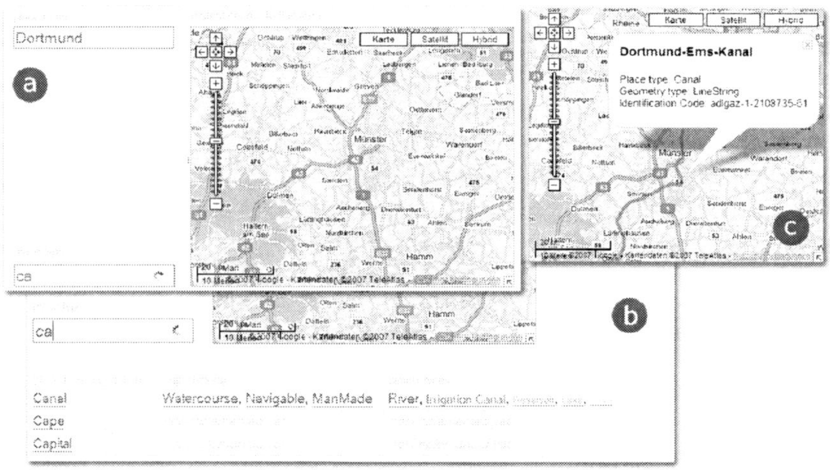


**Fig. 2:** The similarity enabled gazetteer web interface (Janowicz and Keßler, 2008).

## 4 Evaluation

This section presents the human participants test that has been performed to prove the cognitive plausibility of SIM-DL similarity rankings. The test results are evaluated and discussed.

### 4.1 Motivation

SIM-DL is intended to measure similarity between computational representations of concepts. The motivation is to improve the accessibility of tasks such as information retrieval and organization for human users. This can only be achieved if there is a high correlation between the similarity rankings calculated by SIM-DL and human similarity judgments. The

SIM-DL measurement process has been developed based on findings from cognitive science. It takes aspects such as asymmetry, alignment, and context into account which are known to play an important role for human similarity ratings. SIM-DL tries to approximate aspects from the human process of reasoning about similarity to achieve meaningful results. Nevertheless, it is a computational theory for description logics rather than a framework towards understanding cognitive processes. Consequently, we neither claim that SIM-DL models (or even explains) the process of human similarity judgments nor that humans represent concepts (if they do) in any kind of logic based serialization.

Figure 3 illustrates the relation between a similarity reasoning service such as the SIM-DL server and human reasoning about similarity. The box at the top represents the cognitive process (marked as dotted line) of deriving similarity judgments. Without discussing whether there is something such as representation and cognition (or only perception) (Gibson 1977; Markman 2000), up to now no direct mapping to computational representations is possible. Similarity theories developed in cognitive science model (i.e., approximate) this process by partitioning it into observable units. The effect of each unit is studied by changing its settings, while all other units remain stable[10]. Such units include Context, Alignment, Asymmetry, and the Max-Effect (Medin et al. 1993). Each of them is depicted as a box on the dotted process line to indicate that they are fragments of the whole process. Most theories from cognitive science focus on the explanation of human similarity reasoning rather than the development of executable services[11]. Consequently, the chosen representation is more on the informal side. In contrast, information science is interested in computational representations to provide a basis for executable theories. While these theories try to approximate cognitive theories, their goal is not explanatory. Instead, they adopt such units that can be computed with appropriate resources. From this point of view, computational models form a subset of theories established in cognitive science. Typical application areas include human computer interaction and information retrieval.

---

[10] or by studying patients with lesions.

[11] For some exceptions, see SME and MAC/FAC (Falkenhainer et al. 1989, Gentner and Forbus, 1991).

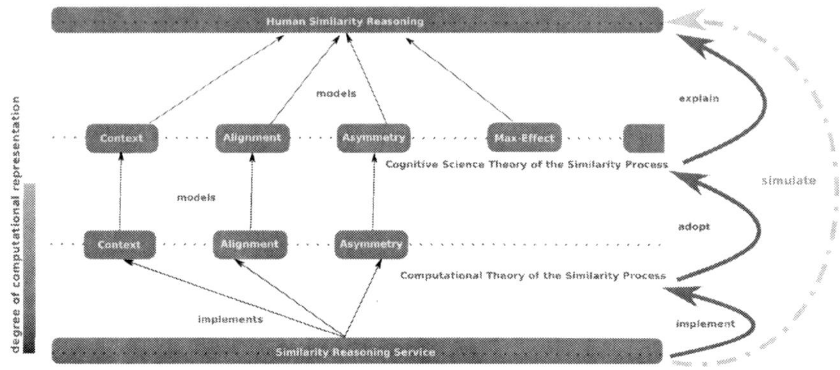


**Fig. 3:** From human similarity reasoning to similarity services such as the SIM-DL similarity server.

The box at the bottom of figure 3 represents concrete similarity reasoning services such as the SIM-DL similarity server. These services implement the computational theories as standalone applications or as parts of a knowledge infrastructure such as the ConceptVISTA[12] toolbox. The motivation for developing similarity-aware applications is to simulate human similarity judgement, thus making tasks such as information retrieval more accessible to the user. It is important to note that not the cognitive process is simulated, but the final similarity ranking, i.e., the reasoning results. The dashed arrow indicates that there is no direct link between the similarity service and human similarity judgments. Computational similarity ratings depend on how compared entities and concepts are represented and which units (parts) of the human similarity process are modeled within the implemented computational theories.

The term *cognitively plausible* will be used, if the similarity rankings produced using SIM-DL correlate with human similarity rankings. In contrast, *cognitively adequate* would require a comparison of the underlying processes (their units) and is out of scope of this work.

## 4.2 Test Setting

A human participants test has been performed to prove that the results calculated by the SIM-DL theory introduced in section 3 correlate with human similarity judgments. Very few objective tests have been carried out so far concerning the usage of similarity measurement in practice. Due to

---

[12] http://www.geovista.psu.edu/ConceptVISTA/

the fact that the participants were all native German speakers, the test was in German, too. In the following, we will translate the German parts of the test into the best-fitting English expressions where it is necessary for the understanding.

28 participants were recruited for the human participants test. The group of participants consisted of 16 males and 12 females. The mean age of the 28 participants was 27.3 with a range from 22 to 31 years. The mean female age was 26.4 and the mean age of the males was 27.8. The questionnaire[13] was distributed randomly among the participants (Montello and Sutton 2006).

The first step for every participant was to read the introduction, consisting of a brief motivation for the test, and instructions on how to fill it in. According to Harrison (1995), written instructions are preferred by participants over spoken instructions. Next, every participant was asked to read the concept descriptions of the given feature types: the named search concept *Canal* (ger.: Kanal) and a set of anonymous target concepts (see figure 4). Every participant was requested to assess the similarity between the description of the search concept and every description of the target concepts by placing a mark between a line ranging from minimum to maximum similarity. Additionally, the participants made a statement how confident they felt when placing the mark using a discrete scale with five classes from not sure (ger.: nicht sicher) to sure (ger.: sicher). It is assumed that a continuous scale for assessing the concept similarity is reasonable due to the provided granularity[14] which is not required for the confidence assessments. The range for the continuous scale went from minimum similarity (ger.: minimale Ähnlichkeit) to maximum similarity (ger.: maximale Ähnlichkeit). The reason for omitting the names of the target concepts was to ensure that the similarity judgments only depend on the concept descriptions and are not biased by the participants' individual conceptualizations.

[13] The questionnaire is available at http://sim-dl.sourceforge.net/downloads/.

[14] For example, to allow statements such as "the similarity between Canal and concept A is almost equal to the similarity between Canal and concept B, but the former seems to be a bit higher".

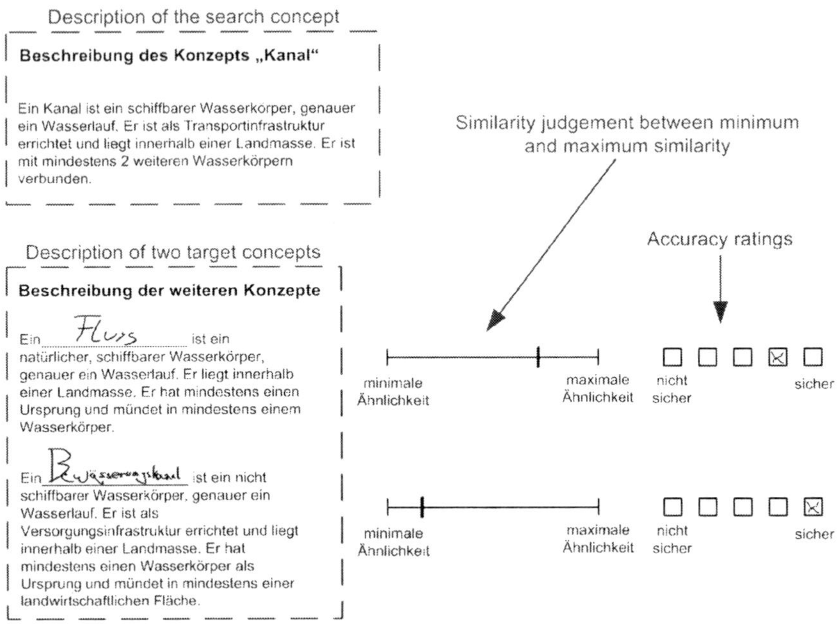


**Fig. 4:** Section of the questionnaire, showing the search concept *Kanal* (eng. Canal: "A canal is a navigable body of water, namely a watercourse. It is constructed as transport-infrastructure, that is inside landmass. It is connected to at least two other bodies of water") and two of the six target concepts, *Fluss* (eng. River: "A river is a natural, navigable body of water, namely a watercourse. It is inside landmass. It is connected to at least one spring as origin and at least one body of water as destination") and *Bewässerungskanal* (eng. Irrigation Canal: "An irrigation canal is a non-navigable body of water, namely a watercourse. It is constructed for supply and infrastructure and is inside landmass. It is connected to at least one body of water as origin and at least one agricultural area as destination.")

In the final step, the participants were asked to assign a given list of (concept) names to the anonymous concept descriptions. This final step was introduced to check whether the presented concept descriptions correspond to the participant's conceptualization. Moreover, wrong assignments of the concept names are a strong hint that the test was filled in randomly and thus useless for the evaluation; this check was considered necessary as there was no financial compensation for the participants' effort.

While a detailed insight into the underlying feature type ontology used for similarity reasoning is out of scope here, the following example demonstrates how the concepts were specified.

$$
\begin{aligned}
Canal \sqsubseteq\, & WaterBody \sqcap Watercourse \sqcap Navigable \sqcap (\exists inside.Landmass) \\
& \sqcap(\exists constructedAs.Transportation) \\
& \sqcap(\geq 2\, connectedTo.Waterbody)
\end{aligned}
$$

While some concepts used to describe *Canal* are primitives (e.g., *Navigable*), other concepts are defined within the ontology. For instance, *WaterBody* is a subconcept of *HydrographicFeature*. Note that for reasons of simplification, and to keep the cognitive load low, the plain text descriptions presented to the participants hide some details. Beside their role in transportation, canals can also be constructed for additional purposes[15].

## 4.3 Results

Out of the 28 questionnaires, 26 were taken for further processing. First, it was checked whether the concept names were properly assigned to the descriptions. All 26 questionnaires satisfy this requirement, however, several participants made updates (changed the names) while performing the test. Next, the similarity values and confidence assessments were transformed to values and weights, respectively, between 0 and 1. Each confidence box corresponds to a weighting step of 0.2. The first box was transformed to 0.2, the second to 0.4, and so on. Table 1 shows the absolute similarity values obtained using the SIM-DL similarity server, the arithmetic mean of the human similarity judgments, and the weighted mean using the confidence assessments.

**Table 1** Mean (absolute) similarity judgments by test subjects, compared to SIM-DL.

| Concept | Fluss (River) | Bewässerungskanal (Irrigation Canal) | Stausee (Reservoir) | See (Lake) | Ozean (Ocean) | Förderplattform (Offshore Platform) |
|---|---|---|---|---|---|---|
| SIM-DL server | 0.75 | 0.67 | 0.58 | 0.5 | 0.38 | 0.08 |
| Arithm. mean | 0.7 | 0.53 | 0.59 | 0.43 | 0.33 | 0.14 |
| Weighted mean | 0.72 | 0.55 | 0.6 | 0.43 | 0.32 | 0.13 |

In a next step, the absolute similarity values from each questionnaire were transformed to ordinal scale, i.e., into a descending similarity ranking. The most similar concept (with respect to Canal) was ranked 6, while the least similar got the rank 1. If two or more concepts had the same absolute similarity values, a mean rank (tie) was chosen (e.g., 4.5). The weights have no influence on the ranking position. Figure 5 shows the resulting box plot for the 26 questionnaires.

[15] This does not influence similarity, as the same mapping was performed for all concepts used for the test and the participants had to compare the descriptions (not knowing which concepts were actually described).

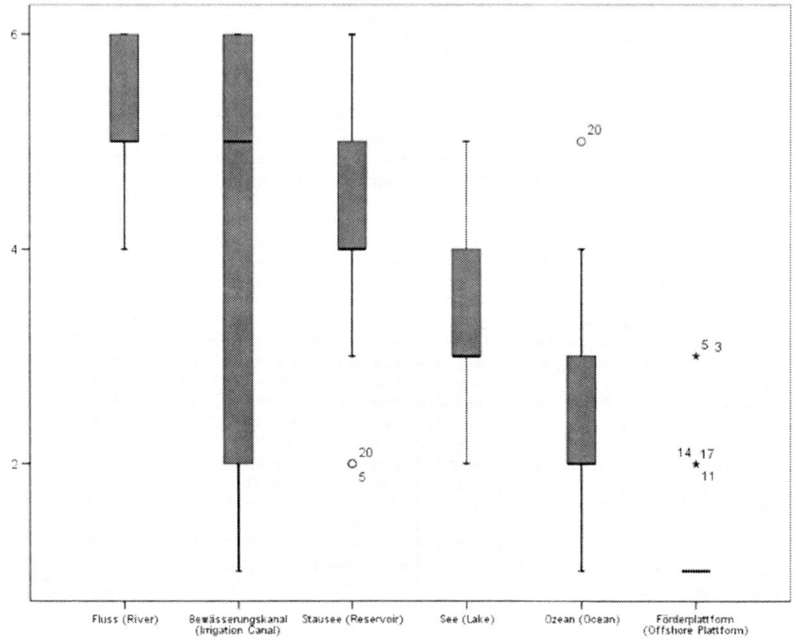


**Fig. 5:** Boxplot showing the human similarity rankings and their dispersion.

It depicts the lowest non-outlier ranking, the lower quartile (25%), the median, upper quartile (75%), and highest non-outlier ranking per target concept. The stars and dots represent mild and extreme outliers. *River*, *Reservoir*, *Lake*, and *Ocean* have a comparable interquartile range, while the boxplot for the *Offshore Platform* is collapsed. In contrast, the *Irrigation Canal* boxplot shows a high distribution among test subjects.

As depicted in table 2, the individual ranking data from each questionnaire was used to compute the median and mode for each target concept. In both cases, the resulting order corresponds to the computed similarity ranking except that *River* and *Irrigation Canal* share the same rank. In terms of frequencies, this means that the majority of test subjects has chosen the same rank as SIM-DL for *Reservoir*, *Lake*, *Ocean*, and *Offshore Platform*. In case of *River*, the same number of participants had chosen the $6^{th}$ and $5^{th}$ rank (12 times), while SIM-DL ranks *River* as most similar concept to *Canal* ($6^{th}$ rank). The remaining two participants selected the $4^{th}$ rank. While the median for *Irrigation Canal* corresponds to the computed $5^{th}$ rank, the mode is 6. This is caused by the high dispersion for this concept. The human rankings range from the first (4 times) up to the sixth rank (8 times).

**Table 2** Median and mode similarity ranks for the target concepts based on the test results.

| | | Fluss (River) | Bewässerungskanal (Irrigation Canal) | Stausee (Reservoir) | See (Lake) | Ozean (Ocean) | Förderplattform (Offshore Platform) |
|---|---|---|---|---|---|---|---|
| N | Valid | 26 | 26 | 26 | 26 | 26 | 26 |
| | Missing | 0 | 0 | 0 | 0 | 0 | 0 |
| Median | | 5.0000 | 5.0000 | 4.0000 | 3.0000 | 2.0000 | 1.0000 |
| Mode | | 5.00[a] | 6.00 | 4.00 | 3.00 | 2.00 | 1.00 |
| frequency (#) | | | | | | | |
| 6th rank | | **12** | **8** | 6 | 0 | 0 | 0 |
| 5th rank | | **12** | 6 | 4 | 1 | 1 | 0 |
| 4.5th rank[b] | | 1 | - | 1 | | - | - |
| 4th rank | | 1 | 2 | **12** | 10 | 1 | 0 |
| 3rd rank | | 0 | 3 | 1 | **14** | 7 | 2 |
| 2nd rank | | 0 | 3 | 2 | 1 | **16** | 3 |
| 1st rank | | 0 | 4 | 0 | 0 | 1 | **21** |

a: Multiple modes exist (5 and 6). The smallest value is shown.
b: This rank is caused by the normalized ranking process of SPSS.

A correlation analysis between the median human similarity ranking and the ranking computed by SIM-DL yields $r_s$ = 0.986 (p = 0.01) using Spearman's $\rho$. As depicted in figure 5, the data is not normally distributed, i.e., skewed. In addition, we cannot assume equi-distance between the ranks. Hence, the correlation was also determined using Kendall's $\tau$ and yields 0.966 (p = 0.01).

To measure the consensus among participants with respect to the chosen rank, Kendall's coefficient of concordance *W* was used. To determine whether an obtained *W* value is significant, chi-square was computed for given degrees of freedom and compared to significance tables for probability. The analysis (taking the ties from the ranking process into account) yields a value of 0.632 for *W* with a Chisq(5) of 82.1. If we hypothesize that the participant's ranks are associated, this corresponds to a proability of $p < 0.001$ that we accept the hypothesis while it is false. Consequently, and with respect to the high number of participants, the results are significant.

## 4.4 Discussion

The test shows a strong and significant correlation between human similarity rankings and those obtained using the SIM-DL similarity server. Based on our previous definition, the computed similarity judgments can be called cognitively plausible. The correspondence between the absolute similarity values is difficult to interpret. Each participant has its own (cognitive) similarity scale and distribution, i.e., the similarity value for the most and least similar concept differs between participants. For instance, the absolute values for the concept *River* range from 0.93 to 0.73 for participants that had chosen *River* to be the most similar concept to *Canal.*

Overall, SIM-DL values are close to the (weighted) mean similarity judgments, but tend to overestimate.

While these results look promising, the interquartile ranges raise some questions. This becomes especially apparent in case of *Irrigation Canal* and partly also for *Reservoir*. In the first case, while most participants had chosen a high similarity ($5^{th}$ or $6^{th}$ rank), several subjects ranked *Irrigation Canal* as very dissimilar. There may be two potential explanations for these results. Out of all compared concept descriptions, *Irrigation Canal* is the only one specified as a non-navigable body of water, while all others (except *Offshore Platform*) are navigable. When subjects compare *Irrigation Canal* to *Canal*, they use the previously made similarity judgments as points of reference. While *Offshore Platform* is too different to serve as a reference, all other concepts share a feature that is missing for *Irrigation Canal*. In this case navigable becomes the characteristic feature of the set of compared concepts and gets a high weighting. This explanation corresponds well to the variability context weighting[16] proposed by Rodrígez and Egenhofer (2004) as well as to Tversky's (1977) notion of diagnosticity. Tversky argues that features which are diagnostic for a particular classification have a disproportionate influence on similarity judgments.

A second explanation could be based on different kinds of information processing and extraction. One has to keep in mind that while the similarity server and the participants share the same information about the presented concepts, their representation is different (plaintext versus description logics). The similarity ranking task involves some deductive reasoning steps. For instance, canals were defined as entities which are connected to at least two bodies of water, while rivers have at least one waterbody as origin and one waterbody as destination. The underlying ontology represents this using the three relations *connectedTo* and its sub-relations *hasOrigin* and *hasDestination*. When searching for entities connected to waterbodies, an entity with a waterbody as origin satisfies this requirement and should be similar. Participants seem to perform this kind of reasoning and therefore assign a high rank to *River*. In contrast, irrigation canals have at least one waterbody as origin and one agricultural area as destination. Instead of judging the origin and destination separately, participants may summarize both to a non matching feature (Tversky 1977).

[16] Up to now, SIM-DL only supports a context weighting comparable to the commonality weighting in MDSM.

## 5 Conclusions and Further Work

Based on the performed human participants test, the SIM-DL theory returns cognitively plausible similarity rankings. To ensure that both the human and the computer similarity judgments are based on the descriptions of the concepts, i.e., their representations, the concept names were left blank during the first step of the test. Accordingly, the participants had to assess the similarity of the concept *descriptions*, instead of their own conceptualizations. The complexity of the ontology used for the test was limited to a small number of concepts that were specified only by their most important characteristics to avoid a cognitive overload for the participants. Future work needs to investigate how more complex ontologies can be presented during a human participants test.

While the test shows that the similarity rankings correlate, it does not answer the question whether their integration and visualization within the proposed gazetteer web interface improves usability. Strictly speaking, one should also not argue that the rankings delivered by the Gazetteer Web Interface correspond to human similarity judgments. The motivation for using a gazetteer might put the focus on other parts of the concept description and hence influence similarity. Consequently, the next step has to be an evaluation of the web interface. Moreover, the feature type ontology needs to be extended by more concepts and more detailed specifications to demonstrate that the developed methods are able to cope with larger information bases.

Parts of the SIM-DL theory have been used within other projects such as a web service for identity assumptions for historical places. As SIM-DL has no own visualization module, an integration within the ConceptVISTA toolkit might be a promising next step.

## Acknowledgments

This work is funded by the project *Semantic Similarity Measurement for Role-Governed Geospatial Categories* (*SimCat*) granted by the German Research Foundation (DFG Ra1062/2-1).

## References

Baader, F., Calvanese, D., McGuinness, D. L., Nardi, D., and Patel-Schneider, P. F., editors (2003). The Description Logic Handbook: Theory, Implementation, and Applications. Cambridge University Press.

Bechhofer, S. (2003). The DIG description logic interface: Dig/1.1. In DL2003 Workshop, Rome, Italy.

Borgida, A.,Walsh, T., and Hirsh, H. (2005). Towards measuring similarity in description logics. In Proceedings of the 2005 International Workshop on Description Logics (DL2005), volume 147 of CEUR Workshop Proceedings. CEUR, Edinburgh, Scotland, UK.

d'Amato, C., Fanizzi, N., and Esposito, F. (2006). A dissimilarity measure for ALC concept descriptions. In Proceedings of the 2006 ACM Symposium on Applied Computing (SAC), pages 1695–1699, Dijon, France.

Falkenhainer, B., Forbus, K., and Gentner, D. (1989). The structure-mapping engine: Algorithm and examples. Artificial Intelligence, 41, 1–63.

Frank, A. U. (2007). Similarity measures for semantics: What is observed? In COSIT'07 Workshop on Semantic Similarity Measurement and Geospatial Applications, Melbourne, Australia.

Gahegan, M., Agrawal, R., Banchuen, T., and DiBiase, D. (2007). Building rich, semantic descriptions of learning activities to facilitate reuse in digital libraries. International Journal on Digital Libraries, 7(1), 81–97.

Gentner, D. and Forbus, K. D. (1991). MAC/FAC: a model of similarity-based retrieval. In Proceedings of the 13th Cognitive Science Conference, pages 504–509, Chicago. Erlbaum, Hillsdale.

Gibson, J. (1977). The theory of affordances. In R. Shaw and J. Bransford, editors, Perceiving, Acting, and Knowing - Toward an Ecological Psychology, pages 67–82. Lawrence Erlbaum Ass., Hillsdale, New Jersey.

Goldstone, R. and Son, J. (2005). Similarity. In K. Holyoak and R. Morrison, editors, Cambridge Handbook of Thinking and Reasoning. Cambridge University Press.

Goodman, N. (1972). Seven strictures on similarity. In N. Goodman, editor, Problems and projects, pages 437–447. Bobbs-Merrill, New York.

Harrison, S. (1995). A comparison of still, animated, or nonillustrated on-line help with written or spoken instructions in a graphical user interface. In Proceedings of the SIGCHI conference on Human factors in computing systems, pages 82–89. ACM Press/Addison-Wesley Publishing Co. New York, NY, USA.

Hill, L. L. (2006). Georeferencing: The Geographic Associations of Information (Digital Libraries and Electronic Publishing). The MIT Press.

Horrocks, I. (2003). The description logic handbook: theory, implementation, and applications, chapter Implementation and optimization techniques, pages 306–346. Cambridge University Press, New York, NY, USA.

Janowicz, K. (2006). SIM-DL: Towards a semantic similarity measurement theory for the description logic ALCNR in geographic information retrieval. In R. Meersman, Z. Tari, P. Herrero, et al., editors, SeBGIS 2006, OTM Workshops 2006 , volume 4278 of Lecture Notes in Computer Science, pages 1681 - 1692. Springer, Berlin.

Janowicz, K. (2007). Similarity-based retrieval for geospatial semantic web services specified using the web service modeling language (WSMLCore). In A. Scharl and K. Tochtermann, editors, The Geospatial Web - How Geo- Browsers, Social Software and the Web 2.0 are Shaping the Network Society, Lecture Notes in Computer Science. Springer, Berlin.

Janowicz, K. (2008; forthcoming). Kinds of contexts and their impact on semantic similarity measurement. In 5th IEEE Workshop on Context Modeling and Reasoning (Co-

MoRea) at the 6th IEEE International Conference on Pervasive Computing and Communication (PerCom'08), Hong Kong. IEEE Computer Society.

Janowicz, K. and Keßler, C. (accepted for publication 2008). The role of ontology in improving gazetteer interaction. To appear in issue 10/2008 of International Journal of Geographical Information Science.

Janowicz, K., Keßler, C., Schwarz, M., Wilkes, M., Panov, I., Espeter, M., and Bäumer, B. (2007). Algorithm, implementation and application of the SIM-DL similarity server. In F. Fonseca and M. A. Rodríguez, editors, Second International Conference on GeoSpatial Semantics, GeoS 2007 , Lecture Notes in Computer Science 4853, pages 128–145. Springer-Verlag Berlin Heidelberg.

Jurisica, I. (1994). Technical report dkbs-tr-94-5: Context-based similarity applied to retrieval of relevant cases. Technical report, University of Toronto, Department of Computer Science, Toronto.

Keßler, C. (2007). Similarity measurement in context. In B. Kokinov, D. C. Richardson, T. R. Roth-Berghofer, and L. Vieu, editors, 6th International and Interdisciplinary Conference CONTEXT 2007, Roskilde, Denmark. Lecture Notes in Artificial Intelligence 4635, pages 277–290. Springer-Verlag Berlin Heidelberg.

Keßler, C., Raubal, M., and Janowicz, K. (2007). The effect of context on semantic similarity measurement. Paper presented at the 3rd International IFIP Workshop On Semantic Web & Web Semantics (SWWS '07). In R. Meersman, Z. Tari, P. Herrero et al.: On The Move – OTM 2007 Workshops Part II, pages 1274-1284. Lecture Notes in Computer Science, Springer Verlag Berlin Heidelberg.

Li, B. and Fonseca, F. (2006). Tdd - a comprehensive model for qualitative spatial similarity assessment. Spatial Cognition and Computation, 6(1), 31–62.

Markman, A. B. and Dietrich, E. (2000). In defense of representation. Cognitive Psychology, 40, 138–171.

Medin, D., Goldstone, R., and Gentner, D. (1993). Respects for similarity. Psychological Review, 100(2), 254–278.

Montello, D. and Sutton, P. (2006). An Introduction to Scientific Research Methods in Geography. Sage Publications Ltd.

Nedas, K. and Egenhofer, M. (2003). Spatial similarity queries with logical operators. In T. Hadzilacos, Y. Manolopoulos, J. Roddick, and Y. Theodoridis, editors, SSTD '03 - Eighth International Symposium on Spatial and Temporal Databases, Santorini, Greece, volume 2750 of Lecture Notes in Computer Science, pages 430–448.

Raubal, M. (2004). Formalizing conceptual spaces. In A. Varzi and L. Vieu, editors, Formal Ontology in Information Systems, Proceedings of the Third International Conference (FOIS 2004), volume 114 of Frontiers in Artificial Intelligence and Applications, pages 153–164. IOS Press, Amsterdam, NL.

Rissland, E. L. (2006). Ai and similarity. IEEE Intelligent Systems, 21(3), 39–49.

Rodríguez, A. and Egenhofer, M. (2004). Comparing geospatial entity classes: an asymmetric and contextdependent similarity measure. International Journal of Geographical Information Science, 18(3), 229–256.

Schwering, A. and Raubal, M. (2005). Spatial relations for semantic similarity measurement. In J. Akoka, S. Liddle, I.-Y. Song, M. Bertolotto, I. Comyn-Wattiau, W.-J. vanden Heuvel, M. Kolp, J. Trujillo, C. Kop, and H. Mayr, editors, Perspectives in Conceptual Modeling: ER 2005 CoMoGIS Workshop, Klagenfurt, Austria., volume 3770 of Lecture Notes in Computer Science, pages 259–269. Springer, Berlin.

Sunna, W. and Cruz, I. (2007). Using the agreementmaker to align ontologies for the OAEI campaign 2007. In The Second International Workshop on Ontology Matching, collocated with the 6th International Semantic Web Conference ISWC, Busan, Korea.

Tversky, A. (1977). Features of similarity. Psychological Review, 84(4), 327–352.

# A Geospatial Implementation of a Novel Delineation Clustering Algorithm Employing the *K-means*

Tonny J. Oyana, Kara E. Scott

Department of Geography and Environmental Resources, Southern Illinois University, 1000 Faner Drive, MC 4514, Carbondale, IL 62901-4514, USA, tjoyana@siu.edu; skara@siu.edu

**Abstract.** The overarching objective of this paper is to introduce a novel Fast, Efficient, and Scalable *k-means* (*FES-k-means**) algorithm. This algorithm is designed to increase the overall performance of the standard *k-means* clustering technique. The *FES-k-means** algorithm uses a hybrid approach that comprises the *k-d* tree data structure, the nearest neighbor query, the standard *k-means* algorithm, and Mashor's adaptation rate. The algorithm is tested using two real datasets and two synthetic datasets and is employed twice on all four datasets. The first trial consisted of previously *MIL-SOM** trained data, and the second was on raw, untrained data. The approach presented with this method enables unfounded knowledge discovery, otherwise unclaimed by conventional clustering methods. When used in conjunction with the *MIL-SOM** training technique, the *FES-k-means** algorithm reduces the computation time and produces quality clusters. In particular, the robust *FES-k-means** method opens doors to (1) *faster* cluster production than conventional clustering methods, (2) *scalability* allowing application in other platforms, and its ability to handle small and large datasets, compact or scattered, and (3) *efficient* geospatial data analysis of large datasets. All of the above makes *FES-k-means** live up to defending its well-deserved name—Fast, Efficient, and Scalable *k-means* (*FES-k-means**). The findings of this study are vital to the relatively new and expanding subfield of geospatial data management.

**Keywords:** spatial data, data mining, clustering techniques, large datasets, k-means

# 1 Introduction

The most substantive and authoritative review on the use and applications of *k-means* clustering by Steinley (2006), a half-century synthesis, observed the following: (1) a continuing demand to design methods that provide optimal partitions within a reasonable amount of computation time; (2) a persistent lack of detailed explanation in many current texts regarding statistical properties of *k-means* and related techniques despite its popularity and widespread use in many domains and applications; and (3) varying research efforts and successes in lieu of the local optima, methods of initialization, methods to estimate *k*, variable selection, and the detection of influential observations. Steinley (2006) made solid, interesting recommendations and also suggested solutions on the most promising efforts and recent findings from his synthesis of past work. In this study, we attempted to address Steinley's first concern, and through searching and tagging together with analytical visualization, we detected some influential observations from the experimental datasets.

The most common grouping schemes are hierarchical and partitional (Jain et al. 1999; Vesanto and Alhoniemi 2000). In addition to hierarchical and partitional, other major clustering techniques include density-based algorithms and model-based algorithms. Clustering algorithms that fall within these groupings include but are not limited to, *k-means* & *k-mediods* (partitioning algorithms); BIRCH (Balanced Iterative Reducing and Clustering using Hierarchies), (Zhang et al. 1996); CURE (Clustering Using Representatives), (Guha et al. 1998); CHAMELEON (hierarchical clustering techniques for large datasets), (Karypis et al. 1999); DBSCAN (density-based clustering algorithm), (Ester et al. 1996); and genetically guided algorithms, neural networks-based algorithms, and vector quantization algorithms using a model-based technique. There are methods that are simulated in conjunction with common clustering techniques used as a preprocessor or initialization to data delineation, which include non-linear projection methods, such as Multi-dimensional Scaling (MDS), (Kruskal 1964), and data reduction techniques, including Principal Component Analysis (PCA); Sammons mapping (Sammon 1969; Jain et al. 1999; Xu and Wunsch II 2005), the Self-Organizing Map (SOM) (Kohonen 1990); among many others.

Partitional clustering methods first partition the data set of *N* objects into an initial set of *k* clusters; second, they improve the clustering by iteratively relocating the objects from one group to another using a cluster quality validating technique.

There are two heuristic approaches to this partitioning. The first approach uses the *k-means*. It is the simplest and most commonly used squared error-based clustering method (MacQueen 1967; Jain et al. 1999; Xu and Wunsch II 2005; Steinley 2006). The *K-means* algorithm is applicable only when the mean of a cluster is defined; it is not, however, suitable for discovering clusters with non-convex shapes nor clusters of size variation. In addition, the standard *k-means* algorithm is sensitive to outliers, commonly referred to as "noise" since a small number of such data can substantially influence the mean value of the points. Another major drawback of the standard *k-means* algorithm is that it relies heavily on the initial starting conditions, which can limit clustering to local versus global optimization (Chinrungrueng and Sequin 1995; Vrahatis et al. 2002; Likas et al. 2003). In such a case, an error may exist in the initial cluster selection (Jain et al. 1999). The *k-means* algorithm has also been found to be computationally expensive regarding the number of partitions and iterations for a given dataset.

The second approach employs the *k-medoids* in which each cluster is represented by one of the objects of the cluster located near the center. It solves the problem of sensitivity to outliers that is present in the well-known *k-means* algorithm. Partitioning around *medoids* (PAM) uses the *k-medoids* method. Due to its slow performance, PAM works well with small datasets. Other algorithms in this category include clustering large applications (CLARA) which applies PAM. Clustering large applications based on randomized search (CLARANS) is an improved *k-medoids* method designed for larger datasets. In summary, partitioning methods work well with finding spherical-shaped clusters in small or medium datasets. In some instances, partitioning methods can be extended for larger datasets. The major setback for any solo *k*-partitioning method is the need for a predetermined value of *k* or the number of clusters to be discovered.

Due to the problems mentioned earlier, several variations of the *k-means* algorithm have been proposed to produce better-quality clusters. To improve computational speed, a filtering *k-means* algorithm using a *k-d* tree structure was proposed by Alsabti et al. (1998) and Kanungo et al. (2002). Ball and Hall (1967) introduced a method that employs a merging and splitting technique to address the problem of initializing *k* to identify the number of clusters and the sensitivity to outliers, namely ISODATA. This technique dynamically addresses the number of clusters according to predefined thresholds. The splitting operation eliminates the clusters that contain minimal patterns or patterns that cause an elongation of clusters that can result from the inclusion of outliers; this technique though valuable in attempt, can prove to be problematic because outlier elimination may not always be desirable. In some cases, identifying and reviewing outliers can assist in revealing valuable information in cluster analysis. To conquer the local minima problem, Likas et al. (2003) proposed the deterministic global *k-means* clustering algorithm, which performs a series of local searches to obtain an optimal global solution for clustering. The claim is that each successive centroid is independent of the previous centroid, which minimizes the clustering error that accompanies the search procedure for the initial centers in the standard *k-means*. For the latter case, although the method is computationally efficient, the performance quality is greatly reduced (Likas et al. 2003). A comparable effort was reported in the work of Tsai et al. (2004)—the efficient ACODF clustering algorithm for data mining in large databases produces smaller error outcomes by combining the *FSOM+k-means* (Su and Chang 2000) and genetic *k-means* algorithm (GKA), (Krishna and Narasimha 1999). One of the drawbacks of these algorithms is that convergence is slow and there is a need to investigate better adaptation rates for the *k-means*.

A detailed review by Steinley (2006) has documented multiple research efforts in *k-means*. Also, recent work is reported in several other articles. For example, Kim and Yamashita (2007) applied the *k-means* method to analyze spatial patterns of pedestrian crashes in Honolulu, Song and Rajasekaran (2005) proposed three constant approximation algorithms for the *k-means* method. A density-sensitive distance metric (DSKM) was proposed for the *k-means* method by Wang et al. (2006) to identify complex non-convex clusters. And Birant and Kut (2007) developed a density-based clustering algorithm ST-DBSCAN, where *k-means* was tested on spatiotemporal data.

## 1.1 Basic Idea for the *FES-k-means** Algorithm

Many attempts have been made to rid the conventional *k-means* algorithm of its limitations: (1) its computationally expensive nature for segregating large-scale datasets; (2) its inaccurate cluster initialization; and (3) the local minima problem. In this study we used the *FES-k-means** algorithm in conjunction with the *MIL-SOM** (Mathematically Improved Learning-Self Organizing Map), (Oyana 2006) algorithm to resolve two of the aforementioned limitations that are commensurate with ideas of Su and Chang (2000), while employing an adaptation rate suggested by Mashor (1998). For the *FES-k-means** algorithm, we have proposed the integration of Kanungo et al.'s 2002 filtering algorithm based on the *k-d* tree data structure, nearest neighbor query, standard *k-means*, and Mashor's adaptability rate. The main goal in doing this is to improve the cost equations involved in *k-means* so as to maintain consistency and accuracy in the results. The improvement will provide a good means of generating almost the same number and correct clusters every time. To achieve the improvement, an adaptation factor is selected that suitably adjusts itself at each learning step to find the winning cluster. The suggested improvements are expected to update the cluster centers by taking time into consideration and also analyzing the cluster center during the previous clustering steps while generating new cluster centers. Figure 1 gives the pseudo code for the proposed *FES-k-means** algorithm.

There is no one clustering techniquc that is universally applicable to uncover the variety of structures: spherical, linear, drawn-out, etc. (Ng and Han 1994), and that can address the multiple-dimensionality of datasets. In most cases, data classification is based on prior knowledge and induces the generation of useful hypothesis concerning structural relationships of data.

## 1.2 The Basic Structure of the *FES-k-means** Algorithm

Let

*u* represents the node, for each node *u* a set of candidate centers is maintained

*k* represents the number of centers and the candidate centers for the root consists of all *k* centers

z represents each candidate or each data point

*z** represents the winning candidate

Recall in Kanungo et al's filtering algorithm for each node *u, C* denotes its cell and *Z* denotes its candidate set. So *z** ▯ *Z* that is closest to the midpoint of *C*

/* Randomly select the initial candidate for *k* centers after initialization */

1. Build k-d tree with the given data
   - Stores weighted centroid of each node *u* and also the max and min coordinates *v* of each cell *C* associated with the node *u*.
   - Computes the nearest neighbor for each candidate
2. For each Iteration
   - Call filter function with arguments node, candidates, and number of candidates
     a. If node is a leaf
        - Select the winning candidate z* from set of candidates nearest to the centroid of the node
        - Update candidate's weighted centroid and count
     b. Else
        - Select the winning candidate *z** from set of candidates nearest to the midpoint of the cell associated with the node
        - For each candidate *z* in the set other than the win-

ning candidate
- Find out the *u* vector as *z* -*z**
- For each coordinate in *u*
  - If coordinate is negative, assign corresponding coordinate in *v* as min coordinate of the cell
  - If coordinate is positive, assign corresponding coordinate in *v* as max coordinate of the cell
- If distance between *v* and *z** < distance between *v* and *z*
  - Prune *z* and update the candidate list

▪ If only one candidate remains in the candidate list
  - Update candidate's weighted centroid and count

▪ Else
  - Call Filter function with arguments—left child, candidates, and number of candidates
  - Call Filter function with arguments—right child, candidates, and number of candidates

• The center is updated by using the weighted centroid and count associated with each center and Mashor's updating principle.

**Fig. 1:** A pseudo code for the proposed *FES-k-means** algorithm

Cluster formation contains implicit or explicit assumptions about cluster shape and multi-cluster configurations based on a variety of any combinations of the following: similarity measure, clustering criteria and algorithm, and proximity measure employed by the chosen algorithm (Jain et al. 1999; Theodoridis and Koutroumbas 2003; Xu and Wunsch II 2005; Steinley 2006). It is, therefore, essential for the user to clearly define the goals set forth for clustering the data to better identify the requirements for analysis. Other key considerations include the use of correct analysis/validation tools and a basic understanding of the data, i.e., what is the clustering tendency in the data—being able to illustrate whether there is even a possible clustering structure within the data.

While the standard *k-means* method, as first proposed by MacQueen (1967), is sufficient for datasets with minimal data, it has been known to have serious limitations when dealing with very large datasets or datasets with outliers. Other drawbacks are reviewed in Steinley (2006). Worth noting though is the complexity of the *k-means* algorithm, which increases linearly with the size of the dataset, causing it to perform poorly with data-rich environments, and therefore it yields poor accuracy for high-volume datasets (Pham et al. 2004). The primary aim of this study is to explore the combination of two methods designed to increase the overall performance of *k-means* clustering using Mashor's updating method. In previous experiments, we found the Davies-Bouldin validity index (DBI) to be quite problematic when it comes to very large datasets with a large number of clusters.

In summary, the more information a user of a clustering technique has about the data and the algorithm, the more the accuracy of the clustering and, henceforth, analysis of the underlying structure of the data stands to be a success! The remainder of this paper is organized into four main sections. The next section deals with the materials and methods used in this study. The results are then presented. A discussion section follows this section. Lastly, conclusions are given that highlight the key benefits/properties of the *FES-k-means** algorithm.

## 2 Materials and Methods

The *FES-k-means** algorithm was tested using two published real datasets and two synthetic datasets. Each dataset comprised an untrained, actual version and its reduced *MIL-SOM** trained counterparts. The *MIL-SOM** trained data optimize cluster quality by providing fore-knowledge of the topological relationships of the data. Due to its dimension-reduction property, the output can be visually inspected and interpreted. The heuristic nature of the *MIL-SOM** training algorithm enables the user to visualize and, thence, provide preliminary information about the parameter *k*, where *k* is the number of clusters. The findings of this study are vital for several areas of data analysis and aid in identifying characteristic similarities using spatial entities resulting in geospatial data management.

Our approach for testing the *FES-k-means** algorithm was to compare it with two of the most commonly used *k-means* methods: the standard *k-means* and MacQueens *k-means* methods. MacQueen's rendition is a convergent subsequence of the standard *k-means* (MacQueen 1967).

### 2.1 Datasets

For impartial examination of our novel method, this study explores a total of four distinct datasets of varying types, sizes, and dimensions. We use two real georeferenced and two synthetic datasets. Each of these datasets comprises a trained and an untrained subset. The first georeferenced dataset earmarks notable housing characteristics and the prevalence of elevated blood lead levels (BLL) in children residing in Chicago, Illinois, and the second one unveils the spatial distribution of childhood asthma in Buffalo, New York. The other two datasets are synthetic computer-generated, denoted as DS1, $n$ = 18,500 data points with 8 clusters, and DS2, $n$ = 32,000 data points clustered into 10 groups.

### 2.1.1 Cases of Elevated BLL Based on Housing Data

The first of four datasets used in this study evaluates whether there is a relationship between housing characteristics and blood lead levels for children diagnosed with elevated BLL residing in Chicago, Illinois. This study focused on regions at the census block group level. According to the U.S. Center for Disease Control (CDC), elevated BLL has been defined as all diagnostic results $\geq 10\mu g/dl$ (micrograms per deciliter). The actual and trained datasets comprise 2,605 records and 260 records, respectively. Both the trained and untrained sets have the following 16 characteristics, formally referred to as dimensions: (dimension 1) child population; (dimensions 2-10) count of homes built per decade (pre-1935 to 1999); (dimension 11) median year of homes built; (dimension 12) elevated BLL prevalence in year 1997; (dimension 13) elevated BLL prevalence in year 2000; (dimension 14) elevated BLL prevalence in year 2003; and finally, (dimensions 15 and 16) geographic location based on $x$- and $y$- geographic coordinates.

### 2.1.2 Childhood Asthma in Buffalo, New York

This dataset is a report of geographic locations of child residences relative to pollution sites in Buffalo, New York. This study was conducted according to individual cases. The untrained set for these data comprises 11,384 records, and the trained, reduced, set contains 253 records. Both sets have five dimensions: geographic location based on $x$- and $y$-coordinates; case control code; distance of site to major road; distance of resident's living quarters to pollution source; and distance to particulate matter. These data were tracked using binary digits (0 and 1), where 1 indicates whether the given location is within 1,000 meters of noted risk element.

### 2.1.3 Synthetic Dataset 1

This randomly generated dataset, DS1, comprises 18,500 records, returning 8 clusters. A pair of $x$-, $y$-coordinates was used as a two-dimensional quantifier for the clusters. The data structure of this dataset is compact and highly dense.

### *2.1.4 Synthetic Dataset 2*

DS2, the second synthetic dataset yielded 10 disjoint subgroups and has a total of 32,000 data points. Similar to DS1, the clusters of the dataset were evaluated using the two-dimensional *x*-, *y*-coordinate pairs of the data points. The data points in this dataset are tightly packed and distinguishable.

## 2.2 Data Analysis

To achieve the goals of this research, we ran several tests employing the new *FES-k-means** clustering method. The testing procedure comprised three major steps: (1) pre-processing, (2) experimentation, and (3) post-processing. These experiments were conducted in SOM Toolbox 2.0 for Matlab (SOM Project, HUT, Finland) and Matlab 7.0 (The MathWorks, Inc., Natick, Massachusetts). These computational environments were used to perform the algorithms because the SOM Toolbox and Matlab enable complex computations. Geographic information systems (GIS) mapping and spatial analysis were conducted using ESRI ArcGIS 9.2 (ESRI, Inc., Redlands, California).

### *2.2.1 Pre-Processing*

The four datasets used in this study were selected during this pre-processing stage. These datasets were selected because of their practicality and ease of evaluation, analysis, and interpretation of results. We relied on previous experience in selecting the published datasets because of our familiarity with their structural characteristics. We then prepared the datasets for modeling by cleaning them and correctly formatting the entities. The data were, finally, imported into the working environment for experimentation.

### *2.2.2 Experimentation*

During experimentation, we compared the *FES-k-means** method with the standard *k-means*, and MacQueen's *k-means* methods. The performance of each algorithm was assessed using three approaches: (1) speed efficiency, (2) cluster quality, and (3) trained and untrained data consistency.

Using run-time, in seconds, speed efficiency was measured against percent of data processed for each of the three aforementioned clustering methods. The percentage of data processed was selected in 10-percent increments from 10 to 100—10%, 20%, 30%, etc.

To test clustering quality of the *FES-k-means** method, we graphically compared the mean square error (MSE) measured in decibels (dB), versus percentage of data processed. The MSE is a function of number of centers ranging from 1 to *k*, where *k* depends on the respective dataset. The percentage of data processed was selected on the basis of percentages that ranged from 10 to 100, increasing in 10-percent increments (10%, 20%, 30%, etc.).

As we expected, the *MIL-SOM** method trained the data by reducing the number of entities and indirectly formalized a visualization tool for the user to initialize the number of clusters *k* in the datasets. From this essential pre-clustering convenience, we delineated the clusters of each dataset using the *FES-k-means** method.

*SOM*, in a geographical context, is used to reduce multivariate spatially referenced data to discover homogeneous regions and to detect spatial patterns (Kohonen 1990; Bação et al. 2004). In *SOM*, a winning neuron is randomly selected to represent a subset of data, while preserving the topological relationships. The algorithm continues until all data are assigned to a neuron. Assignments are based on characteristic similarities using distance as a determinant, whereby similar data are grouped together and dissimilar data are clustered separately. Resulting clusters may be visualized using a multitude of techniques, including the *U*-matrix, histograms, or scatter plots, among others available within the SOM Toolbox. For the purposes of testing, the *U*-matrix shows distances between neighboring units and displays cluster structure of data in a grid-like fashion.

The *k*-value for all trained datasets was initialized to 10. Although all trained sets were initialized to 10, each dataset returned different values for the actual number of centers. The BLL housing data had six centers; the childhood asthma data had eight clusters; and both synthetic datasets had 10 clusters. The number of clusters were determined via the *MIL-SOM** training, which was used to determine the number of clusters for the datasets and was fused into visualizing the actual full versions of each dataset. The clusters formed from the *MIL-SOM** training were then delineated using the *FES-k-means** algorithm. There were 20 iterations for each experiment on each dataset.

### 2.2.3 Post-Processing

For post-processing and validation, we compared the *FES-k-means** with the standard *k-means* algorithm using SPSS 14.0 for Windows (SPSS, Inc., Chicago, IL) and found that the resulting clusters using either method were comparable; however, interestingly, our method produced outliers that were otherwise disregarded by traditional methods. Further analyses were conducted in this study to investigate these outlier findings.

The clusters containing the most records are referred to as major clusters. Box plots were used to identify the major clusters. Exploratory analyses were conducted using SPSS, and we employed ESRI ArcGIS 9.2 (ESRI, Inc., Redlands, California) to perform GIS mapping and spatial analysis.

We digitally mapped the clusters using ArcGIS to visualize, compare, and confirm cluster formation and point distributions for the *MIL-SOM** trained data as well as the untrained versions for each of the four datasets. However, due to the limitation of space we have focused our illustrations only on the *MIL-SOM** trained data. The *MIL-SOM** algorithm was used in conjunction with *FES-k-means** algorithm to visually explore and exploit the relational tendencies that exist within the dataset in a geographic form and delineated clusters.

To further explore clusters and outliers especially in our untrained datasets, we complemented theory with field testing using communal/housing investigations in Chicago, Illinois. Photos were provided to confirm findings for the elevated BLL linked with the housing dataset in a separate communication.

## 3 Results

Initial observation is that the reduced *MIL-SOM** trained datasets contain fewer clusters than the untrained datasets. While all the data points of the trained datasets fall within a cluster, the untrained datasets contain outlying anomalies not grouped into clusters. Although the clusters in both categories were visually similar, the clusters of the trained datasets were small and less dense than their untrained counterparts.

Figure 2 (*a* through *d*) displays the spatial distribution of the four datasets. Figure 2*a*, DS1 illustrates eight clusters, seven of which are connected at the edges in a linear fashion. In Figure 2*b*, DS2 reflects 10 cluster groupings, each connected to the other at the edges and having a linear tendency. Figure 2*c* shows the spatial distribution of childhood asthma with a greater concentration towards the upper mid-center area and gradually disperses out from the center to the surrounding edges. Figure 2*d* shows a highly dense clutter of points that are very concentrated along the point distribution line.

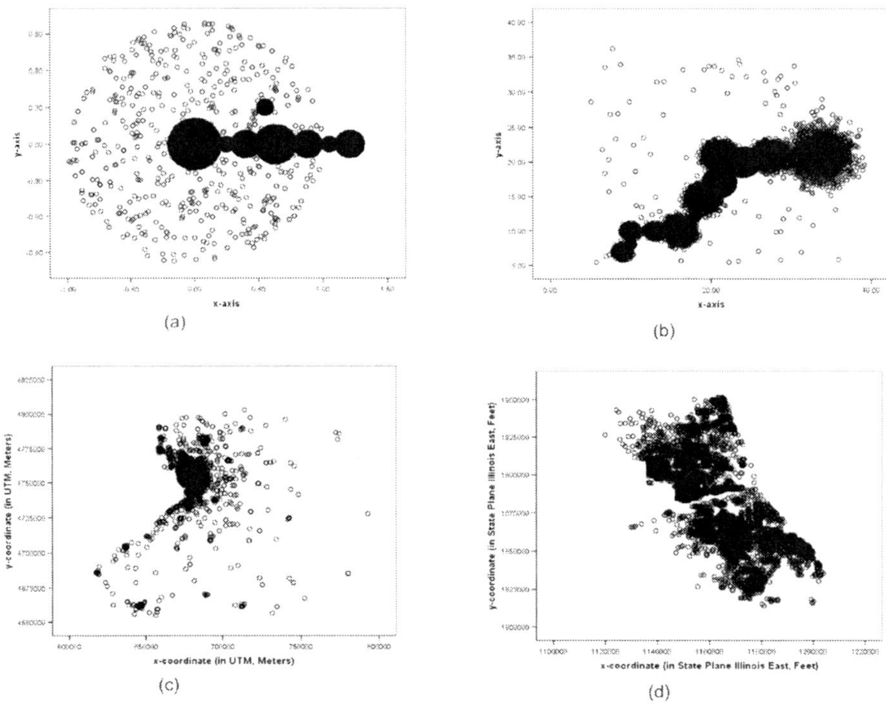


**Fig. 2:** The spatial distribution of original datasets: (*a*) synthetic data I; (*b*) synthetic data II; (*c*) childhood asthma; and (*d*) elevated BLL

Figure 3 (*a* through *d*) shows comparison plots of clustering results using three distinct adaptation rates of three variations of the *k-means* algorithm: the standard *k-means*, MacQueen, and *FES-k-means**. These plots compare the runtime, in seconds, to the percentage of data processed. In Figure 3*a*, the standard *k-means* and MacQueen methods took longer to process than the *FES-k-means** method. Figure 3*b* is a plot of the second synthetic dataset. While all three methods began at the same runtime, less than 0.5 seconds for 10 percent of the data, they all finish separately, from slowest to fastest, in their respective order MacQueen, the standard *k-means*, and *FES-k-means**. Figure 3*c* shows line curves of runtime versus percent of data for childhood asthma. The standard *k-means* and MacQueen both show a steep upward slope, reaching 100 percent at approximately 2.7 seconds, while the *FES-k-means** produces a plot that reflects a slower rise, indicating greater consistency, reaching 100 percent at just below 0.5 seconds. Figure 3*d* shows runtime versus the percentage of data processed for the fourth dataset, elevated childhood BLL. This plot shows a relatively rapid increase in runtime as percentage of data increases. The increase in runtime for *FES-k-means** is not as extensive as that of the others, terminating at 0.7 seconds for 100 percent of data versus just below 1 second for the standard *k-means* and MacQueen methods.

In Figure 4, we report the cluster performance using MSE versus the percent of data. The four test plots (*a* through *d*) reveal similar characteristics for all three methods: linearity of the data processed and mean square error, an increasing relationship between the percentage of data processed and mean square error, and parallel similarities between the three tested methods.

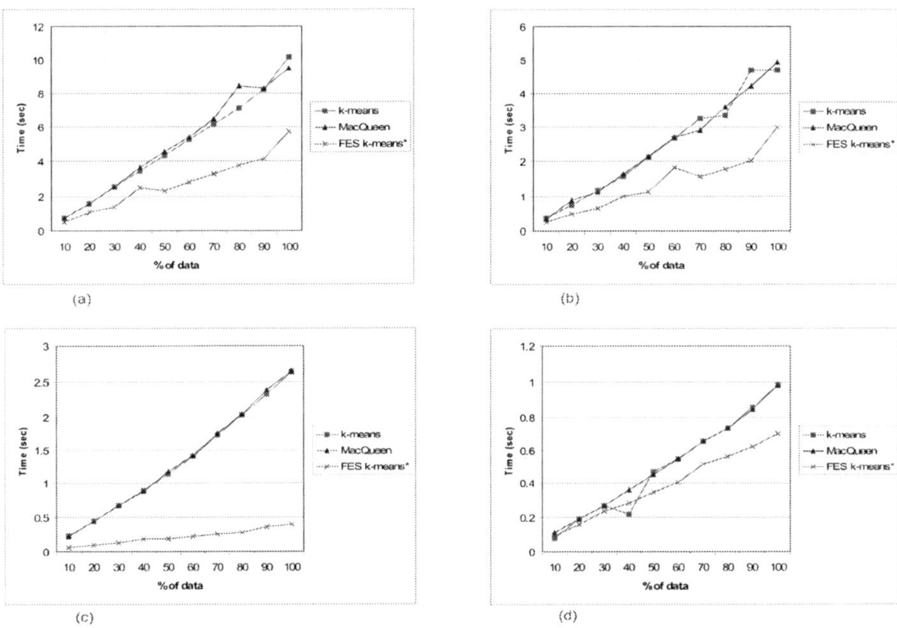


**Fig. 3:** A comparison of three *k-means* algorithms using runtime versus percent of data processed: (*a*) synthetic data I; (*b*) synthetic data II; (*c*) childhood asthma; and (*d*) elevated BLL

Figure 4*a*, shows the cluster performance of the three methods on DS1. Similarities are apparent with each plot beginning at an MSE of less than 9 dB for 10 percent of data processed and ending at an MSE of less than 11dB for 100 percent of data processed. As percentage of data increases, the performance error increases as well. Figure 4*b* reveals that as the number of data increases, the mean square error also increases. The cluster performance error for this dataset begins at 5 dB for 10 percent of data and maximizes at less than 8 dB for 100 percent of data for each of the three methods. In Figure 4*c*, the line curves of the child asthma dataset show increasing performances for the three methods from start to finish of the data processing. The rank order from lowest MSE to highest is standard *k-means*, MacQueen, and *FES-k-means**. The plot for the standard *k-means* and MacQueen methods shows an initial MSE of roughly 15.3 dB and a final MSE of about 17.6 dB at 10 percent of data and 100 percent of data, respectively. The MSE of *FES-k-means**, on the other hand, at 10 percent is greater than 15.5 dB and, at 100 percent, approximately 17.6 dB. Figure 4*d* is the plot of the elevated blood lead level dataset. As the percentage of data increases, the mean square error also increases. All three methods are relatively the same, each producing an upward trend and comparable clustering performance. The standard *k-means* and MacQueen algorithms produce identical curves, and the curve of *FES-k-means** is within close range. Starting at an MSE of 15.5 dB, all three methods cluster 100-percent data at a slightly different MSE. Errors are as follows: standard *k-means* at just below 18 dB, MacQueen exactly 18 dB, and *FES-k-means** just above 18 dB.

Overall, the clustering performance for all three methods is not significantly different. For all four datasets, synthetic and real, the difference in mean square error from 0 to 100 percent of data is rather small at less than 4 decibels.

The test results further prove that the cluster performance of *FES-k-means**, based on MSE, is within range of and has a more enhanced runtime than the standard *k-means* and MacQueen methods. From these plots, it is clear that *FES-k-means**, which employs Mashor's updating rate, has a faster convergence than do the standard *k-means* and MacQueen's methods.

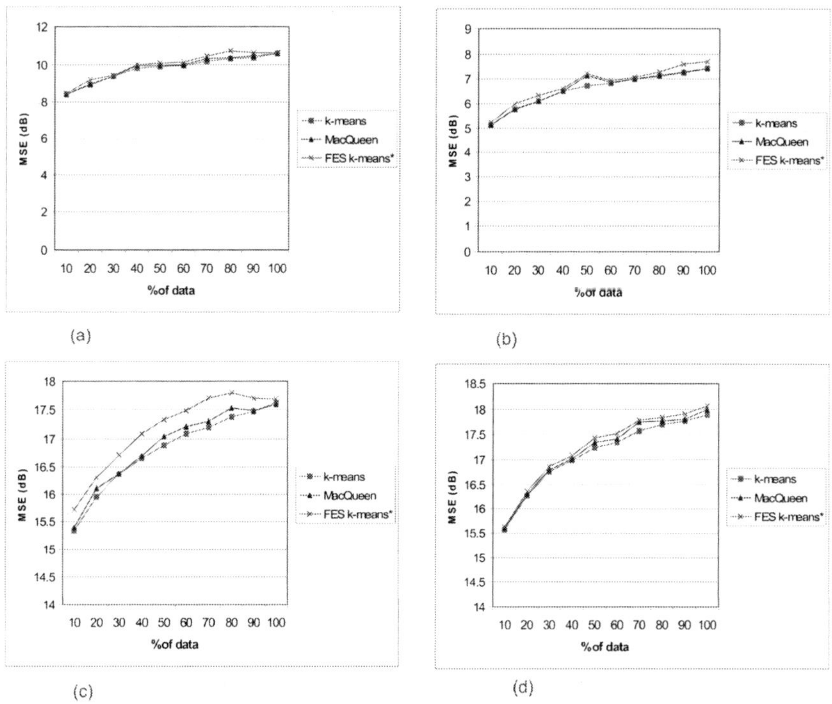


**Fig. 4:** A comparison of three *k-means* algorithms using MSE versus the percent of data: (a) synthetic data I; (b) synthetic data II; (c) childhood asthma; and (d) elevated BLL

Figure 5 (*a* through *d*) presents *FES-k-means** delineated boundaries of *MIL-SOM** trained datasets for two synthetic and two real ones. Figures 5*a* and 5*b* illustrate DS1 and DS2, while Figures 5*c* and 5*d* illustrate childhood asthma and elevated BLL, respectively. The new spatial configurations achieved through *MIL-SOM** training allow for better data visualization simply because the trained datasets are now reduced in size and shape from original datasets. Also clusters are now much smaller in area and less dense than original datasets to help facilitate their easier identification and delineation using the *FES-k-means** algorithm. Though reduced, the trained datasets are still representative of the original datasets. We can deduce this fact from the four scatter plots since they contain similar clusters. The only obvious differences are in ranges, which now are more compressed than in the original datasets due to the normalization process. In addition, outlying anomalies have either been completely eliminated or reduced into different clusters. Below we explore the other two real datasets.

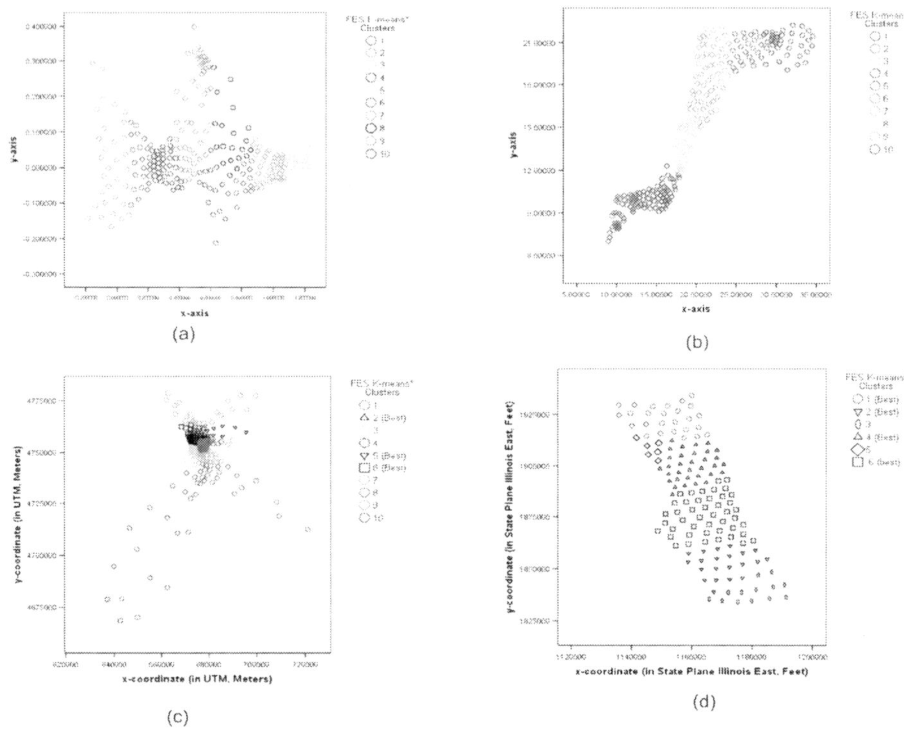


**Fig. 5:** *FES-k-means** delineated boundaries of *MIL-SOM** trained data for the following: (a) synthetic data I; (b) synthetic data II; (c) childhood asthma; and (d) elevated BLL

Data exploration of elevated blood lead prevalence in the city of Chicago was conducted using multiple scatter and box plots within ArcGIS. Four major clusters (1, 2, 4, and 6) were identified, and according to the statistical summaries following the *FES-k-means** method, the averages for the BLL prevalence rates of the major clusters were 95.91257 per 1,000 children, 119.1012 per 1,000 children, 154.9350 per 1,000 children and 180.2174 per 1,000 children for cluster numbers 4, 6, 2, and 1, respectively. The aforementioned clusters are located in four distinct geographical areas, the west, south, far south, and far north. As found in previous studies, cluster 2, with the highest prevalence rate, is located in the far southern region of the city followed by cluster 4 on the western side of Chicago. Cluster 1, which is located on the northern section of the study region, has the lowest prevalence of BLL in children. The north is considered the reference area of this study and previous studies alike. These results are consistent with the findings of previous studies of BLL prevalence in the city of Chicago.

The childhood asthma in Buffalo, New York, was also explored using scatter and box plots. In this dataset, we identified three best clusters (2, 5, and 7) as shown in Figure 5c. Three distinct geographic regions in the western region of Buffalo, New York, were identified as major clusters; each cluster is exclusive to the northern, central, or southern parts of the city. We learned that the clustering results found using the *FES-k-means** cluster on *MIL-SOM** trained and raw data are comparable. The general findings of geospatial location and clustering patterns for the childhood asthma data for the three major clusters are clearly defined and well segregated, and in general, the data within them are tightly compact.

## 4 Discussions

We have found that with the combination of several prominent components to achieve optimum data segregation our *FES-k-means** algorithm has shown remarkable clustering results. This novel approach shows promise to address several key issues pertaining to clustering, data mining, and knowledge discovery. We propose that this method is versatile in data analysis with its ability to (1) optimize clustering by choosing to employ the *MIL-SOM** to initialize clusters, (2) enable handling of large geospatial data, (3) detect outlier anomalies that reveal new information, and 4) formulate meaningful hypotheses concerning data structures, including outliers, for further investigation.

The *k-d* tree data structure, nearest neighbor query, standard *k-means*, and Mashor's adaptability rate, when implemented and utilized together, provide a very fast, computationally efficient, and scalable environment in which to segregate very large geospatial datasets. Preliminary results lend credibility to this original research idea.

The untrained data appear to be plagued with noise and outliers. Providing the user with the option to initialize and reduce the size, by employing the *MIL-SOM** algorithm, enables effective management of these outliers, thus facilitating visual ease of cluster distinction. Moreover, the *MIL-SOM** algorithm is used not only to train but also to create an imagined two-dimensional geography. The trained datasets are presented as a means to eliminate the complexities associated with visual analysis of untrained datasets. For a more thorough evaluation, the trained and untrained analyses can prove to be fruitful in data analysis as determined by the DOD (details on demand) or need-to-know principles. So, according to what the end user is seeking in the analysis, either or both techniques may be favorable.

Thus far, this study has demonstrated that even with a slightly higher MSE, when one is comparing all test methods, including runtime and MSE, the *FES-k-means** is fairly significant to the performance of the other tested methods. An interesting observation, though, is that the curves for the synthetic datasets differ from the real datasets in evaluating the MSE. The MSE values for the synthetic datasets are almost exact for each of the three methods, whereas the MSE for the real datasets reveals slight variation in between 40 and 50% of the data. This finding should be further investigated to quantify and confirm a noted difference in analyzing real and synthetic datasets.

We used two measures (runtime and MSE) to explore the attributes of the *FES-k-means** algorithm. For the runtime, we compared the *FES-k-means** processing speed based on Mashor's adaptation rate with the processing speeds of the standard *k-means* and MacQueen's methods. All test plots revealed the same general characteristics: an increasing linear relationship of the data processed and runtime and similar positioning of the standard *k-means* and MacQueen methods for all three tests. From these plots, it is clear that *FES-k-means**, which employs Mashor's updating rate, has a faster convergence than do the standard *k-means* and MacQueen's methods. Next, we evaluated the quality of the clusters using MSE based on percentage of data processed. Each test plot yielded similar characteristics for the three methods: linearity of the data processed and mean square error, an increasing relationship between the percentage of data processed and mean square error, and parallel similarities between the three tested methods—the standard *k-means*, MacQueen, and *FES-k-means**. All three methods illustrate consistency according to error measures for multiple datasets. It turns out that the cluster performances of the three methods are not significantly different. This further illustrates that in spite of its complexity, *FES-k-means** returns desired outputs efficiently. These results show that the *FES-k-means** method is compatible and scalable with traditional methods.

From the results, we find that the suggested compilation of the conventional clustering and pruning methods has resolved many problems that exist within the individual methods. The two new clever methods, referenced as *MIL-SOM** (Oyana et al. 2006) followed by *FES-k-means** algorithms, have shown ground-breaking improvements in their ability to handle varying data, to segregate very large geospatial data, to efficiently transform high-dimensional datasets to low dimensions while preserving topological relationships, to minimize the number of identified cluster fluctuations, and to adequately analyze largely scattered datasets. Lastly, a very prominent benefit is its ability to perform the aforementioned tasks at an efficient speed.

We believe that modifying the adaptation rate of the *k-means* algorithm with that of Mashor (1998) improves the overall clustering performance. The results from Mashor's study show that his proposed adaptation rate,

$$\eta(t) = \eta(t-1) / e^{[1/r]} \tag{1}$$

is superior to other updating methods: MacQueen, the Square Root method, and Chen's method. In his study, his method produced the best overall clustering performance (Mashor 1998). During experimentation, we compared the *FES-k-means** with the standard *k-means* and MacQueen's *k-means* methods. We used two measures, the runtime and MSE. Our findings report that the *FES-k-means** has a faster runtime, but the MSE in our results are higher than but comparable with MacQueen and the standard *k-means*. For each of these methods on all datasets, the MSE decreases as the number of clusters increase. The lack of better performance here for *FES-k-means** could be a result of the complexity of the *FES-k-means** algorithm. Nevertheless, the advantage of Mashor's method is that by evaluating each updated cluster at its previous step, we produce efficient clusters at a faster runtime, which leads us to ideal clustering conditions—optimal updating at a fast convergence rate and a small steady state value at the end of training time (Mashor 1998). The movement of centers is also minimized during updating, further lessening overall clustering fluctuation. The results reveal that the *FES-k-means** method is prematurely deemed to be more robust than the standard *k-means* and MacQueen's clustering methods.

While the engaging implementation and performance of the *FES-k-means** clustering algorithm undoubtedly reveals great potential, we acknowledge that further research is required to investigate how to effectively integrate the montage of these methods, while achieving an overall minimal MSE. This study is, however, continuing to explore these discovered potentials, along with other properties, while anticipating additional benefits.

Since we are primarily concerned with the clustering of large geospatial data, according to Roussopoulos et al. (1995), the efficient implementation of nearest neighbor queries is of a particular interest in GIS. They further state that nearest neighbor query is useful when the user is not familiar with the layout of the spatial objects. Clustering structured data improves overall clustering performance in terms of total distance calculations and computation time. Alsabti et al. (1998) report improvements by one to two orders of magnitude when compared to clustering data with no structure. The *k-d* tree and nearest neighbor query do not fall short on any of these expectations for our *FES-k-means** method. In fact, it produces the same or comparable clustering results as the direct *k-means* algorithm.

Although the *FES-k-means** algorithm can be applied broadly to any application domain, clustering results still have key implications and could provide preliminary data regarding some of the spatial aspects of clustering for knowledge generation and discovery. More significantly, with the identification of clusters, we can separately study the spatial configurations of each cluster and gain additional insights. Also the resulting spatial information developed from visualizing and classifying these clusters can be valuable in further establishing key characteristics of subgroups. For instance, we can obtain answers to pertinent key questions such as what are the key characteristics or geographic profiles of identified clusters and what makes them similar or different from one another? The exploratory knowledge derived from these questions could support meaningful variable analysis and the formulation of superior study hypotheses. As has been clearly demonstrated by the experimental data, the use of the *MIL-SOM** algorithm for data training together, with the *FES-k-means** algorithm for delineation of clusters, allows for simultaneous assessment of spatial changes in locations being investigated.

Not only can *FES-k-means** be utilized in the SOM Toolbox, the efficiency and robustness of the algorithm enables it to be used on multiple platforms and program applications. For example, the original code was written in C, and then it was exported to Matlab and *SOM* neuron-computational environments. Also, due to its scalability we find that *FES-k-means** yields equally successful clustering results on both small and large datasets, unlike the fast global *k-means* algorithm in which the quality begins low and gradually increases as the number of clusters increase with larger cluster separation and smaller dimensionality (Likas et al. 2003), indicating that it performs poorly on small datasets. Furthermore, *FES-k-means** revealed sub-clusters within clusters that were not otherwise determined by conventional methods hence yielding robust results when compared to standard *k-means*.

We recommend five areas for additional research in data clustering: (1) how to explore and handle outliers; (2) how to evaluate resulting clusters; (3) how to measure reliability of results; (4) how to effectively display more than three dimensions; and (5) how to minimize error. In the future, we hope to pursue some of these ideas. Presently, a major challenge of multi-dimensional analysis is with the visual display of datasets containing more than three dimensions.

The success of data clustering techniques will provide better visual exploration and data mining tools for a range of disciplines that rely on clustering methods for managing, exploring, and visualizing large datasets.

## 5 Conclusions

The overarching properties of *FES-k-means** algorithm are: (1) production of clusters similar to the standard *k-means* method at a much faster rate; (2) scalability, measured by its transferability to other platforms and its innateness to adequately handle small and large datasets—compact or scattered; (3) efficient analysis of large geospatial data; and (4) production of stable clusters illustrating that the algorithm is robust. In summary, we find that with the hybrid approach of the *FES-k-means** method, we are able to enjoy the benefits of all of the aforementioned properties.

In terms of its efficacy and remarkable clustering results, the novel *FES-k-means** clustering method in conjunction with *MIL-SOM** trained data shows promise for a better clustering solution than does just clustering with *FES-k-means** on raw untrained data. With the marrying of these two fresh approaches, we are able to immediately observe the detection of outliers and confirm results from previous studies.

### Acknowledgments

This study was supported by a summer grant from the Responsive and Reflective University Initiative (RRUI) Fund, Southern Illinois Unversity. Dharani Babu Shanmugam and Haricharan Padmanabana Computer Programmers assisted with program and code implementation. Professor Jamson S. Lwebuga-Mukasa provided the asthma datasets, and Anne Evens and Patrick MacRoy provided child lead poisoning data from their Chicago Department of Public Health unit. Special thanks to Professors Luke E. K. Achenie and Patrick A. Rivers for their useful advice, comments, and suggestions. We thank Dr. Ernesto Cuadros-Vargas and several anonymous reviewers for offering their insightful and critical reviews. These timely critiques of the *FES-k-means** algorithm facilitated its design, implementation, and overall quality. All research was approved by the Southern Illinois University Carbondale Human Investigation Review Board in accordance with national and institutional guidelines for the protection of human subjects.

## References

Alsabti, K., S. Ranka, V. Singh. (1998). An efficient k-means clustering algorithm. Proceedings in IPPS: 11th International Parallel Processing Symposium Workshop on High Performance Data Mining, IEEE, Computer Society Press.

Bação, F., Lobo, V., and M. Painho. (2004) Geo-Self-Organizing Map (Geo-SOM) for Building and Exploring Homogeneous Regions. In Egenhofer, M.J., Freskes, C. and Miller, H.J., (eds.) Geographical Information Science. Proceedings of Third International Conference, GIScience. Adelphi, MD, USA. October 20–23, 2004. Springer-Verlay Berlin Heidelberg.

Ball, G. H., and D. J. Hall. (1967) A clustering technique for summarizing multivariate data. Behavioral Sciences. 12(2):153–55.

Birant, D. and A. Kut. (2007) ST-DBSCAN: An algorithm for clustering spatial-temporal data. Data & Knowledge Engineering. Elsevier. 601(1): 208–221.

Chinrungrueng C., and C. H. Sequin. (1995) Optimal adaptive k-means algorithm with dynamic adjustment of learning rate. IEEE Transactions on neural networks. 6(1): 157–69.

Ester, M., H.P. Kriegel, J. Sander, and X. Xu. (1996). A density based algorithm for discovering clusters in large geospatial databases with noise. In Proceedings of the Second International Conference on Knowledge Discovery and Data Mining, Portland, Oregon.

Guha, S., R. Rastogi, and K. Shim. (1998) Cure: An efficient clustering algorithm for large databases. In Proceedings of the ACM SIGMOD Conference on Management of Data (SIGMOD, Seattle, WA, June). ACM Press, New York, NY.

Jain, A.K., M.N. Murty, and P.J. Flynn. (1999). Data clustering: a review. ACM Computing Surveys. 31(3): 264–323.

Kanungo, T., D.M. Mount, N. S. Netanyahu, et al. (2002). An efficient k-means clustering algorithm: analysis and implementation. IEEE Transactions on pattern analysis and machine intelligence. 24(7): 881–92.

Karypis, G., E.H Han, V. Kumar. (1999) Chameleon: Hierarchical clustering using dynamic modeling. IEEE Computer Special Issue on Data Analysis and Mining. 32(8): 68–75.

Kim, K. and E. Yamashita (2007) Using a k-means clustering algorithm to examine patterns of pedestrian involved crashes in Honolulu, Hawaii. Journal of Advanced Transportation. 41(1): 60–89.

Kohonen, T. (1990) The self-organizing map. Proceedings of the IEEE. 78(9): 1464–80.

Krishna, K. and M. Narasimha. (1999) Genetic k-means algorithm. IEEE Transactions on Systems, Man and Cybernetics, Part B. 29(3): 433–439.

Kruskal, J.B. (1964) Non-metric multidimensional scaling: a numerical method. Psychometrika 29: 115–29.

Likas A., N. Vlassis, and J.J. Verbeek. (2003) The global k-means clustering algorithm. Pattern Recognition. 36: 451–61.

MacQueen, J.B. (1967) Some methods for classification and analysis of multivariate observations. Proceedings of the Fifth Berkeley Symposium on Mathematical Statistics and Probability, Statistical Laboratory, University of California. 282–97.

Mashor, M.Y. (1998) Improving the performance of k-means clustering algorithm to position the centers of RBF network. International Journal of the Computer, The Internet and Management. 6(2).

Ng, R. and J. Han. (1994) Efficient and Effective Clustering methods for spatial data mining. In Proceedings of the 20th International Conference of Very Large Data Bases (VLDB), September 1994, Santiago, Chile. 144–55.

Oyana TJ, Achenie LEK, Cuadros-Vargas E, Rivers PA, and Scott KE. (2006) A Mathematical Improvement of the Self-Organizing Map Algorithm. Chapter 8: ICT and

Mathematical Modelling (pp 522–531). In Mwakali J.A. and Taban-Wani G. (eds.): Advance in Engineering and Technology, London: Elsevier Ltd. 847.

Pham D.T., S.S. Dimov, C.D. Nguyen. (2004) A two-phase k-means algorithm for large datasets. Proc. Instn Mech. Engrs, Part C: Journal Mechanical Engineering Science. 218(C). 1269–73.

Roussopoulos N., S. Kelley, F. Vincent. (1995) Nearest Neighbor Queries. In Proceedings of Special Interest Group on Management of data (SIGMOD), San Jose, CA, USA. 71–79.

Sammon, J. W. (1969) A nonlinear mapping for data structure analysis. IEEE Transactions on Computers. 18(5): 401–09.

Song, M. and S. Rajasekaran. (2005) Finding frequent item sets by transaction mapping. Symposium on Applied Computing. 488–492.

Steinley, D. (2006) K-means clustering: A half-century synthesis. British Journal of Mathematical and Statistical Psychology. 59: 1–34.

Su, M. and H. T. Chang. (2000) Fast self-organizing feature map algorithm. IEEE transactions on neural network. 11(3): 721–33.

Theodoridis, S. and K. Koutroumbas. (2003) Pattern Recognition (2ed). Elsevier Science (USA). China machine press. 431–535.

Tsai, C.F., C.W. Tsai, H.C. Wu, and T. Yang. (2004) ACODF: a novel data clustering approach for data mining in large databases. Journal of Systems and Software. 73(1): 133–145.

Vesanto, J. and E. Alhoniemi. (2000) Clustering of the Self-Organizing Map. IEEE Transactions on Neural Networks. 11(3): 586–600.

Vrahatis, M. N., B. Boutsinas, P. Alevizos, and G. Pavlides. (2002) The new k-windows algorithm for improving the k-means clustering algorithm. Journal of Complexity. 18. 375–91.

Wang, L., L. Bo, L. Jiao. (2006) A modified k-means clustering with a density-sensitive distance metric. Lecture notes in Computer Science. Springer Berlin / Heidelber. 4062: 544–551.

Xu, R and D. Wunsch II. (2005) Survey of Clustering Algorithms. IEEE Transactions on Neural Networks. 16(3): 645–78.

Zhang T., R. Ramakrishnan, M. Livny. (1996) BIRCH: An efficient data clustering method for very large databases. Proc. ACM SIGMOD Conf. Management of Data. 103–14.

# DBSCAN-MO: Density-Based Clustering among Moving Obstacles

Emmanuel Stefanakis

Department of Geography, Harokopio University of Athens
70 El.Venizelou Ave, 17671 Kallithea Athens, Greece
estef@hua.gr

**Abstract.** This paper introduces *DBSCAN-MO*, an algorithm for density-based clustering of point objects on a planar surface with moving obstacles. This algorithm extends a well known spatial clustering method, named *DBSCAN*, which has been initially proposed to cluster point objects in a static space. DBSCAN-MO is able to form a set of *spatio-temporal clusters* and may be readily customized to complex dynamic environments. A prototype system, which implements the algorithm, developed in Java and tested through a series of synthetic datasets, is also presented.

**Keywords:** moving objects, spatio-temporal data, data mining, clustering techniques, DBSCAN

## 1 Introduction

Nowadays, there is an increasing need for efficient spatial data mining methods to extract useful information contained implicitly in large collections of geographical data sets. *Spatial clustering* is one of the basic data mining tasks applied in geography, and may lead to the grouping of geographical units (entities or objects) into meaningful classes (i.e., clusters) so that the members of a cluster are as similar as possible, whereas the members of different clusters differ as much as possible from each other (Fayyad et al. 1996; Miller and Han 2001). The similarity above is based on their spatial or thematic attributes; whereas in dynamic applications the values of these attributes are subject to change over time.

Computer scientists have developed a wide collection of methods for clustering objects with alphanumeric attributes during the last couple of decades (Goebel and Gruenwald 1999). These methods can be separated into four general categories: (a) *partitioning*, (b) *hierarchical*, (c) *density-based*, and (d) *grid-based*. Most of these methods have also been adopted to support the clustering of spatial (or geographic) objects (Miller and Han 2001). Which method is the most appropriate depends heavily on the application goal, the trade-off between quality and speed, and the characteristics of data (Han et al. 2001).

In this study, we introduce a method for clustering a set of point objects $P$ that lie on a two-dimensional (plane) surface $S$, which comprises a set of moving obstacles $MO$. For simplicity, the surface is orthogonal – with its borders parallel to the $X,Y$-axes – and described through two pairs of $(x,y)$ coordinates, the lower left (or south west – $X_{LL}$, $Y_{LL}$) and the upper right (or north east – $X_{UR}$, $Y_{UR}$) corners (Figure 1). The space is considered during a temporal interval $[T_{from}, T_{to}]$, defined by a pair of time instances, the $T_{from}$ and $T_{to}$, where $T_{to}$ is subsequent to $T_{from}$. We call this period of time as *space life*. Hence, a *spatio-temporal cube C* defined by the triples ($X_{LL}$, $Y_{LL}$, $T_{from}$) and ($X_{UR}$, $Y_{UR}$, $T_{to}$) is considered.

The surface $S$ comprises a set of *moving obstacles* ($MO$). In other words, a set of objects, which are moving in the spatio-temporal cube $C$. For simplicity, in this study we consider that each obstacle $MO_i$ has a constant circular shape with a radius $r_i$, and carries out a straight route with a constant velocity $v_i$. Specifically, each obstacle $MO_i$ is defined by the following set of parameters (more complex configurations may be readily applied):

$$( r_i , x_{from\text{-}i} , y_{from\text{-}i} , t_{from\text{-}i} , x_{to\text{-}i} , y_{to\text{-}i} , t_{to\text{-}i} ) \tag{1}$$

where the triples ($x_{from\text{-}i}$ , $y_{from\text{-}i}$ , $t_{from\text{-}i}$) and ($x_{to\text{-}i}$ , $y_{to\text{-}i}$ , $t_{to\text{-}i}$) correspond to the starting and ending locations of the obstacle $MO_i$ in space-time.

Figure 2 presents an example moving obstacle $MO_i$ with radius $r_i$ on the surface $S$ (a projective view), which travels from point $A(x_A,y_A)$ to point $B(x_B,y_B)$, during the temporal interval $[t_{A\text{-}i},t_{B\text{-}i}]$ . The object velocity is constant and equal to:

$$v_i = \frac{\sqrt{(x_A - x_B)^2 + (y_A - y_B)^2}}{(t_{B-i} - t_{A-i})} \tag{2}$$

A simplified example scenario is given in Figure 3. The five points (i.e., $P = \{A,B,C,D,E\}$) considered for clustering are located on the planar surface. Among them there is a set of moving obstacles $MO$ (of circular shape). Our scope is to form a set of clusters and group appropriately the points $P$ in the spatio-temporal cube $C$.

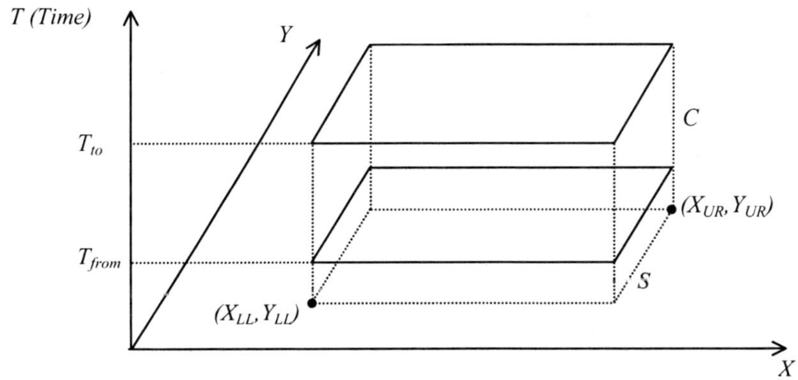


**Fig. 1:** The space-time.

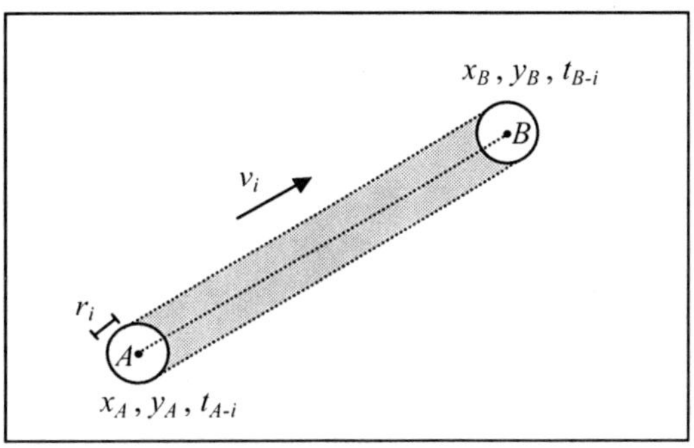


**Fig. 2:** An example of a moving obstacle (projective view).

Obviously, the clusters of these points are *dynamic*, i.e., they change over time and depend on the location of the obstacles in the space-time. A *spatio-temporal cluster* is valid for a specific period of time (temporal interval) as will be shown in Section 3. For instance, points *A*, *B* and *D* may form two clusters during the temporal intervals $[t_1,t_2]$ and $[t_3,t_4]$, respectively; while points *A*, *C* and *D* may form another cluster in the meantime, i.e., the temporal interval $[t_2,t_3]$.

To determine the temporal clusters, we attempt to extend existing (static) spatial clustering algorithms, and make them capable to handle dynamic spaces. Specifically, we focused our efforts to extend an efficient and widely recognized spatial clustering algorithm, named *DBSCAN* (Ester et al. 1996; Sander et al. 1998).

*Density-Based Spatial Clustering of Applications with Noise* (DBSCAN) is a clustering algorithm, which generates the clusters based on the notion of *density* and meets three basic requirements of large spatial databases (Han et al. 2001): (a) minimal requirements of domain knowledge to determine the input parameters; (b) discovery of clusters with arbitrary shape; and (c) good efficiency. We named the extended version as *DBSCAN-MO* (DBSCAN among Moving Obstacles). It is worthy to mention that, although this study is focused on the extension of DBSCAN algorithm, a similar approach can be applied to lead on the extension of other spatial clustering algorithms to handle sets of points located on a planar surface with moving obstacles.

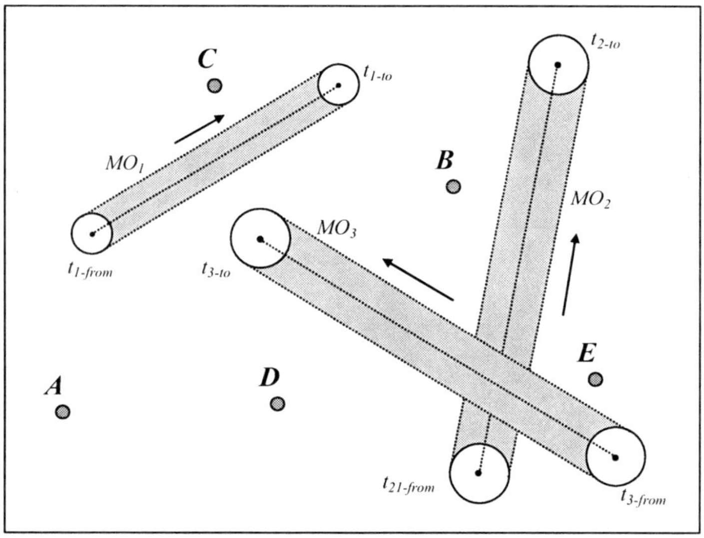


**Fig. 3:** A simplified scenario (projective view). Five points in space (i.e., *A*, *B*, *C*, *D* and *E*) and three moving obstacles (each arrow depicts the direction of movement; $t_{i\text{-}from}$ and $t_{i\text{-}to}$ are the temporal instants when moving obstacle $MO_i$ appears and disappears).

To the best of our knowledge no method has been proposed so far to solve a similar problem. The new algorithm combines the ideas published recently at Stefanakis (2006) and Stefanakis (2007). The former paper provides a method for scheduling trajectories on a planar surface with moving obstacles; while the latter introduces *NET-DBSCAN*, an algorithm for clustering the nodes of a dynamic linear network. The scope of this paper is to introduce the new method. Therefore no special focus is given on scalability issues or the application of this method in a real-world situation. These are directions for our future research (see Section 5).

The discussion is organized as follows: Section 2 briefly presents the DBSCAN clustering algorithm. Section 3 describes DBSCAN-MO, an adapted algorithm to support the clustering of point objects in a dynamic space with moving obstacles. Section 4 presents a prototype system, which implements the algorithm as well as the experiments performed on simplified synthetic data sets. Finally, Section 5 concludes the discussion by summarizing the contribution of the paper and giving several hints for future research.

## 2 DBSCAN Algorithm

*Density Based Spatial Clustering of Applications with Noise* (DBSCAN, Ester et al. 1996) is a density-based clustering method initially developed to cluster point objects. The associated algorithm requires two input parameters: $\in$ and *MinPts*. The neighborhood of a given point *p* is examined and judged to be sufficiently dense, if the number of data points within a distance $\in$ from *p* is greater than *MinPts*. If so, *p* is called a *core point* and forms an initial cluster for the data set. The neighborhood within radius $\in$ from the point *p* is called the *$\in$-neighborhood* of *p*. All point objects of the data set that lie within the *$\in$-neighborhood* of *p* are called *non-core points* of the initial cluster of *p*. The clustering formed from DBSCAN follows the steps below:

Step 1: Give the parameter values for $\in$ and *MinPts*.
Step 2: Form the initial clusters by applying the following rule: For each point *p* of the data set, examine if it is a *core point*. If so, generate a new cluster with *core point* the point *p* and *non-core points* all other points of the set that lie within the *$\in$-neighborhood* of *p*.
Step 3: Compare repetitively the initial clusters in pairs; if the *core point* in the one cluster is a *non-core point* in the other cluster, merge the two clusters in one.
Step 4: If a point is assigned as a *non-core point* to more than one cluster, remove it so that it is assigned to only one of them.

At the end of this process each data point is assigned to one or none cluster. In the latter case, the point is considered to be *noise*.

Figure 4 presents an example of the DBSCAN algorithm. As shown in Figure 4g, *MinPts* is assigned the value of 3 and $\in$ the radius of the circle. Figure 4a shows the point data set. In Figure 4b, a circle is drawn from each point and all *core points* are shown in dark black dots. Notice that the circle from each dark dot contains at least 3 (*MinPts*) other points of the set. Figure 4c shows how two initial clusters are compared and finally merged as of Figure 4d. Figure 4e presents the result of merging the initial clusters. It consists of two clusters, and two isolated points, which are considered to be noise. Notice that the two clusters in Figure 4e share one point. This point is decided to being removed from the top cluster. The decision to which cluster the point will be assigned is arbitrary. However, the application domain may impose the rule. Figure 4f shows the final two clusters discovered by DBSCAN algorithm.

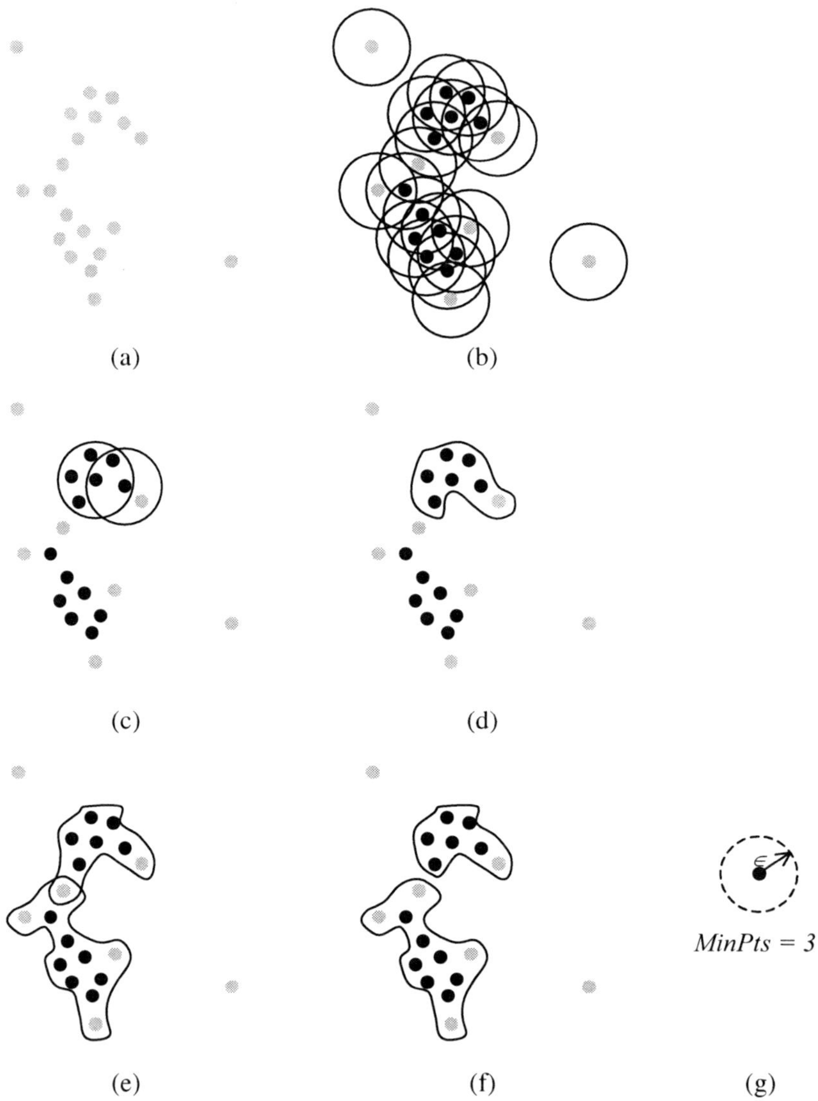


**Fig. 4:** An example of the clustering process adopted by DBSCAN algorithm.

## 3 DBSCAN-MO: The Extended Algorithm

The scope of this paper is to show how DBSCAN algorithm may be adapted to cluster a set of points *P* that lie on a planar surface *S*, which in turn comprises a set of moving obstacles *MO*. Obviously, the algorithm needs to be modified in two aspects to consider the dynamic space. First, the clusters to be formed are *dynamic*, i.e., they change over time and depend on the location of the obstacles in the space-time. A *spatio-temporal cluster* is valid for a specific period of time (temporal interval). Second, due to the existence of the moving obstacles, the notion of $\in$-neighborhood must be extended with a temporal description, i.e., its density is judged for specific temporal intervals. The $\in$-neighborhood is no longer of circular shape; instead it comprises all individual locations that may be reached through an accessible (and not necessarily straight path as in pure DBSCAN) of total length less than $\in$ from point *p* during the corresponding temporal interval (Figure 5). Hence, in order to examine how close two points of the set *P* are, we need to consider a free movement in space (on the plane) and not only their Euclidean (straight) distance.

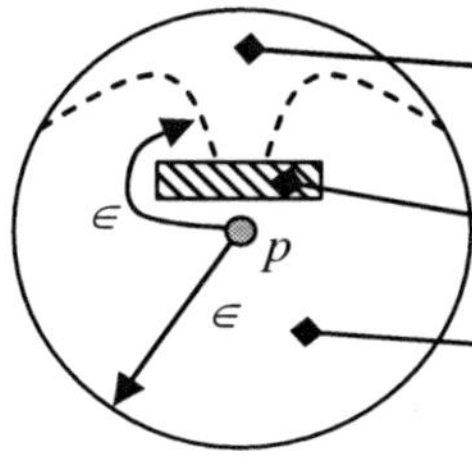


**Fig. 5:** The $\in$-neighborhood is no longer of circular shape. The presence of an obstacle (here static) may render part of the circle not being accessible by point p with a cost less than $\in$.

As stated already in Section 1, these new restrictions have been examined and published recently in Stefanakis (2006) and Stefanakis (2007) to solve two different but relevant problems. The former paper provides a method for scheduling trajectories on a planar surface with moving obstacles; while the latter introduces *NET-DBSCAN*, an algorithm for clustering the nodes of a dynamic linear network. In the sequel, the ideas introduced in these papers have been combined and adapted to lead to *DBSCAN-MO*, the new method for clustering a set of points that lie on a space with moving obstacles.

Assuming a set of points *P* and a set of moving obstacles *MO* on a planar space *S* as in Figure 3, DBSCAN-MO method consists of eight steps, which are discussed in the Subsections that follow:

Step 1: Assign values to the adapted parameters of $\epsilon$ and *MinPts*.
Step 2: Establish a network in space.
Step 3: Formulate the travel cost model.
Step 4: Compute the temporal intervals during which the network nodes and edges are not accessible (due to the presence of the obstacles).
Step 5: Compute the virtual edges for each point of the data set *P*.
Step 6: Compute virtual edges accessible temporal intervals.
Step 7: Form the initial clusters for each point of the data set *P*.
Step 8: Merge the clusters based on the notion of density.

## 3.1 Assign the Parameter Values

The DBSCAN-MO, as an extension of DBSCAN, inherits two basic parameters: $\epsilon$ and *MinPts*. However, these parameters take an adapted meaning.

Specifically, $\epsilon$ is tight to the notion of the cost (*C*) of traversing a path in space, which consists of a set of network edges (the network is established at Step 2). On the other hand, *MinPts* is related to the number of points in *P*, which share a virtual edge (see Section 3.5) with a potential core point $p_i$. The role of these basic parameters will be clarified in the following Subsections.

## 3.2 Establish a Network in Space

As stated previously, in order to examine the nearness ($\epsilon$-neighborhood criterion) of two points in the dynamic space, we need to consider alternative paths in space and not only their linear connection (straight Euclidean distance) as in the static DBSCAN algorithm. This raises the problem of the free movement in space (here, on a planar surface).

The inconvenience of movement in space is the infinite number of spots (i.e., point locations or nodes), involved in the determination of a path. The proposed solution (Stefanakis and Kavouras 1995, 2002) to overcome this problem is based on the technique of discretization of space. *Discretization* (Laurini and Thompson 1992; Worboys 1995) is the process of partitioning the continuous space into a finite number of disjoint areas or volumes (cells), whose union results in the whole space. By representing each of these cells with one node (e.g., its center point), a finite set of nodes is generated.

Obviously, the number of nodes depends on the size of the cell. If these nodes are interconnected through edges, a linear network is established, and appropriate algorithms available in the graph theory and artificial intelligence can be applied to support the navigation. How nodes are interconnected is related to the degrees of freedom characterizing the movement. In this study, we adopt a common scheme, which is based on the *regular grid* tessellation. More details can be found in Stefanakis and Kavouras (2002).

Specifically, a regular grid is superimposed on the plane surface. A network node is then located on the centroid of each cell (Figure 6). Then, a set of network edges are established to connect the network nodes. These edges are driven from the regular grid as follows. Each cell has three types of neighbor cells (Figure 7): (a) *direct*, i.e., neighbors with shared edges; (b) *indirect*, i.e., neighbors with common vertices; and (c) *remote* neighbors. The level of proximity to the cell of reference characterizes remote neighbors.

For instance, level-one (level-two) remote neighbors are the cells, which are direct or indirect neighbors of the direct or indirect neighbors of the cell of reference (of the level-one remote neighbors of the cell of reference). Interconnecting the direct neighbors leads to a set of four directions of movement from each node (*rook's move* is allowed – Figure 8a). Interconnecting the indirect neighbors adds another set of four directions (*queen's move* is allowed – Figure 8b). Interconnecting the level-one remote neighbors adds another set of eight directions of movement (*queen's+knight's moves* are allowed – Figure 8c). An exhaustive network would interconnect all direct, indirect and remote (of any level) neighbors.

We assume, for simplicity, that all points in *P* are located on network nodes. Figure 9 presents an example of four points on a plane surface *S* (Figure 9a) and the established network with four (Figure 9b) and eight directions of movement (Figure 9c).

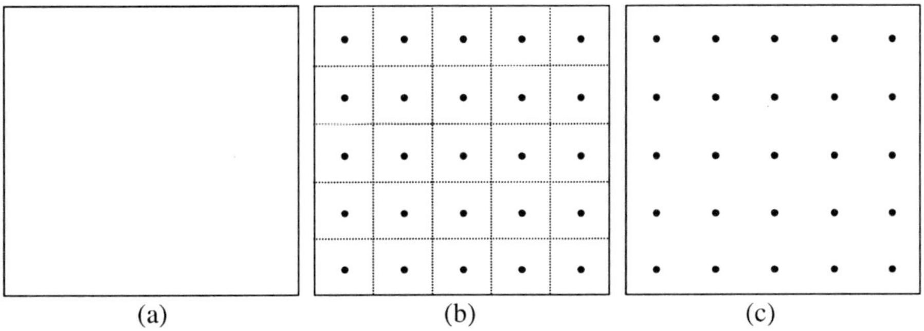


**Fig. 6:** Establishing the network nodes. The space (a), the tessellation superimposed on the space (b), and the resulting nodes (c).

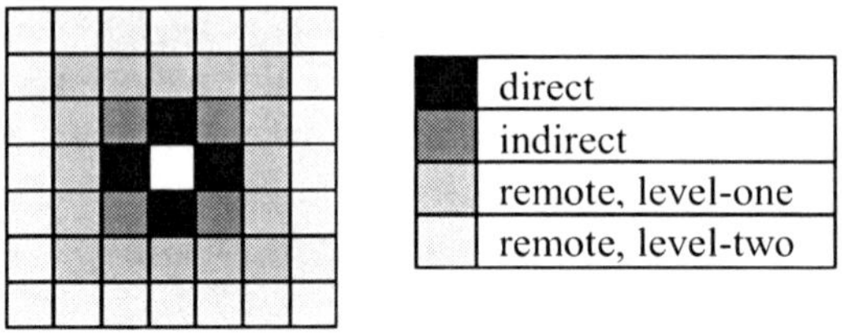


**Fig. 7:** Types of cell neighbors in a regular grid.

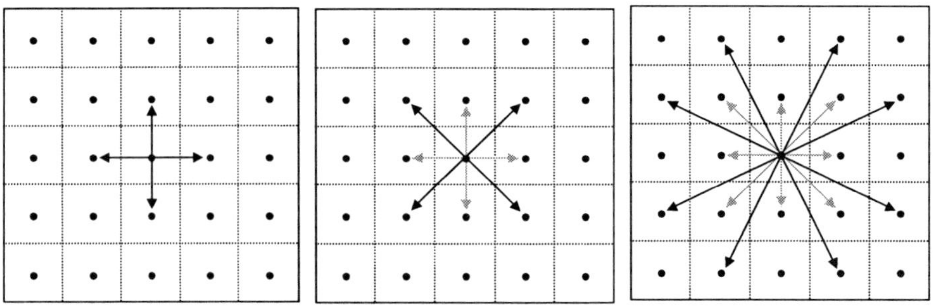

(a) (b) (c)

**Fig. 8:** Four (a), eight (b) and sixteen (c) directions of movement.

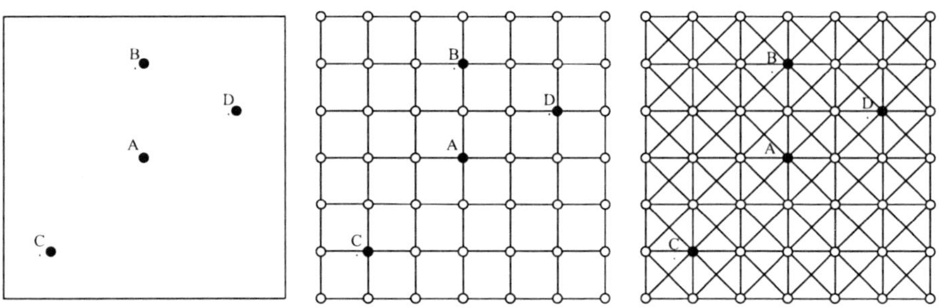


(a) (b) (c)

**Fig. 9:** Four points on a plane surface (a) and the established network with four (b) and eight directions of movement (c).

Consider the network (or graph) *G(N,E)* as established in this step. It consists of *N* nodes, which are connected through *E* bi-directional edges. Each edge $E_i$ of the network must be assigned two parameters: $(C_i,D_i)$. $C_i$ accommodates the *cost of traversing* the edge (either direction; this is an assumption for simplicity); while $D_i$ accommodates the *duration of traversing* the edge (either direction; this is also an assumption for simplicity). Additionally, each edge $E_i$ must be assigned a set of *temporal intervals* $\{TI_{ij}; j=1, \ldots, k\}$, which correspond to the periods of time the edge is *not accessible*, due to the presence of a moving obstacle. Hence, the structure of a network edge $E_i$ is as follows:

*The network edge structure:* [ $E_i$, $N_{i1}$, $N_{i2}$, $C_i$, $D_i$, $\{TI_{ij}; j=1, \ldots, k\}$ (3)

where $E_i$ is the edge identifier; $N_{i1}$ and $N_{i2}$ are the two network nodes connected by the edge.

The parameters $C_i$ *and* $D_i$ will be assigned in the following Step (Step 3); while the temporal intervals $TI_{ij}$ will be computed at Step 4 of DBSCAN-MO method.

## 3.3 Formulate of the Travel Cost Model

The *travel cost model* assigns weights to the edges of the network established in the previous step. Its form depends on both the space under study and the application needs. Some representative examples of travel cost models are:

- the model of *distance* (it assigns the overall distance)
- the model of *time* (it assigns the overall time)
- the model of *expenses* (it assigns the overall expenses)
- the model of *risk* (it assigns a measure for the overall risk)

In each case, the space under study consists of areas that are characterized by a weight, which indicates the cost of movement across them per unit of movement; and depends on the travel cost model in use. A detailed analysis is can be found in Stefanakis and Kavouras (2002).

In this paper, we consider the cost model of *Euclidean distance*. Hence, the cost parameter $C_i$, as defined in Section 3.2, will be assigned the length of the edge segment for both directions of movement along each edge $E_i$ of the network. Additionally, the duration parameter *Di* of traversing the network edge will be assigned a value derived from the division of $C_i$ over a constant velocity of movement *U*. Obviously, more sophisticated models can be easily applied.

## 3.4 Compute the Temporal Intervals during Which Nodes and Edges are Not Accessible

After the network has been established, the obstacles are considered in order to compute all those temporal intervals during which nodes and edges are not accessible. All nodes and edges are compared against the moving obstacles locations in time. At the end of this comparison, each individual node and edge of the network is assigned a list of temporal intervals during which it is not accessible, because an obstacle intersects it.

Figure 10 presents an example of two nodes *A, B* and the edge *A_B* connecting them. An obstacle moves from point *K* to point *L*. As it is shown, the obstacle covers node *A* during the temporal interval $[t_2,t_3]$ and intersects the edge *A_B* during the temporal interval $[t_2,t_4]$. During these temporal intervals the corresponding node and edge are not accessible.

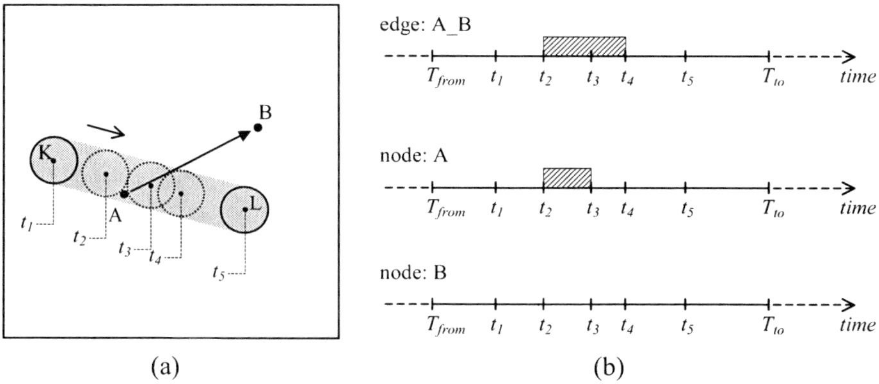


**Fig.10:** An example of a moving object (a), and the temporal intervals during which nodes *A,B* and edge *A_B* are not accessible (b).

Notice that, if a moving obstacle *MO* crosses a network node where a point of the data set *P* is located, the point will not be available for clustering during the corresponding temporal interval. In the sequel, we make the assumption that never an obstacle crosses the points in *P*; hence, all the points in *P* are available for clustering during the whole space life.

## 3.5 Compute the Virtual Edges for Each Point

In this Step, we compute all distinct paths over the established network between the points of the set *P* with an accumulated cost less than $\in$. Those paths coming out of point $p_i$ are called *virtual edges* ($VE_i$) of $p_i$ over the set *P*. Obviously, all virtual edges satisfy the *$\in$-neighborhood* criterion; i.e., their network length is less than or equal to $\in$.

Figure 11 shows the virtual edges coming out from point *A* to point *B* in Figure 9b. Assuming that the cost of traversing each network edge is equal to 10 units, seven (7) virtual edges come out of *A* for a value of $\in$ equal to 40 units (Figure 11a). On the other hand, only one (1) edge for a value of $\in$ equal to 20 units (Figure 11b).

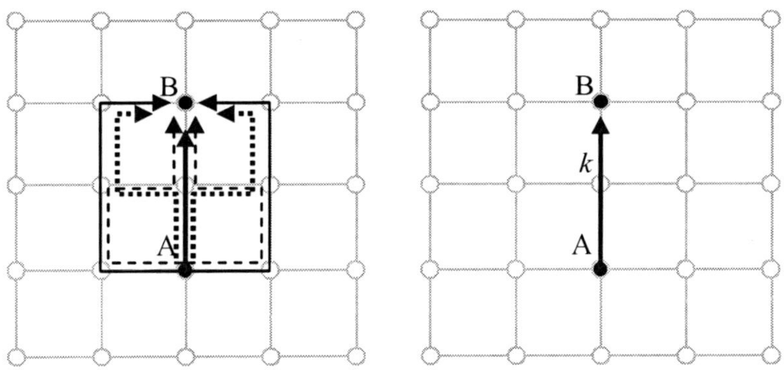


(a) (b)

**Fig. 11:** The virtual edges from point A to point B in Figure 9b. For $\epsilon = 40$, 7 virtual edges are available (a). For $\epsilon = 20$, only one virtual edge is available (b). The cost of each network edge is equal to 10 units.

The discovery of the virtual edges coming out from each point $p_i$ of the set $P$ may be supported by any *breadth-first search algorithm* (Sedgewick 1990). Each individual path coming out of $p_i$ should be recorded and assigned the accumulated cost. The recursion is halted when either an end node is reached or the passing to a neighbor node leads to an accumulated cost that exceeds $\epsilon$.

Obviously, the complexity of such an algorithm is high, especially for dense networks and a high $\epsilon$ value. Nevertheless, it is worthy to emphasize that (a) only the edges coming out from the points of $P$ and satisfy the *$\epsilon$-neighborhood* criterion are computed; and (b) the computation can be performed in a *pre-processing mode* (with various $\epsilon$ values) so that the execution of the DBSCAN-MO method is not delayed. Notice that, from all the computed paths we maintain only those that end up to another point $p_j$ of the set $P$.

At the end of this process, the set of virtual edges coming out from each point $p_i$ of the set $P$ are computed. Each virtual edge consists of one or more network edges and is assigned the accumulated cost of traversing it (which is always less than or equal to $\epsilon$).

Apart from the accumulated cost, each virtual edge is assigned the accumulated duration for traversing it and a set of temporal intervals during which the edge is not accessible. As for the duration, it is simply the sum of durations assigned to the network edges composing the path. For instance, the accumulated duration of the *virtual edge A_B* in Figure 9b, which corresponds to the *path A→k→B*, is equal to the sum of durations assigned in the two network edges A_*k* and *k_B*. Adopting the simple model described in Section 3.3 and a constant velocity of movement $U$ which is equal to 1 unit per minute, the traversing lasts for 10+10=20 min.

As for the temporal intervals, in here we make the simple assumption that each virtual edge is assigned the *union* of all temporal intervals assigned to the network edges composing the corresponding path. For instance, considering again the virtual edge *A_B* of the *path A→k→B* (Figure 11b); if *edge A_k* is not accessible during the temporal intervals { [20, 30], [80,100] }, and *edge k_B* is not accessible during the temporal intervals { [30,40], [90,110], [150,170]}; then the *virtual edge A_B* will be assigned the temporal intervals { [20,40], [80,110], [150,170] }.

## 3.6 Compute Virtual Edges Accessible Temporal Intervals

Each virtual edge $ve_i$ is assigned (Step 3.5) a duration for traversing $D_i$, and a set of temporal intervals during which this edge is not accessible $\{TI_i; i=1, \ldots, n\}$. Based on these values, we can compute the set of temporal intervals during which the edge is accessible, named *accessible temporal intervals* $\{ATI_j; j=1, \ldots, m\}$. In order to accomplish that, all temporal intervals of the set $\{TI_i; i=1, \ldots, n\}$ must be extended (on their left side; see next paragraph) by the duration of traversing the edge $D$ and then subtracted from the space life ($[T_{from}, T_{to}]$). The result of this process is the set $\{ATI_j; j=1, \ldots, m\}$.

Consider the example situation in Figure 10. Assume that the duration of traversing the *edge A_B* from node $A$ to node $B$ is equal to $D_{AB}$. Provided that the edge is not accessible during the temporal interval $[t_2,t_4]$, a vehicle traveling from $A$ to $B$ is restricted to depart from $A$ during the temporal interval $[t_2\text{-}D_{AB},t_4]$. By subtracting this temporal interval from the space life, the following set of accessible temporal intervals is derived: $[T_{from}, t_2\text{-}D_{AB})$ and $(t_4, T_{to}]$.

## 3.7 Form the Initial Clusters

In this step of the algorithm the initial clusters of the point set $P$ are generated. Provided that both the space and the established network are dynamic in nature – due to the presence of the moving obstacles – the derived clusters are valid for specific temporal intervals. Before we examine how the initial clusters of the point set $P$ are discovered, we need to define the structure of a (temporal) cluster. In here, we assume that a cluster is valid during only one temporal interval and hence its structure is as follows:

*The temporal cluster structure:* [ $id$, $p_c$, {$NP_i$; $i$=1, …, $n$}, [$t_f$,$t_t$] ] (4)

where $id$ is a unique identifier for the cluster; $p_c$ is the *core point* of the cluster; {$NP_i$; $i$=1, …, $n$} is the set of *non-core points* assigned to the cluster; and [$t_f$,$t_t$] is the temporal interval during which the cluster is valid.

The structure above can be explained as follows. During the temporal interval [$t_f$,$t_t$], there is a virtual edge (path) connecting point node $p_c$ to each one of the nodes {$NP_i$; $i$=1, …, $n$} and this edge is accessible. Additionally, the cost $C$ of traversing assigned to all virtual edges is less than or equal to $\epsilon$, while the number of non-core points $n$ is greater than or equal to *MinPts*.

A detailed example of discovering the initial clusters in a simple dynamic network can be found in Stefanakis (2007). The algorithm for generating the initial clusters is given in Figure 12. For a network *G(N,E)* generated over a plane surface *S* and a set of points *P* (1), the algorithm scans all the points in *P* (2). For each point $p_i$ in *P* it computes its virtual edges and assigns them their parameters (accumulated cost, accumulated duration, and union of temporal intervals; see Sections 3.5 and 3.6) (3). This is accomplished by traversing recursively the network starting from that point. The recursion stops when the accumulated cost becomes greater than $\epsilon$. Notice that, all virtual edges have an accumulated cost that is less than or equal to $\epsilon$.

```
(1)  Graph: G(N,E) ; Set of points: P
(2)  for each point p_i of P
(3)    VE_pi: the set of all virtual edges coming out of p_i
           (and satisfy the ϵ-neighborhood criterion)
(4)    if sizeof(VE_pi) >= MinPts
(5)      for the temporal instant T running from T_from to T_to
(6)        NP_i: the set of all opposite points of edges in
                VE_pi, which are accessible at T
(7)        If sizeof(NP_i) >=  MinPts
(8)          new_cluster[id,p_i,{NP_i;i=1,…,sizeof(NP_i)},[T,T]]
(9)          merge_with previous_cluster()
```

**Fig. 12:** An algorithm for discovering the initial clusters on the network nodes.

If the number of the virtual edges is smaller than *MinPts*, the process continues to the next point of the set *P*. Otherwise (4), for each temporal instant *T* of the space life (with a step equal to the granularity of time) (5), the algorithm generates the set of the opposite points ($NP_i$) for all out coming virtual edges, which are accessible at the temporal instance *T* (6). If their number is greater than or equal to *MinPts* (7), a new cluster is generated (8). The new cluster is assigned a unique identifier (*id*); it has $p_i$ as core point; all points of the set $NP_i$ as non-core points; and a zero duration

[*T*,*T*]. This new cluster is compared with the one generated in the previous loop (if any; function: *merge_with_previous_cluster()*) (9). If all the parameters in both clusters, except the temporal interval, are assigned the same values and their temporal intervals differ by the time granularity, the two clusters are merged into one with the union of their temporal intervals.

## 3.8 Merge the Clusters

After the initial clusters are generated, a process similar to the one performed in DBSCAN takes place in repetition to lead in the generation of the final clusters. Specifically, clusters are compared repetitively in pairs. If the *core point* in one of them is assigned as a *non-core point* to the other the two clusters are merged in one. After an exhaustive comparison, which leads to no more left merging processes, the final clusters are generated.

In contrary to the original DBSCAN version, when two (temporal) clusters merge, they may generate one to three new clusters depending on the topological relations of their temporal intervals (Allen 1983). This is shown through the three examples in Figure 13.

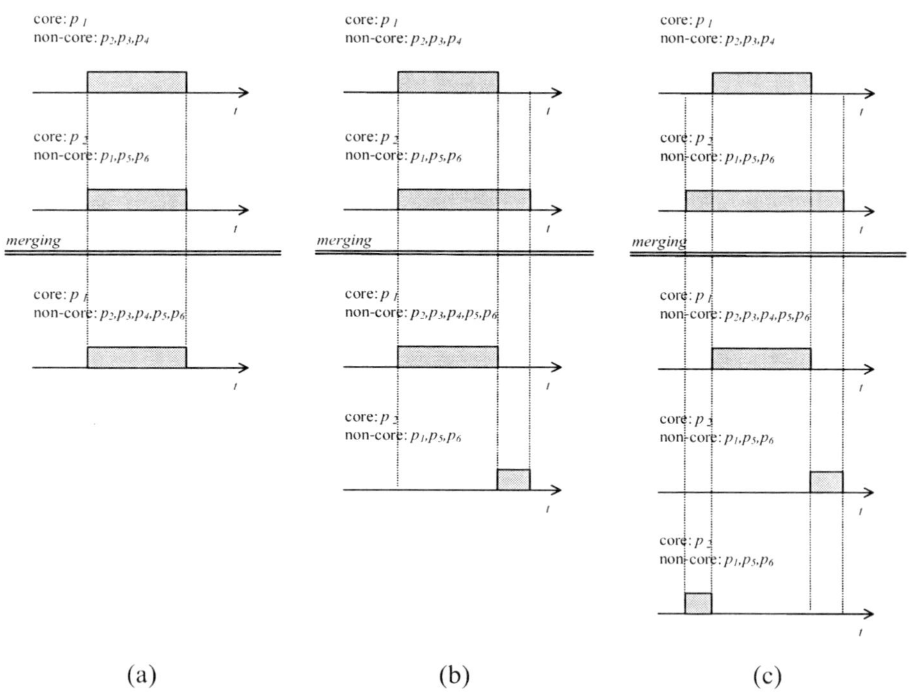


**Fig. 13:** Three examples for Step 8 of the DBSCAN-MO method.

The core point of the top cluster ($p_l$) is assigned as a non-core point on the bottom cluster. Additionally, the two clusters are contemporary, i.e., they share in all three examples a temporal interval. In Figure 13a, the temporal intervals of the two clusters coincide. Hence, they merge into one cluster. In Figure 13b, only the one edge of their temporal intervals is the same. This situation leads to two new clusters; one cluster for the common part of the temporal interval and one cluster for the non-common part. In Figure 13c, the temporal interval of the first cluster is entirely included in the temporal interval of the second cluster. This is the case where three new clusters are generated.

## 4 The Prototype System – Experimentation

A prototype system has been developed in Java to implement the DBSCAN-MO method as described in the previous Section. Figure 14 presents the system interface. The prototype displays the point set $P$ (in red triangles), the underlying network nodes and edges (in red squares and green line segments), as well as the moving obstacles $MO$ in an animation mode (in purple circles). The points forming a temporal cluster – after the execution of the DBSCAN-MO algorithm – are shown to the user by flashing in the same color, while the temporal details (the temporal intervals during which they are valid) are listed in the result window (middle frame at the right side of Figure 14).

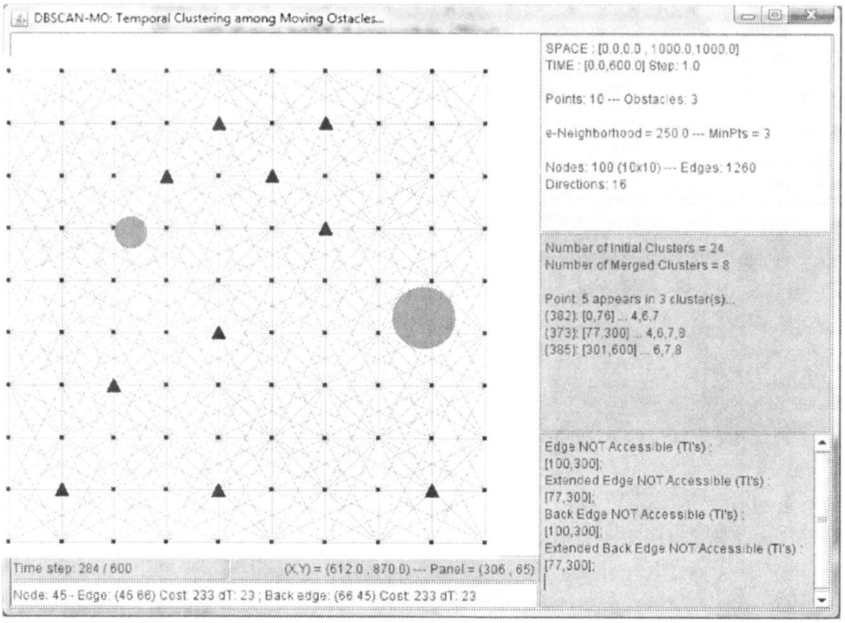


**Fig. 14:** The prototype system implementing DBSCAN-MO method. The three frames on the right side (from top to bottom) are: (i) *parameter values window* (it lists the parameter values of the current session); (ii) *results window* (it reports the clusters of a point after clicking on it); and (iii) *edges window* (it provides a description of edges after clicking on them).

The prototype has been tested in numerous synthetic datasets. Figure 14 presents a snapshot of the situation in space *S* [(0,0), (1000, 1000)] at time instant 284 (see at the bottom left side) for a space life of [0, 600] and a granularity of time equal to 1. Ten points comprise the set *P*, while three obstacles appear and move in the space *S* during the space life; two of them are present at the snapshot instant. The established network consists of 100 nodes and 1260 edges derived from the 16 directions of movement. 24 clusters have been formed initially (Step 7 of the algorithm), which have been merged to 8 final clusters. As it appears in Figure 14, one of the points in *P* (point 4) is a member in three temporal clusters during the space life (see the results window in Figure 4) for the values of 250 and 3 assigned to the parameters $\in$-neighborhood and *MinPts*, respectively.

As mentioned already, the scope of this paper has merely been to present the DBSCAN-MO method. Currently, we are working on an extended experimentation involving both synthetic and real data sets. Additionally, we are considering optimizing the algorithm by supporting its individual steps through the adoption of appropriate spatial index structures (Samet 1990, Manolopoulos et al. 2005).

## 5 Conclusion

In this study, we introduce a method for clustering a set of point objects *P* that lie on a two-dimensional (plane) surface *S*, which comprises a set of moving obstacles *MO*. Obviously, the clusters of these points are *dynamic*, i.e., they change over time and depend on the location of the obstacles in the space-time. To determine the *spatio-temporal clusters*, we extend an existing (static) spatial clustering algorithm, the *DBSCAN* (Ester et al. 1996; Sander et al. 1998) and make it capable to handle dynamic spaces. We named the extended version as *DBSCAN-MO* (i.e., DBSCAN among Moving Obstacles).

The new algorithm combines the ideas published recently at Stefanakis (2006) and Stefanakis (2007). The former paper provides a method for scheduling trajectories on a planar surface with moving obstacles; while the latter introduces *NET-DBSCAN*, an algorithm for clustering the nodes of a dynamic linear network.

The scope of this paper is to introduce the new method. Therefore no special focus is given on scalability issues or on the application of this method in a real-world situation. These are directions for our future research. Specifically, our future research will be focused on the following issues.

Firstly, we strongly believe that DBSCAN-MO method is potentially useful to support real-world situations. We plan to work on this direction by discussing with geographers and geoscientists who are involved in the management of real-world problems. An extended experimentation on real-world data sets will help significantly on the customization of the DBSCAN-MO method to the needs of specific application domains in geography.

Secondly, we plan to eliminate the assumptions that have been made in the current version of DBSCAN-MO algorithm. For instance, in this version we assumed that the point set *P* to be clustered is static. Additionally, the travel cost model in the established network is rather simple. A first attempt to eliminate the former assumption can be found in Stefanakis (2005). Additionally, more sophisticated cost models and methods for free movement in space have been reported in Stefanakis (2006) and Stefanakis and Kavouras (2002).

Last but not least, we are currently focusing on the optimization of the algorithm by supporting its individual steps through the adoption of appropriate spatial index structures (Samet 1990; Manolopoulos et al. 2005) and the application of the algorithm in spaces of higher dimensionality (e.g., 3D space with moving polyhedrons).

## References

Allen, J.F., (1983) Maintaining Knowledge about Temporal Intervals. Communications of the ACM, 26(11), pp. 832-843.

Ester, M., Kriegel, H.P., Sander, J., and Xu, X. (1996) A Density-Based Algorithm for Discovering Clusters in Large Spatial Databases. In the Proceedings of the International Conference on Knowledge Discovery and Data Mining (KDD'96). Portland, Oregon, pp. 226-231.

Fayyad, U.M., Piatetsky-Shapiro, G., Smyth, P., and Ulthurusamy, R. (Eds) (1996) Advances in Knowledge Discovery and Data Mining. MIT Press.

Goebel, M., and Gruenwald, L. (1999) A Survey of Data Mining and Knowledge Discovery Software Tools. SIGKDD Explorations (1), ACM, pp. 20-33.

Han, J., Kamber, M., and Tung, A.K.H. (2001) Spatial Clustering Methods in Data Mining: A Survey. In Miller and Han 2001.

Laurini, R., and Thompson, D. (1992) Fundamentals of Spatial Information Systems. Academic Press Ltd.

Manolopoulos, Y., Nanopoulos, A., Papadopoulos A.N., and Theodoridis, Y. (2005) R-trees: Theory and Applications. Series in Advanced Information and Knowledge Processing, Springer.

Miller, H.J and Han, J. (Eds) (2001) Geographic Data Mining and Knowledge Discovery. Taylor & Francis.

Samet, H. (1990) The Design and Analysis of Spatial Data Structures. Addison-Wesley, Reading MA.

Sander, J., Ester, M., Kriegel, H.P., and Xu, X., (1998) Density-Based Clustering in Spatial Databases: The Algorithm GDBSCAN and Its Applications. Data Mining and Knowledge Discovery, 2(2), pp. 169-194.

Sedgewick, R. (1990) Algorithms. Addison-Wesley.

Stefanakis, E. (2004) Navigating among Moving Obstacles. In the Proceedings of the 3rd International Conference on Geographic Information Science (GIScience 2004), Adelphi, MD, Oct. 20-23, 2004.

Stefanakis, E. (2005) Clustering Dynamic Map Objects Based on Density Measures. In Proceedings of the 22nd International Cartographic Conference. A Coruna, Spain, July 2005.

Stefanakis, E. (2006) Scheduling Trajectories on a Planar Surface with Moving Obstacles. Informatica. Vol. 17(1), pp. 95-110.

Stefanakis, E. (2007) NET-DBSCAN: Clustering the Nodes of a Dynamic Linear Network. International Journal of Geographical Information Science. Taylor & Francis. Volume 21(4), 427-442.

Stefanakis, E., and Kavouras, M. (1995) On the Determination of the Optimum Path in Space, In Frank, A., and Kuhn, W., (Ed's.), Spatial Information Theory: A Theoretical Basis for GIS (COSIT 95). Springer-Verlag, pp. 241-257.

Stefanakis, E., and Kavouras, M. (2002) Navigating in Space under Constraints, International Journal of Pure and Applied Mathematics (IJPAM), Vol. 1(1), Academic Publ., pp. 71-93.

Worboys, M.F. (1995) GIS: A Computing Perspective. Taylor & Francis.

# A Metric of Compactness of Urban Change Illustrated to 22 European Countries

Alex Hagen-Zanker[1,2], Harry Timmermans[1]

[1]Urban Planning Group, Eindhoven University of Technology, PO Box 513, 5600 MB Eindhoven, The Netherlands, a.h.hagen.zanker@tue.nl
[2]Research Institute for Knowledge Systems, PO Box 463, 6200 AL Maastricht, The Netherlands

**Abstract.** Most metrics of urban spatial structure are snapshots, summarizing spatial structure at one particular moment in time. They are therefore not ideal for the analysis of urban change patterns. This paper presents a new spatio-temporal analytical method for raster maps that explicitly registers *changes* in patterns. The main contribution is a transition matrix which cross-tabulates the distance to the nearest urbanized location at the beginning and end of the analyzed period. The transition matrix by itself offers a powerful description of urban change patterns from which further metrics can be derived. In particular, a metric that is an indicator of the compactness of urban change is derived. The new metric is applied first to a synthetic dataset demonstrating consistency with existing classifications of urban change patterns. Next, the metric is applied country by country on the European CORINE land cover dataset. The results indicate a striking contrast in change patterns between Western and Eastern European counties. The method can be further elaborated in many different ways and can therefore be the first in a family of spatio-temporal descriptive statistics.

**Keywords:** urban change, spatial patterns, CORINE, metric of compactness, spatial planning

## 1 Introduction

Over the years, a variety of models of urban patterns and urban dynamics has been suggested. In this context, a distinction between descriptive models and explanatory models is highly relevant. Descriptive models summarize data. Well-known examples are the rank size distribution of city populations (Ioannides and Overman 2003; Chen and Zhou 2008), the cluster size distributions of urban areas (Benguigui et al. 2006) and fractal relations in urban form (Batty and Longley 1996; Shen 2002; Thomas et al. 2008). Furthermore, several metrics of spatial clustering and diffusion such as enrichment factors (Verburg et al. 2004) and transiograms (Li 2006) have been proposed.

In contrast, explanatory models go beyond mere description by focusing on processes underlying spatial patterns, thereby offering an interpretation of reality. Explanatory models of urban areas are typically based on the dynamic interactions between actors and their relative geographic position. Such relations can include for instance network effects, benefits of scale and spatial externalities, and lead for instance to buffers, segregation, agglomeration and sprawl effects. Especially, Cellular Automata modeling (White and Engelen 1993; Clarke et al. 1997; Couclelis 1997) and Agent Based Modeling (Parker et al. 2003), based on the assumed process of self-organization (White and Engelen 1993; Irwin and Geoghegan 2001) have become quite popular.

Because of this difference in focus, a gap can be identified between descriptive models and the inherent dynamic nature of urban change. Dynamic descriptive models would constitute a building block in the sense that such models support empirical evidence of dynamic processes and could also serve as a basis for theory development. An examination of the existing literature however suggests that such descriptive models have been used in rather limiting ways. First, explanatory models have been fit to historical data (Silva and Clarke 2004; Straatman et al. 2004) and their predictive capacity has been investigated (Pontius 2004; Hagen-Zanker et al. 2005), these studies have evaluated dynamic models only in terms of the static end-situation of a simulation run as opposed to focusing on the dynamic process. Secondly, although there is some recent evidence of spatio-temporal analysis of urban form (e.g., Tao et al. 2004; Herold et al. 2005; Seto and Fragkias 2005; Xiao et al. 2006), mirroring earlier developments in the field of landscape ecology (Turner 1989), these studies first apply a spatial analysis to summarize the structure of maps in multiple metrics for multiple moments in time and then conduct a temporal analysis to summarize the trajectories in time. Although this is a valid approach, it does have some clear disadvantages. The most important of these is the fact that already in the first step of the analysis all spatial information is lost, making it problematic to interpret the trajectory of individual metrics over time in terms of processes or patterns of change. For instance, one of the most commonly used metrics is patch size. Increasing patch size over time can be the consequence of disappearing small patches, but also of appearing large patches or of expansion of existing patches. Which of these change patterns occurred can only be conjured from other metrics, for instance the number of patches. If mixed change patterns occur (e.g. some patches appearing and some disappearing), it becomes impossible to untangle different change patterns.

A notable exception to this approach can be found in Wilson et al. (2003) whose spatial analysis is based on patterns of change in spatial structure. They identified five different patterns of urban change, presented in an urban change map: *Infill, Expansion, Isolated growth, Linear branch*, and *Clustered branch*. Xu et al. (2007) used a similar classification of *Infill, Edge-expansion* and *Spontaneous growth*.

Other spatio-temporal analyses describing land use change are based on the land use transition matrix (Debussche et al. 1977; Muller and Middleton, 1994). This matrix cross-tabulates land use categories of locations (cells) at the beginning and end of the analyzed period. However, these land use transition matrices typically do not consider spatial structure, except for cell-to-cell overlap, and therefore are of limited interest when investigating the link between pattern and process.

The goal of the present paper therefore is to suggest a new method to alleviate these limitations of existing approaches. More specifically, a distance class transition matrix is suggested to capture descriptively processes of land use change. The method can be viewed as an extension of the traditional land use transition matrix for only two classes; *Urban* and *Non-urban*, and builds on the concept of urban change maps (Wilson et al. 2003; Xu et al. 2007). It does not arbitrarily break down the spatial and temporal analysis and is also not based on the spatial configuration of land use pertaining to the end situation only, as in the calibration process.

The paper is organized as follows. First, we will introduce the method that we propose. Next, we will illustrate and apply the method to two data sets: synthetic data and European land cover data. The paper is completed by discussing the results of the analysis and reflecting on possible elaborations of the suggested method.

## 2 Method

### 2.1 Distance to Urban

Input is a pair of binary (*Urban*, *Non-urban*) raster maps that delineates the urban area at the beginning and end of the study period. From these two maps, indicator maps are derived that express for every cell the distance to the nearest cell of class *Urban*.

To allow cross-tabulation, distance classes are defined. These distance classes are simple bins with a lower (included) and upper (excluded) boundary. The upper boundary of one distance class is the lower boundary of the next. In the equations that follow, the distribution over distance classes is used as an approximation of the distribution over distances. The precision of this approximation depends on the number and size of the bins. In the current application, increasingly broader bins are applied to larger distances. The rationale for this choice is that at larger distances the required (absolute) precision is lower.

The first bin is always for the cells at distance 0 to *Urban*, i.e. cells that are *Urban* themselves. Table 1 gives the general form of a transition matrix for two classes *Urban* and *Non-urban*; it illustrates that the distance classes are a further specification of the class *Non-urban,* and that the class *Urban* is identical to the first distance class ($D_1$).

**Table 1** Generic distance class transition matrix.

| | | | Final | | | | | |
|---|---|---|---|---|---|---|---|---|
| | | | Urban | Non-urban | | | | |
| | | | $D_1$ | $D_2$ | $D_3$ | ... | $D_n$ | Sum |
| Initial | Urban | $D_1$ | $t_{1,1}$ | $t_{1,2}$ | $t_{1,3}$ | ... | $t_{1,n}$ | $t_{1,+}$ |
| | Non-urban | $D_2$ | $t_{2,1}$ | $t_{2,2}$ | $t_{2,3}$ | ... | $t_{2,n}$ | $t_{2,+}$ |
| | | $D_3$ | $t_{3,1}$ | $t_{3,2}$ | $t_{3,3}$ | ... | $t_{3,n}$ | $t_{3,+}$ |
| | | : | : | : | : | | : | : |
| | | $D_n$ | $t_{n,1}$ | $t_{n,2}$ | $t_{n,3}$ | ... | $t_{n,n}$ | $t_{n,+}$ |
| | | Sum | $t_{+,1}$ | $t_{+,2}$ | $t_{+,3}$ | ... | $t_{+,n}$ | $t_{+,+}$ |

*Urban* and *Non-urban* are land use classes. $D_1$, $D_2$, $D_3 \ldots D_n$ are distance classes. $t_{i,j}$ is the number of cells changing from class $D_i$ to $D_j$. $t_{i,+}$ is the number of cells originally in class $D_i$. $t_{+,i}$ is the number of cells finally in class $D_j$. $t_{+,+}$ is the number of cells in the map.

Even though the distance to cells with an urban land use class is a simple concept, the calculation of these distances is not straightforward. Naive implementations will demand prohibitively long calculation times on substantial datasets. This problem is known in computer science as the Euclidean Distance Transform and over the years many algorithms have been proposed. Typically, these algorithms trade accuracy of the distance estimates for calculation time. We settled for the exact algorithm of Felzenszwalb and Huttenlocher (2003) which is reasonably fast and does not introduce errors into the analysis. The algorithm approaches the Euclidean Distance Transform as a minimization problem and applies dynamic programming to solve it. The execution time of the algorithm is proportional to the number of cells and it manages 18 million cells in 5 seconds on a 1.6 GHz AMD Turion processor.

Fig. 1 illustrates the relation between urban land use, distance to urban and distance classes for the case of Luxembourg. Note that for legibility less distance classes are displayed than in the results section. The distance classes that are used and their total presence on the maps are tabulated in table 2. The distance class transition matrix is given as table 3.

**Table 2** Distance classes of the Luxembourg example

| Class | From | To | Area in 1991 | Area in 1991, cumulative | Area in 2000 |
|---|---|---|---|---|---|
| $D_1$ | 0 | 1 | 20839 | 20839 | 22591 |
| $D_2$ | 1 | 6 | 70634 | 91473 | 73331 |
| $D_3$ | 6 | 20 | 130649 | 222122 | 128802 |
| $D_4$ | 20 | 82 | 37439 | 259561 | 34837 |

Distance and area are measured in cell units; the cell size is 100 m

**Table 3** Transition matrix of the Luxembourg example

| | $D_1$ | $D_2$ | $D_3$ | $D_4$ |
|---|---|---|---|---|
| $D_1$ | 20810 | 29 | 0 | 0 |
| $D_2$ | 1303 | 69305 | 26 | 0 |
| $D_3$ | 458 | 3687 | 126504 | 0 |
| $D_4$ | 20 | 310 | 2272 | 34837 |

## 2.2 Summary Metric

The distance class transition matrix itself can be interpreted in terms of change patterns by visual inspection. It is clear that when urban growth takes place (and no shrinking) all non-zero values are found below or on the diagonal of the matrix. If the growth pattern is compact, transitions are found close to the diagonal (indicating that urban areas are only encroaching slowly) and towards the upper left corner (indicating that cells close to urban areas are affected, but those far away from urban areas are not). If loss of urban area takes place, which is not common given the irreversible nature of urbanization, the transitions are registered above the diagonal. In this case, compactness is gained when transitions are found away from the diagonal (creating large non-urban areas) and in the upper right corner (affecting those areas at great distances from urban cells).

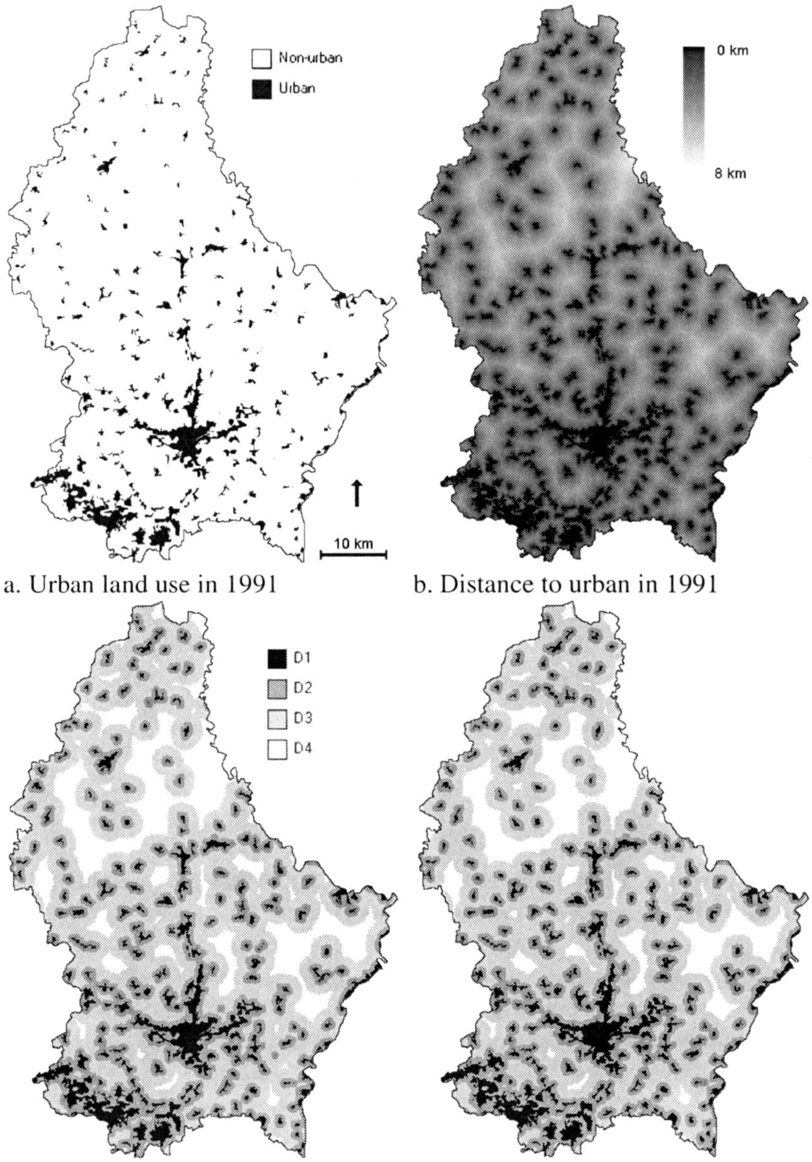


a. Urban land use in 1991 b. Distance to urban in 1991

c. Distance classes in 1991 d. Distance classes in 2000

**Fig. 1:** Urban land use and distance classes in Luxembourg

Different summary metrics can be envisioned on the basis of the transition matrix. We will introduce only one here, focusing on the relative loss of compactness, normalized to the total area of change.

The relative loss in compactness of a cell on the map is calculated as the drop in the cumulative distribution of distance to urban areas on the map, using the following equation:

$$L\left(d_{before,cell}, d_{after,cell}\right) = \frac{F_{before}\left(d_{after,cell}\right) - F_{before}\left(d_{before,cell}\right)}{\frac{1}{2}\left(F_{before}\left(d_{after,cell}\right) + F_{before}\left(d_{before,cell}\right)\right)} \tag{1}$$

where the function $L(d_{before,cell}, d_{after,cell})$ yields the loss of compactness associated to the transition from distance $d_{before,cell}$ to $d_{after,\ cell}$ at the location of *cell*. $F_{before}$ is the cumulative distribution of distance to urban area of all *nCells* cells, it is defined as follows:

$$F_{before}\left(d^*\right) = \frac{\sum_{cell=1}^{nCells}\left[d_{cell,before} \le d^*\right]}{nCells} \tag{2}$$

where square brackets are Iverson brackets; $[P]$ returns 1 if proposition $P$ is true and 0 otherwise. Thus, $F_{before}(d^*)$ is the proportion of all cells that lie at distance $d^*$ or closer to urban areas in the initial situation.

Using distance classes means that information on the precise distance is lost. Therefore, the loss of compactness cannot be calculated exactly on the basis of the transition matrix. However an approximation can be made following:

$$F_{before}\left(d^*\right) \approx \frac{\sum_{ii=1}^{i} t_{ii,+}}{t_{+,+}} \tag{3}$$

where distance $d^*$ is within distance class $D_i$. Yielding the following for each pair of distance classes:

$$l_{i,j} = \frac{\sum_{ii=1}^{i} t_{ii,+} - \sum_{jj=1}^{j} t_{jj,+}}{\frac{1}{2}\left(\sum_{ii=1}^{i} t_{ii,+} + \sum_{jj=1}^{j} t_{jj,+}\right)} \tag{4}$$

where $l_{i,j}$ expresses the loss of compactness related to cells that are originally in distance class $i$ and finally in distance class $j$. The indices $ii$ and $jj$ iterate over all distance classes equal to or smaller than $i$ resp. $j$. Note that the value $l_{i,j}$ solely depends on the distribution over distance classes in the initial situation.

Using the cumulative areas per distance class of table 2 the loss of compactness associated to each element of the transition table of the Luxembourg example can be calculated. For instance the loss associated to the transition from $D_3$ to $D_1$ is calculated according to eq. 5. The outcomes for all combinations of distance classes are presented in table 4. The spatial distribution of the different degrees of loss is presented in fig. 2.

$$l_{3,1} = \frac{(222122 - 20839)}{\frac{1}{2}(222122 + 20839)} \approx 1.7 \tag{5}$$

**Table 4** Loss of compactness ($l_{i,j}$) for the Luxembourg example

| | D1 | D2 | D3 | D4 |
|---|---|---|---|---|
| D1 | 0 | -1.3 | -1.7 | -1.7 |
| D2 | 1.3 | 0 | -0.8 | -1.0 |
| D3 | 1.7 | 0.8 | 0 | -0.2 |
| D4 | 1.7 | 1.0 | 0.2 | 0 |

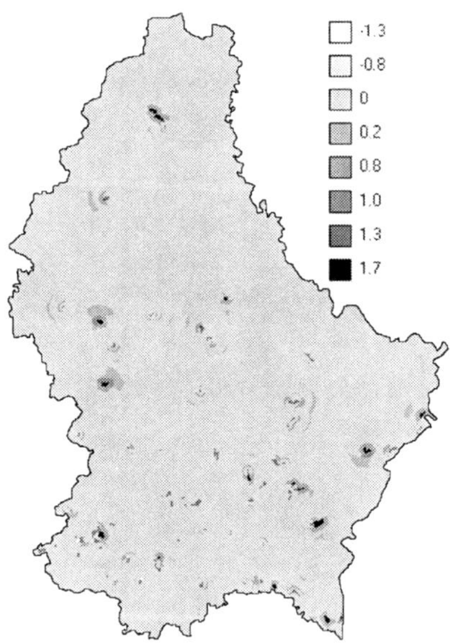


**Fig. 2:** Spatial distribution of loss of compactness (1991-2000) in the Luxembourg example. Most cells display no loss of compactness.

As indicated, the effect of this formulation is that the change in compactness is weighted relative to the distribution of distances to urban areas in the initial situation. This introduces a scale-independency which means that a loss in cells lying within (say) 5 km of urban areas is registered as a strong loss in compactness in densely and scattered built-up countries (e.g., Belgium), and as only a mild loss in countries with vast open areas (e.g., Spain). Likewise, a change pattern that is considered compact relative to the whole country may not be compact relative to a region.

The overall loss of compactness is calculated as the area weighted mean. Therefore it is equivalent to the mean over the map presented in fig. 2 and it is calculated as follows:

$$l_{mean} = \frac{\sum_{i=1}^{n}\sum_{j=1}^{n} t_{i,j} * l_{i,j}}{t_{+,+}} \tag{6}$$

The metric presented here is a measure of spatial structure, but there are also non-spatial metrics of compactness of urban change, such as the overall increase in urban area and the increase in population density. In order to express the spatial structure component independently of the other non-spatial metrics of compactness of change, the total loss in compactness is normalized in such fashion that the resulting metric can be interpreted as a measure of elasticity: the relative loss in compactness per relative increase in urban area. It is calculated as follows:

$$l_{region} = \frac{l_{mean}}{\frac{u_{after} - u_{before}}{\frac{1}{2}\left(u_{after} + u_{before}\right)}} \tag{7}$$

where $l_{region}$ is the normalized loss in compactness of the studied region; $u_{before}$ and $u_{after}$ are the total urban area in the initial and final situation and can be read from the transition matrix, since distance class $D_1$ corresponds to *Urban*.

$$\begin{aligned} u_{before} &= t_{1,+} \\ u_{after} &= t_{+,1} \end{aligned} \tag{8}$$

Eq. 9 integrates eqs. 4, 6, 7 and 8; it expresses the loss of compactness as a function of the transition matrix only.

$$l_{region} = \frac{t_{1,+} + t_{+,1}}{t_{1,+} - t_{+,1}} * \frac{1}{t_{+,+}} \sum_{i=1}^{n} \sum_{j=1}^{n} \left( t_{i,j} * \frac{\sum_{ii=1}^{i} t_{ii,+} - \sum_{jj=1}^{j} t_{jj,+}}{\sum_{ii=1}^{i} t_{ii,+} + \sum_{jj=1}^{j} t_{jj,+}} \right) \tag{9}$$

For the Luxembourg example, this yields: $l_{Luxembourg} = 0.29$. This number differs from the result presented further on, because the number of classes in the example is (too) small.

# 3 Data

The method is tested on two datasets. First, it is applied to a synthetic dataset of which the loss in compactness is well understood. The results for that dataset serve as a verification of the method. Secondly, the method is used to analyze Pan-European land cover data. The interpretation of the results of this application uses the first dataset as reference levels.

## 3.1 Synthetic Dataset

The synthetic dataset (fig. 3) consists of six maps. The first three maps represent urban growth patterns as identified by Wilson et al. (2003) and Xu et al. (2007). These are *Infill*, *Expansion* (also called *Edge-expansion*) and *Isolated* also called *Spontaneous growth*). In reality, urban areas will not develop exclusively according to one of these patterns, but in fact there may be combinations or in-between forms. Three more maps (fig. 4) give in-between patterns of *Infill-expansion*, *Infill-isolated* and *Expansion-isolated.* The maps do not refer to an actual situation and have no particular scale. The map size is 50 by 50 pixels.

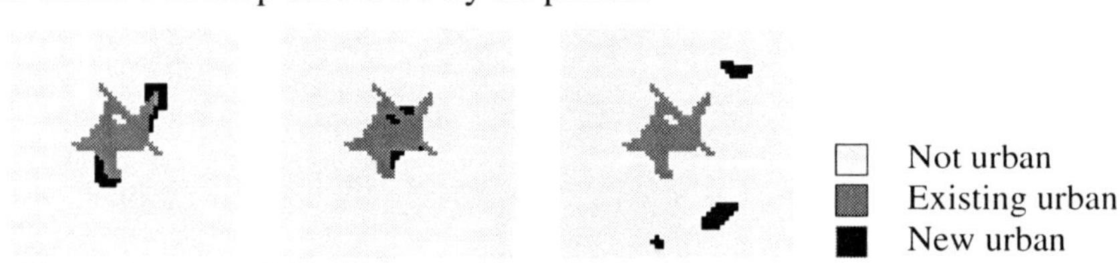


a. Expansion b. Infill c. Isolated

**Fig. 3:** Archetypical growth patterns

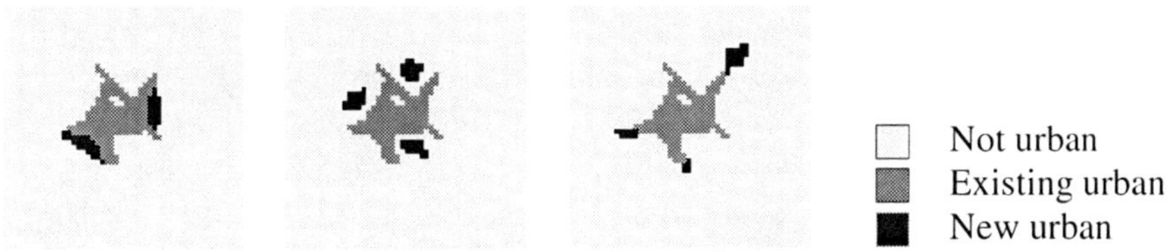


a. Infill-expansion b. Infill-isolated c. Expansion-isolated

**Fig. 4:** Mixed type growth patterns

## 3.2 CORINE Land Cover

CORINE is a Pan-European land cover map produced by the Environmental Assessment Agency (EEA) It recognizes 44 types of land cover, however we only consider one main category and its complement. This category is Artificial Surfaces and includes the following sub-classes:

- Continuous urban fabric
- Discontinuous urban fabric
- Industrial or commercial units
- Road and rail networks and associated land
- Port areas
- Airports
- Mineral extraction sites
- Dump sites
- Construction sites
- Green urban areas
- Sport and leisure facilities

There are some particularities to the CORINE dataset; it is available as a 100m raster dataset, but the classification procedure in fact is based on recognition of objects rather than fields. Homogenous objects are functional objects (i.e. the garden belonging to a house is classified as *Urban fabric* and a farmhouse may be classified as *Arable land*). The objects are recognized with a minimum mapping unit of 15 ha, thus one must be careful not to interpret resolution as precision.

The CORINE dataset is available for two moments in time: 1990 and 2000. The map of 2000 is actually based on imagery of 2000, but the 1990 map in fact is based on imagery ranging from 1985 to 1996. The year 1990 is only the median of the dataset. Individual countries are based on data from one year only and when we present the results further on, we will also indicate the period between the initial and final year. The mapping procedure, and in particular application of the minimal mapping unit, implies that differences between the 1990 and 2000 dataset do not all represent changes that took place in reality. Therefore, EEA has performed an elaborate analysis and produced an additional data layer which is the layer of land cover changes. This layer is the most reliable source of spatially explicit pan-European land use/ land cover change.

The dataset that we have is CORINE 2000 overlaid with the 1990 and 2000 exponent of changes to obtain consistent maps for 1990 (median) and 2000. The two CORINE land cover maps are not available for all countries. The CORINE project is ongoing however and over time more countries may become available. A release of 2005 data is pending.

The data has been cleaned by EEA before releasing it to the public. Nevertheless, visual inspection of the maps indicated several differences between 1990 and 2000 that should possibly be attributed to data errors. It is beyond the scope of this project to redo the data cleaning work of EEA. Instead we assume that over time artificial surfaces do not change to non-artificial. This is put into effect by only considering those values in the transition matrix below or on the diagonal.

# 4 Results

## 4.1 Synthetic Examples

The distance class transition matrices are given in table 5. The visual interpretation of the transition matrices confirms our expectations. The *Infill* pattern affects only the smaller distance classes, the cells in distance class $D_7$ or further are unaffected, i.e. lay at the diagonal. The *Expansion* pattern affects cells in all distance classes (off-diagonal values are found for all distance classes), but the magnitude of the effect is small (the off-diagonal values are found close to the diagonal). The *Isolated* pattern only affects cells at larger distances and these distance classes are severely affected.

The transition matrices of the mixed change patterns present a balance of the mixed patterns. *Infill-expansion* has positive values in the same cells as *Expansion*, but the values in the further distance classes are smaller. The matrix of *Infill-isolated* shows that the distance classes at the mid-range are most affected ($D_4$ to $D_6$). It thereby takes the middle of the *Infill* and *Isolated* patterns. The *Expansion-isolated* pattern affects the small as well as the mid-range distances. The larger numbers in the matrix are found in the mid-range ($D_4$ to $D_7$).

The results for the metric of compactness of urban change are listed in table 7. The results are ordered by compactness. The ranking of the change patterns in terms of compactness is fully according to our a priori expectations: *Infill* is the most compact, *Isolated* the least compact and mixed patterns are ranked in between the two patterns that they mix. Only for the mutual ranking of the patterns *Infill-isolated* and *Expansion* we had no a priori expectation.

**Table 5** Distance class transition matrices for the synthetic dataset

**Table 5a** Infill

| | $D_1$ | $D_2$ | $D_3$ | $D_4$ | $D_5$ | $D_6$ | $D_7$ | $D_8$ | $D_9$ | $D_{10}$ | $D_{11}$ |
|---|---|---|---|---|---|---|---|---|---|---|---|
| $D_1$ | 164 | | | | | | | | | | |
| $D_2$ | 15 | 61 | | | | | | | | | |
| $D_3$ | 1 | 4 | 31 | | | | | | | | |
| $D_4$ | 1 | 3 | 1 | 74 | | | | | | | |
| $D_5$ | | | 1 | 2 | 115 | | | | | | |
| $D_6$ | | | | | 1 | 216 | | | | | |
| $D_7$ | | | | | | | 219 | | | | |
| $D_8$ | | | | | | | | 437 | | | |
| $D_9$ | | | | | | | | | 606 | | |
| $D_{10}$ | | | | | | | | | | 479 | |
| $D_{11}$ | | | | | | | | | | | 69 |

**Table 5b** Expansion

| | $D_1$ | $D_2$ | $D_3$ | $D_4$ | $D_5$ | $D_6$ | $D_7$ | $D_8$ | $D_9$ | $D_{10}$ | $D_{11}$ |
|---|---|---|---|---|---|---|---|---|---|---|---|
| $D_1$ | 164 | | | | | | | | | | |
| $D_2$ | 20 | 56 | | | | | | | | | |
| $D_3$ | 8 | 3 | 25 | | | | | | | | |
| $D_4$ | 16 | 8 | 2 | 53 | | | | | | | |
| $D_5$ | 4 | 23 | 3 | 11 | 77 | | | | | | |
| $D_6$ | | | 4 | 29 | 44 | 140 | | | | | |
| $D_7$ | | | | | 7 | 79 | 133 | | | | |
| $D_8$ | | | | | | 14 | 99 | 324 | | | |
| $D_9$ | | | | | | | | 141 | 465 | | |
| $D_{10}$ | | | | | | | | | 134 | 345 | |
| $D_{11}$ | | | | | | | | | | 21 | 48 |

**Table 5c** Isolated

| | $D_1$ | $D_2$ | $D_3$ | $D_4$ | $D_5$ | $D_6$ | $D_7$ | $D_8$ | $D_9$ | $D_{10}$ | $D_{11}$ |
|---|---|---|---|---|---|---|---|---|---|---|---|
| $D_1$ | 164 | | | | | | | | | | |
| $D_2$ | | 76 | | | | | | | | | |
| $D_3$ | | | 36 | | | | | | | | |
| $D_4$ | | | | 79 | | | | | | | |
| $D_5$ | | | | | 118 | | | | | | |
| $D_6$ | | 3 | 3 | 5 | 5 | 201 | | | | | |
| $D_7$ | 9 | 4 | | 4 | 6 | 34 | 162 | | | | |
| $D_8$ | 13 | 18 | 8 | 26 | 38 | 56 | 31 | 247 | | | |
| $D_9$ | 36 | 21 | 11 | 27 | 41 | 77 | 61 | 66 | 266 | | |
| $D_{10}$ | | 2 | 4 | 10 | 21 | 52 | 70 | 120 | 79 | 121 | |
| $D_{11}$ | | | | | | | 1 | 12 | 37 | 19 | |

**Table 5d** Infill-expansion

| | $D_1$ | $D_2$ | $D_3$ | $D_4$ | $D_5$ | $D_6$ | $D_7$ | $D_8$ | $D_9$ | $D_{10}$ | $D_{11}$ |
|---|---|---|---|---|---|---|---|---|---|---|---|
| $D_1$ | 164 | | | | | | | | | | |
| $D_2$ | 19 | 57 | | | | | | | | | |
| $D_3$ | 6 | 4 | 26 | | | | | | | | |
| $D_4$ | 13 | 4 | 3 | 59 | | | | | | | |
| $D_5$ | 7 | 9 | 4 | 10 | 88 | | | | | | |
| $D_6$ | | 3 | 2 | 11 | 26 | 175 | | | | | |
| $D_7$ | | | | | 2 | 40 | 177 | | | | |
| $D_8$ | | | | | | 1 | 43 | 393 | | | |
| $D_9$ | | | | | | | | 55 | 551 | | |
| $D_{10}$ | | | | | | | | | 37 | 442 | |
| $D_{11}$ | | | | | | | | | | 15 | 54 |

**Table 5e** Infill-isolated

| | $D_1$ | $D_2$ | $D_3$ | $D_4$ | $D_5$ | $D_6$ | $D_7$ | $D_8$ | $D_9$ | $D_{10}$ | $D_{11}$ |
|---|---|---|---|---|---|---|---|---|---|---|---|
| $D_1$ | 164 | | | | | | | | | | |
| $D_2$ | | 76 | | | | | | | | | |
| $D_3$ | | 8 | 28 | | | | | | | | |
| $D_4$ | 12 | 10 | 1 | 56 | | | | | | | |
| $D_5$ | 24 | 8 | 4 | 7 | 75 | | | | | | |
| $D_6$ | 18 | 14 | 5 | 13 | 18 | 149 | | | | | |
| $D_7$ | | 3 | 2 | 10 | 15 | 36 | 153 | | | | |
| $D_8$ | | | | | 5 | 29 | 58 | 345 | | | |
| $D_9$ | | | | | | | 6 | 97 | 503 | | |
| $D_{10}$ | | | | | | | | | 54 | 425 | |
| $D_{11}$ | | | | | | | | | | 14 | 55 |

**Table 5f** Expansion-isolated

| | $D_1$ | $D_2$ | $D_3$ | $D_4$ | $D_5$ | $D_6$ | $D_7$ | $D_8$ | $D_9$ | $D_{10}$ | $D_{11}$ |
|---|---|---|---|---|---|---|---|---|---|---|---|
| $D_1$ | 164 | | | | | | | | | | |
| $D_2$ | 4 | 72 | | | | | | | | | |
| $D_3$ | 4 | | 32 | | | | | | | | |
| $D_4$ | 5 | 8 | 3 | 63 | | | | | | | |
| $D_5$ | 7 | 8 | 2 | 12 | 89 | | | | | | |
| $D_6$ | 11 | 9 | 4 | 15 | 29 | 149 | | | | | |
| $D_7$ | 5 | 5 | 3 | 10 | 20 | 49 | 127 | | | | |
| $D_8$ | | 3 | 1 | 8 | 16 | 60 | 94 | 255 | | | |
| $D_9$ | | | | | 2 | 18 | 49 | 179 | 358 | | |
| $D_{10}$ | | | | | | | | 33 | 149 | 297 | |
| $D_{11}$ | | | | | | | | | 1 | 22 | 46 |

**Table 6** Distance classes in cell units

| | From(included) | To(excluded) |
|---|---|---|
| $D_1$ | 0 | 1 |
| $D_2$ | 1 | $\sqrt{2}$ |
| $D_3$ | $\sqrt{2}$ | 2 |
| $D_4$ | 2 | $2\sqrt{2}$ |
| $D_5$ | $2\sqrt{2}$ | 4 |
| $D_6$ | 4 | $4\sqrt{2}$ |
| $D_7$ | $4\sqrt{2}$ | 8 |
| $D_8$ | 8 | $8\sqrt{2}$ |
| $D_9$ | $8\sqrt{2}$ | 16 |
| $D_{10}$ | 16 | $16\sqrt{2}$ |
| $D_{11}$ | $16\sqrt{2}$ | 32 |

**Table 7** Compactness of change of the synthetic dataset

| Change pattern | Loss of compactness | A priori expected rank |
|---|---|---|
| Infill | 0.046 | 1 |
| Infill-expansion | 0.19 | 2 |
| Infill-isolated | 0.31 | 3 or 4 |
| Expansion | 0.38 | 3 or 4 |
| Expansion-isolated | 0.80 | 5 |
| Isolated | 1.1 | 6 |

## 4.2 Patterns of Urban Change across Europe

The transition matrices and the derived loss of compactness metric are calculated for all countries in the CORINE dataset. The results are presented in table 8. The results of the synthetic dataset are used as reference levels.

The results indicate that the loss of compactness for all countries has been in between that of the *Infill* change pattern and the *Expansion-isolated* pattern.

The analysis is performed twice, once with and once without the filter that ignores loss of urban area. The filter corrects one outlier (Slovakia) that without filtering registers a growth even less compact than *Isolated growth*.

**Table 8** Loss of compactness

| | | | Without filtering | | |
|---|---|---|---|---|---|
| Rank | Country | Loss of C. | Loss of C. | Rank | Period (years) |
| *** | *Infill* | *0.05* | *** | *** | *** |
| **1** | **Estonia** | **0.10** | **0.13** | **1** | **6** |
| **2** | **Slovenia** | **0.11** | **0.13** | **2** | **5** |
| **3** | **Bulgaria** | **0.16** | **0.16** | **3** | **10** |
| **4** | **Romania** | **0.18** | **0.18** | **4** | **8** |
| *** | *Infill-expansion* | *0.19* | *** | *** | *** |
| **5** | **Lithuania** | **0.20** | **0.18** | **5** | **5** |
| **6** | **Poland** | **0.21** | **0.23** | **8** | **8** |
| 7 | The United Kingdom | 0.21 | 0.21 | 7 | 10 |
| 8 | Austria | 0.22 | 0.21 | 6 | 15 |
| 9 | Belgium | 0.25 | 0.25 | 9 | 10 |
| 10 | Spain | 0.26 | 0.26 | 10 | 14 |
| 11 | Portugal | 0.27 | 0.27 | 11 | 14 |
| 12 | France | 0.29 | 0.28 | 12 | 10 |
| *** | *Infill-isolated* | *0.31* | *** | *** | *** |
| **13** | **Slovakia** | **0.31** | **3.71** | **22** | **8** |
| 14 | Ireland | 0.33 | 0.33 | 13 | 10 |
| 15 | The Netherlands | 0.35 | 0.36 | 14 | 14 |
| 16 | Luxembourg | 0.36 | 0.36 | 15 | 11 |
| 17 | Greece | 0.37 | 0.37 | 16 | 10 |
| *** | *Expansion* | *0.38* | *** | *** | *** |
| **18** | **Hungary** | **0.39** | **0.41** | **17** | **8** |
| 19 | Germany | 0.41 | 0.47 | 20 | 10 |
| 20 | Denmark | 0.41 | 0.42 | 18 | 10 |
| 21 | Italy | 0.46 | 0.46 | 19 | 10 |
| **22** | **Latvia** | **0.72** | **0.72** | **21** | **5** |
| *** | *Expansion-isolated* | *0.80* | *** | *** | *** |
| *** | *Isolated* | *1.11* | *** | *** | *** |

Eastern European countries in **bold**

## 5 Discussion

The six urban growth patterns of the synthetic dataset fit well with our intuitive understanding of compactness of urban change. It is therefore comforting that the value of the metric confirms expectations. Moreover, the range of values found on the basis of the CORINE dataset is similar to that of the synthetic dataset. This means that the synthetic dataset provides a useful frame of reference and the results can be well interpreted.

A striking distinction emerges between Western and Eastern European countries. It appears that Eastern countries as a whole have more compact urban change patterns. It is difficult to attribute this difference to one or the other process, particularly because these two regions are distinct in so many aspects. Nevertheless, we like to speculate that in the young Eastern European economies social and economic opportunities primarily occur in the (large) cities. Spatial developments in the countryside are limited and as a consequence there is limited fragmentation. In the Western European countries, rural and peri-urban development is taking place, the contrast between rural and urban areas diminishes, and so does the compactness of the urban areas. Note that the United Kingdom, with London as its strong urban magnet, is the most compact of Western European countries.

One avenue of further investigating the hypothesis that the contrast between rural and urban development explains the distinction between Western and Eastern European countries is to apply the proposed method at a finer scale, for instance European NUTS3 administrative regions. The expectation would then be that in Eastern Europe the compactness at the regional level will be higher, since the regional urban-rural contrast will not longer contribute to the compactness.

Another somehow surprising result is the lack of any clear evidence of the effect of national spatial planning strategy. In particular, the well-known contrast between Belgian (liberal) and Dutch (strict) spatial planning does not materialize. A possible explanation may be that the urban landscape of these countries is the effect of a longer history of spatial planning. It may well be that the recent history breaks that trend. Another explanation may be the role of the initial situation in the sense that in a highly fragmented landscape there may be more possibilities for compact development than in a more compact landscape.

# 6 Conclusion

This paper set out to develop a method for the description of urban change patterns. The method that is introduced centers on a distance class transition matrix and a metric of compactness of urban change is derived from this matrix. Application of the newly developed method to a synthetic dataset confirms the descriptive power of the transition matrix as well as the derived statistic. It could there be applied with confidence on a real dataset of land use / land cover patterns in Europe.

The results of this analysis present a strong contrast between Eastern and Western European countries, and perhaps surprisingly do not demonstrate a clear link between spatial planning practice and patterns of change. We speculate that this may be the effect of the scale of the analysis and plan to investigate urbanization patterns at finer scales in particular European NUTS3 regions.

The metric and transition matrix presented in this paper are not the ultimate tool of describing urban areas. Instead, they offer a novel approach that may be extended and modified in many ways. Even though the link is not explicit, the analysis of transitions in distance to urban areas relates to fractal analysis. There are several methods to calculate the fractal dimension of urban areas. A common approach is based on (erosion-)dilation. The urban area with a dilation of radius $r$, is identical to the area of all locations where distance to urban area is smaller or equal to $r$. This line has not been pursued in the present paper, but the distance class transition matrix may be a useful instrument to derive metrics of fractal change.

The inclusion of other structure indicators than distance to urban area can be readily implemented. A straightforward extension of the transition matrix would be to include distance to non-urban areas in a similar fashion. Other likely candidates are patch size, built up density, population density and edge. Multidimensional transition matrices would allow the evaluation of multiple indicators in a single metric. Further summary metrics can refer to other aspects of urban structure, such as specialization, segregation, accessibility, self-sufficiency, disturbance, and exposure.

## Acknowledgements

This work is part of the project "*Instruments for monitoring and analysis of urban change*" supported by the Netherlands Organisation for Scientific Research (NWO) under the CASIMIR programme. It is an exchange project between the Research Institute for Knowledge Systems (RIKS bv) in Maastricht and the Urban Planning Group of Eindhoven University of

Technology (TU/e). The suggestions of two anonymous reviewers helped improving the paper and directing future work.

## References

Batty M, Longley PA (1996) Fractal cities. Academic press, London and San Diego.

Benguigui L, Blumenfeld-Lieberthal E, Czamanksi D (2006) The dynamics of the Tel Aviv morphology. Environment and Planning B: Planning and Design 33(2):269 – 284.

Chen Y, Zhou Y (2008) Scaling laws and indications of self-organized criticality in urban systems. Chaos, Solitons and Fractals 35(1):85-98.

Clarke KC, Gaydos LJ (1998) Loose-coupling a cellular automaton model and GIS: long-term urban growth prediction for San Francisco and Washington/Baltimore. International Journal of Geographical Information Science 12(7):699-714.

Couclelis H (1997) From cellular automata to urban models: New principles for model development and implementation. Environment and Planning B-Planning and Design 24(2):165-174.

Debussche DM, Godron M, Lepart J, Romane F (1977) An account of the use of a transition matrix. Agro-Ecosystems 3:81-92.

Felzenszwalb PF, Huttenlocher DP (2004) Distance transforms of sampled functions, pp 15. Cornell Computing and Information Science, Ithaca.

Hagen-Zanker A, Straatman B, Uljee I (2005) Further developments of a fuzzy set map comparison approach. International Journal of Geographical Information Science 19(7):769-785.

Herold M, Couclelis H, Clarke KC (2005) The role of spatial metrics in the analysis and modeling of urban land use change. Computers, Environment and Urban Systems 29(4):369-399.

Ioannides YM, Overman HG (2003) Zipf's law for cities: an empirical examination. Regional Science and Urban Economics 33(2):127-137.

Irwin EG, Geoghegan J (2001) Theory, data, methods: developing spatially explicit economic models of land use change. Agriculture Ecosystems and Environment 85(1-3):7-23.

Lambin EF, Turner BL, Geist HJ, Agbola SB, Angelsen A, Bruce JW, Coomes OT, Dirzo R, Fischer G, Folke C, George PS, Homewood K, Imbernon J, Leemans R, Li X, Moran EF, Mortimore M, Ramakrishnan PS, Richards JF, Skanes H, Steffen W, Stone GD, Svedin U, Veldkamp TA, Vogel C, Xu J (2001) The causes of land-use and land-cover change: moving beyond the myths. Global Environmental Change 11(4):261-269.

Li W (2006) Transiogram: A spatial relationship measure for categorical data. International Journal of Geographical Information Science 20(6):693-699.

Muller MR, Middleton J (1994) A Markov model of land-use change dynamics in the Niagara Region, Ontario, Canada. Landscape Ecology 9(2):151-157.

Parker DC, Manson SM, Janssen MA, Hoffmann MJ, Deadman P (2003) Multi-agent systems for the simulation of land-use and land-cover change: A review. Annals of the Association of American Geographers 93(2):314-337.

Pontius RG, Huffaker D, Denman K (2004) Useful techniques of validation for spatially explicit land-change models. Ecological Modelling 179(4):445-461.

Seto KC, Fragkias M (2005) Quantifying Spatiotemporal Patterns of Urban Land-use Change in Four Cities of China with Time Series Landscape Metrics. Landscape Ecology 20(7):871-888.

Shen G (2002) Fractal dimension and fractal growth of urbanized areas. International Journal of Geographical Information Science 16(5):419-437.

Silva EA, Clarke KC (2002) Calibration of the SLEUTH urban growth model for Lisbon and Porto, Portugal. Computers, Environment and Urban Systems 26(6):525-552.

Straatman B, White R, Engelen G (2004) Towards an automatic calibration procedure for constrained cellular automata. Computers, Environment and Urban Systems 28(1-2):149-170.

Tao Z, Jiyuan L, Xiangzheng D (2004) Spatial patterns of urban land expansion of super-cities of China in 1990s. Paper presented at the 2004 IEEE International Geoscience and Remote Sensing Symposium.

Thomas I, Frankhauser P, Biernacki C (2008) The morphology of built-up landscapes in Wallonia (Belgium): A classification using fractal indices. Landscape and Urban Planning 84(2)99-115.

Turner MG (1989) Landscape Ecology: The Effect of Pattern on Process. Annual Review of Ecology and Systematics 20(1):171-197.

Verburg PH, de Nijs TCM, Ritsema van Eck J, Visser H, de Jong K (2004) A method to analyse neighbourhood characteristics of land use patterns. Computers, Environment and Urban Systems 28(6):667-690.

White R, Engelen G (1993) Cellular-Automata and Fractal Urban Form - a Cellular Modeling Approach to the Evolution of Urban Land-Use Patterns. Environment and Planning A 25(8):1175-1199.

Wilson EH, Hurd JD, Civco DL, Prisloe MP, Arnold C (2003) Development of a geospatial model to quantify, describe and map urban growth. Remote Sensing of Environment 86(3):275-285.

Xiao J, Shen Y, Ge J, Tateishi R, Tang C, Liang Y, Huang Z (2006) Evaluating urban expansion and land use change in Shijiazhuang, China, by using GIS and remote sensing. Landscape and Urban Planning 75(1-2):69-80.

Xu C, Liu M, Zhang C, An S, Yu W, Chen J (2007) The spatiotemporal dynamics of rapid urban growth in the Nanjing metropolitan region of China. Landscape Ecology 22(6):925-937.

# Advanced Data Mining Method for Discovering Regions and Trajectories of Moving Objects: "Ciconia Ciconia" Scenario

Claudio Carneiro[1], Arda Alp[1], Jose Macedo[2], Stefano Spaccapietra[3]

[1]Geographic Information Systems Laboratory (AGILE Member), EPFL
[2]Artificial Intelligence Laboratory, EPFL
[3]Database Laboratory, EPFL
{Claudio.Carneiro; Arda.Alp; Jose.Macedo; Stefano.Spaccapietra}
@epfl.ch

**Abstract.** Trajectory data is of crucial importance for a vast range of applications involving analysis of moving objects behavior. Unfortunately, the extraction of relevant knowledge from trajectory data is hindered by the lack of semantics and the presence of errors and uncertainty in the data. This paper proposes a new analytical method to reveal the behavioral characteristics of moving objects through the representative features of migration trajectory patterns. The method relies on a combination of Fuzzy c-means, Subtractive and Gaussian Mixture Model clustering techniques. Besides, this method enables splitting the analysis into sections in order to differentiate the whole migration into i) migration-to-destination, ii) reverse-migration. The method also identifies places where moving objects' cumulate and increase in number during the moves (bottleneck points). It also computes the degree of importance for a given point or probability of existence of an object at a given coordinate within a certain confidence degree, which in turn determines certain zones having different degrees of importance for the move, i.e. critical zones of interest. As shown in this paper, other techniques are not capable to elaborate similar results. Finally, we present experimental results using a trajectory dataset of migrations of white storks (Ciconia ciconia).

**Keywords:** moving objects, trajectories, regions, spatial patterns, spatio-temporal dataset, data mining, clustering techniques

## 1 Introduction

The analysis of the interaction of moving objects and geographical space is of crucial importance for explaining important natural phenomena and related worldwide events. For example, animal migration analysis, which researches on animal habits, is quite useful for understanding disease spread, animals' extinction threats, global ecological equilibrium, etc. With the advent of low cost global positioning system (GPS) devices, collecting time positional data from moving objects became an easy task. Integration of GPS data and recent technological developments may help us to capture the semantics data. Such ability may let the user to integrate the analysis results for entertainment or marketing purposes. However, deriving relevant knowledge from these data may be a complex task because (1) there is a lack of semantics in the collected data; (2) sensor devices errors and uncertainties are propagated to the trajectory data; and (3) trajectory data analysis is highly dependent on the application domain. Thus, some new methods must be developed in order to provide means for discovering relevant knowledge taking into account those problems.

Global positioning system devices capture a moving object's trajectory only from a geometric point of view, storing geodetic positions (latitude, longitude and altitude: geodetic coordinates referring to the WGS84 system) of the moving object at specific time intervals. As a result, geographic background information that is fundamental for trajectory data analysis is neglected by such devices. Therefore, the lack of semantics in trajectory data representation obstructs trajectory data analysis, avoiding making intelligent use of this data. Although recent researches have responded by developing data mining algorithms that can improve the prospects for uncovering interesting and useful patterns from such large trajectory data collections, this is not sufficient for overcoming the lack of semantics in trajectory data. Trajectory semantics is application dependent and requires to be interpreted in the light of the application domain.

The elements of error and uncertainty that are present in trajectory data are caused by non-ideal reception conditions that deceive current GPS receivers. The position calculated by a GPS receiver requires the measurement of the current time, the position of the satellite and the measured delay of the received signal. The position accuracy is primarily dependent on the satellite position and signal delay. Thus, corrections must be made on trajectory data based on background knowledge. This can only be done by specific methods.

Thus, there is a need for more adequate analytical methods that may be applied over trajectory data in order to disclose hidden knowledge. In this paper, we propose a new analytical method combining Fuzzy c-means, Subtractive and Gaussian Mixture Model clustering techniques. This method allows revealing the characteristics of the representative migration's trajectory patterns. Using this representation we can conclude about the object's movement behavior. Besides, this method enable us characterize the direction of the movement by splitting the analysis into sections in order to differentiate the whole migration as two different habits: i) migration-to-destination ii) reverse-migration. Also, we are able to define places that our experimental moving objects cumulate and increase in number during the move (bottleneck points). We can also determine the degree of importance for a given point, or probability of existence of an object at a given coordinate within a certain degree of confidence, which specifies certain zones as having different degrees of importance for the move. Moreover, we investigate other techniques as to show that they are not adequate for the study of this case. In order to show the usefulness of our method, we present the experimental results using a trajectory dataset of migrations of white storks (Ciconia ciconia).

This paper is organized as follows. The 'Related Work' section overviews several data models that have been proposed for mining moving objects. The following 'White Storks Scenario' section overviews our particular moving object scenario: the migration of white storks. In this section it is presented the motivation of our research and an in-depth on how the migration data is handled. The 'Trajectory Data Analysis Requirements section discusses our advanced data mining method in order to analyze annual migration habits and to discover regions and trajectories of a group of white storks (Ciconia ciconia) based on real trajectory data. The 'Experimental Study' section discusses the experimental study setup and our findings on the behavior of the methods *Fuzzy c-means Clustering* and *Subtractive Clustering*.

## 2 Related Work

Several data models have been proposed for efficiently querying on moving objects (Wolfson *et al.* 1998), (Forlizzi *et al.* 2000), (Brakatsoulas *et al.* 2004), (Mouza 2001). In Wolfson *et al.* (1998) and Forlizzi *et al.* (2000), in which the main focus relies on the geometric properties of trajectories, while in Brakatsoulas *et al.* (2004) and Mouza (2005) both semantics and background geographic information are considered. In Mouza (2005) moving patterns are extracted from data by defining the pattern in advance, for instance, by finding all trajectories that move from zone A to zone B and cross zone Z, and thus, moving patterns are in fact the trajectories that follow the given pattern

From the data mining perspective, many trajectory pattern mining algorithms have been developed (Tsoukatos and Gunopulos 2001), (Laube *et al.* 2005), (Verhein and Chawla 2006), (Gudmundsson *et al.* 2004). Some of these approaches find dense patterns, where moving objects are in the same region and move in the same direction (Cao *et al.* 2006). Other approaches consider spatially and temporally geographically referenced events, instead moving objects (Iyengar 2004)

## 3 White Storks Scenario

### 3.1 Overview

Annual migrations of white storks (Ciconia ciconia) have been thoroughly analyzed by several research groups[17]. The main reason for the migration of this species is the same as for the other species: search of better food availability. This migration starts each autumn from the north hemisphere (e.g. Europe) to reach the south hemisphere (e.g. Africa).

Giannotti *et al.* (2007) states that in many applications the mining problem comes with an inherent knowledge of the regions-of-interests manually obtained by experts through the application domain or simply by commonsense. Both origins and destinations are usually given as background knowledge or as results of some preliminary study. However, like in the authors' case, in our case we do not have this information and we have to derive it somehow. It has to be automatically computed using an appropriate technique. Thus we need to analyze the migration habits of the birds to improve our knowledge about many open questions on their behavior. At this point we consider the expertise of two biologists[18] from the University of Lausanne on bird migrations and the habits of white storks. Some of our questions are critical for biologists and Spaccapietra *et al.* (2007) also underscored them in their research. We selected the following questions as the motivation of our research.

Question 1: What route do these birds use while they are migrating and is it possible to define our characterization of the whole migration path?

[17] http://www.storchenhof-loburg.info and http://www.fr.ch/mhn/

[18] Alan Juilland, and Reto Burri, Doctoral Students of Département d'écologie et evolution, +41 21 692 41 74, Alan.Juilland@unil.ch; Reto.Burri@unil.ch

- Question 2: What is the direction of move?
- Question 3: Do they all choose same migration path?
- Question 4: Are there any places where these birds cumulate and increase in number during the migration (these places are highly critical for biologist to protect the birds from being hunted)?
- Question 5: Where these critical (bottleneck) points or areas (zones) are located on this route?

### 3.2 Data Set

In order to observe migration habits of white storks and to detect flocks, (i.e. storks flying close to each other) the spatio-temporal position of the storks is analyzed. We try to simplify the migration habits of our observation group to a simple trajectory so that we can find the graph belonging to this trajectory. It is possible to organize the trajectory data in three ways: i) according to years (starting year of the migration), ii) according to individual white storks, iii) according to the direction of migration (i.e. to-south, to-north). The first classification is useful in terms of analyzing flock migrations habits and comparing migration routes of each year within the observation period (1999-2006). This classifying constitutes '*time*' dimension. The second classification is useful in terms of analyzing the individual migration habits of these birds and comparing migration routes of each individual. This classifying constitutes '*bird*' dimension. The third and final classification can be constituted according to the definition of direction. Considering the trajectory it is possible to consider several subparts of it and therefore to define several directions respectively.

## 4 Trajectory Data Analysis Requirements

### 4.1 Presentation

In this section we evaluate different approaches for analyzing annual migration habits of a group of white storks (Ciconia ciconia) based on real trajectory data. Our objective is to relate the migration habits of our experiment group to a representative trajectory of the birds' paths. It is basically an aggregating of movement behavior. We aim to find a new pattern, aggregated trajectory pattern, which describes the movement of white storks, and represents a set of individual trajectories that inherits the property of visiting the same sequence of places in the future. This section of the research may be separated into two major interests:

*1. aggregated trajectory pattern,*
*2. regions of interest (critical bird zones)*

We start by analyzing the spatio-temporal position of the storks by classifying GPS data (latitude-geodetic coordinate, longitude-geodetic coordinate) and the date of when the data was captured according to years. We realize that for some years the date of the first migration data is the same for almost all birds (1998, 1999, and 2001). However, for the rest of the years, the date of first migration data is different for all birds (2000, 2002, 2003, 2004, and 2005). Therefore, it is somewhat hard to say that these birds started to migrate at the same day. Also, regarding the first migration data, location of each bird is apart from each other.

Thus, solely by considering each year individually it is not feasible to conclude that these storks migrate in flocks. Also, only a small percentage of storks are equipped with transmitters, and hence the flock to which a stork belongs for a given trajectory may be in fact, unknown. Moreover, considering each year, after tracking the dates of GPS data for each bird, we find out that these dates are inconsistent with each other. Namely for each bird, the GPS data collection date is different, as well as the number of data points that highly differs among the birds for a given year.

## 4.2 Study of Data Analysis Techniques

In order to come to a conclusion with regards to future migration habits of these birds, one possible option may be 'Time Series Prediction' which is used to model or predict future events based on knowledge of past events. This method allows us to predict future data points before they are measured. The sequence of data points must be captured at successive (often uniform) time intervals. However, the nature of our data set does not satisfy the basic condition of time series analysis: uniform time intervals.

Even it was possible to run some time series analysis on individual birds this would not yield satisfactory results for group habits of the birds. It is therefore rather hard to come up with a general conclusion on the migration habits of these birds for different time periods of the migration period. Moreover, this handicap hinders us to drive to a more general conclusion among the years. However it is still possible to run a non-time based statistical analysis on this data set at least to drive a conclusion on the migration coordinates (data points on the map) aiming to determine strategies for creating decision rules and surrogate constraints based on the analysis of scatter of the migration data points.

One of the simplest solutions arose from the knowledge interchange with the above mentioned biology experts. It is possible to devise a simplified and representative migration path by aggregating data points in a systematic way. This can be achieved by dividing the migration map into wards and aggregating the data points in each ward. Aggregation itself is performed by taking the mean value of the considered values of data points. We can create these wards in three ways:

I. By use of horizontal lines, which divides the map into horizontal wards parallel to latitudes;
II. By use of vertical lines, which divides the map into vertical wards parallel to longitudes and each vertical line couple creates a vertical ward;
III. By use of grids, which divides the map into grids. Namely this is the combination of using the vertical and horizontal lines to create grid shaped wards.

Using grids and aggregating the data points enables a simplified representative summary of the all migration data points. No conclusions can be extracted nevertheless due to uniqueness and systematic nature of the latter method. In fact, the quality and reliability of the representative data points is questionable, since these two criteria are highly dependent on how the grids are selected, namely on how the distance between two parallel lines (vertical and horizontal) is determined. Furthermore, this aggregation is done simultaneously for each wand and as a matter of fact aggregation is independent from any relation among close points. Specifically, no interrelation among the neighboring points is considered. On the other hand, during an iterative process it is possible to consider any existing neighboring relations. During the iterations, it is possible to consider different neighborhood relations among close data points. It is clear that an iterative aggregation process yield more room to consider alternative data clustering and to come up with a reasoning based on the performance of past iterations. Thus, an iterative procedure might be assumed to be more realistic compared to a single step clustering approach. In the case where an iterative process is not used (if all the aggregation is done in one step among the grids), the resulting aggregated data points can be reused as an input and several number of runs can be carried out in order to refine the very first initial aggregation of data points. Naturally, the degree of quality of the solution is dependent not only on the selection of grid sizes but also on the selection of iteration number.

## 4.3 Clustering Related Techniques and Statistical Pattern Recognition

All these observations indicate the necessity of an intelligent way of simplifying the migration data points. There may be different research directions depending on what we want to find and how we want to proceed. Briefly, a reasonable idea is centered around 'Vector quantization' which was proposed by Gray (1984) is a quantization technique often used in loss data compression – the main goal is to code or replace with a key, values from a multidimensional vector space into values from a discrete subspace of lower dimension. In this respect, it is representing the groups of interest by finding centroid point which is similar to k-means and other clustering algorithms. Due to ours reasons explained in previous section, now we will focus on clustering methods.

The purpose of clustering is to identify natural groupings of data from a large data set in order to produce a concise representation of a system's behavior. There are many clustering methods available, and each of them may give a different grouping of a dataset. The choice of a particular method will depend on the type of output desired. In our case, our main objective is to use the appropriate method for pattern recognition based on data that is subject to error and uncertainty. Fuzzy classification techniques can deal with the spectral and spatial vagueness, and can be used to model the uncertainty in remote sensing classification. In fuzzy clustering, each point has a degree of belonging to clusters, as in fuzzy logic, rather than belonging completely to a single cluster. Thus, points on the edge of a cluster may be in the cluster to a lesser degree than points in the center of cluster.

If we assume that birds stop in places where some facilities exist (such as water resource, food, protection, etc), our problem seems to be similar to facility location problems – in this case, the Fuzzy c-means (FCM[*]) algorithm gives better results than other clustering algorithms (Žalik 2006). Another important feature of fuzzy c-means algorithm is membership function. In this case an object can belong to several clusters at the same time but with different degrees. This is a useful feature for a facility location problem.

*Fuzzy c-means* is a data clustering technique wherein each data point belongs to a cluster to some degree that is specified by a membership grade.

---

[*] FCM is abbreviated by MATLAB 'Fuzzy Logic Toolbox For Use with MATLAB, User's Guide' Version 2, January 1999. This has no analogy to "facility location problem".

This technique was originally introduced by Jim Bezdek (1981) as an improvement on earlier clustering methods. It provides a method of how to group data points that populate some multidimensional space into a specific number of different clusters.

The Fuzzy C-means algorithm is very similar to the k-means algorithm (MacQueen 1967):

- Choose a number of clusters.
- Assign randomly to each point coefficients for being in the clusters.
- Repeat until the algorithm has converged (that is, the coefficients' change between two iterations is no more than ε, the given sensitivity threshold).
- Compute the centroid for each cluster, using the formula above.
- For each point, compute its coefficients of being in the clusters, using the formula above.

The algorithm minimizes intra-cluster variance as well, but has the same problems as k-means, i.e. the minimum is a local minimum, and the results depend on the initial choice of weights. The Expectation-maximization algorithm is a more statistically formalized method which includes some of these ideas partial membership in classes. It has better convergence properties and is in general preferred to fuzzy-k-means.

Fuzzy c-means is very sensitive to its initial value. It will fall into local optimum solution if the enactment of initial value is inadequate, and it requires the number of clustering before we use it. Therefore, a possible approach to solve this problem is to use subtractive clustering in order to initialize the initial value of FCM before its use. In this manner, we will gain the optimum solution, speed up the rate of convergence without the need to give the cluster number beforehand. *Subtractive Clustering*, is a fast, one-pass algorithm for estimating the number of clusters and the cluster centers in a set of data.

In the case of an analysis that demands focusing on more neighboring structures of the data points and some future estimates, then the following key issues should be considered:

- *Statistical Machine Learning, Neural and Statistical Classification:* The aim of statistical machine learning is to construct systems able to learn to solve tasks given a set of examples of those tasks and some prior knowledge about them. This includes tasks such as image classification, handwritten or speech recognition, time series prediction, etc.
- *Gaussian Density Estimation:* Gaussian density function is placed at each data point, and the sum of the density functions is computed over the range of the data.

## 5 Experimental Study

Real life data is transferred to computer environment and actual data (projected coordinates) is modeled using Clustering GUI toolbox of MATLAB (Figure 1). As explained in the previous section, the *Fuzzy c-means Clustering* and *Subtractive Clustering* allow us to find clusters within input-output training data. Subtractive clustering feature provides estimate for the cluster centers and the number of cluster. For example as an initial trial, considering the default experimentation values, the clustering toolbox yielded three cluster centers (Figure 2). However, in order to have different granularities regarding our objectives, *Fuzzy c-means Clustering (FCM)* should be used. Depending on parameter selection it is possible to have misleading clustering results with underestimated or overestimated number of cluster centers (Figure 3). *Cluster number, Maximum Iteration Number, Min Improvement, Exponent* are the input variables for FCM and *Objective Function* is the output variable.

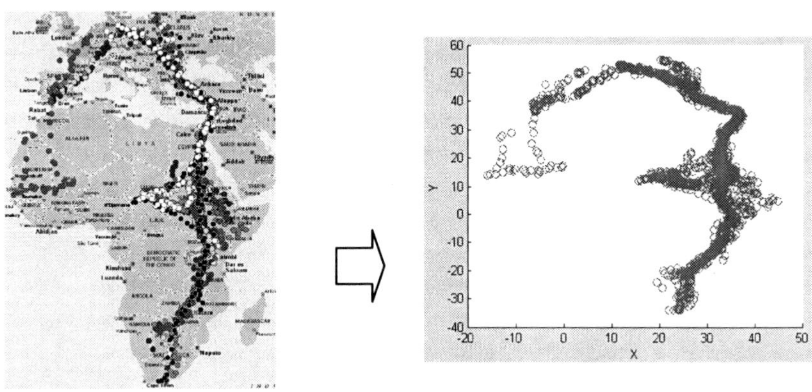


**Fig. 1:** Transfer of actual data into Clustering GUI toolbox, MATLAB

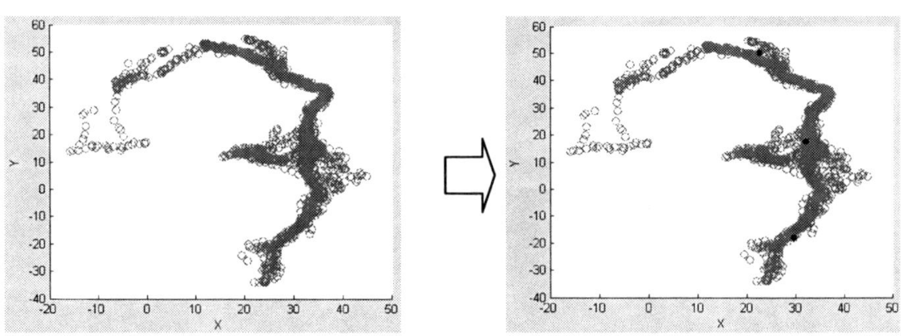


**Fig. 2:** Subtractive Clustering, initial trial

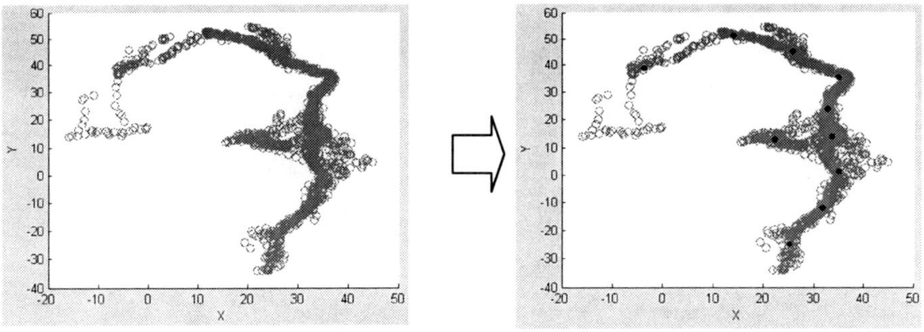

**Fig. 3.1.** Fuzzy C-means Clustering, Initial Trial: Underestimate

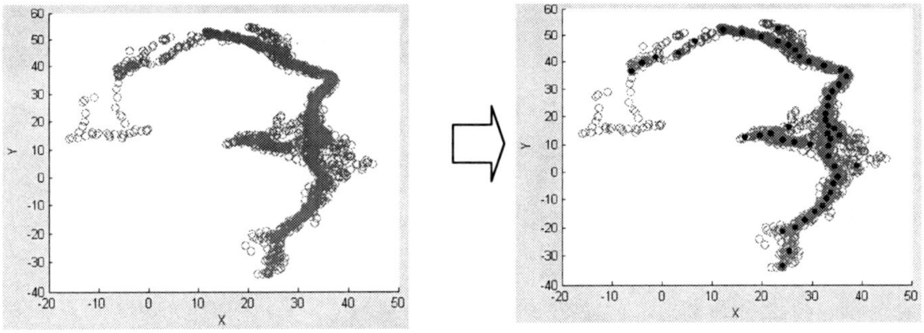

**Fig. 3.2.** Fuzzy C-means Clustering, Second Trial: Exact estimate

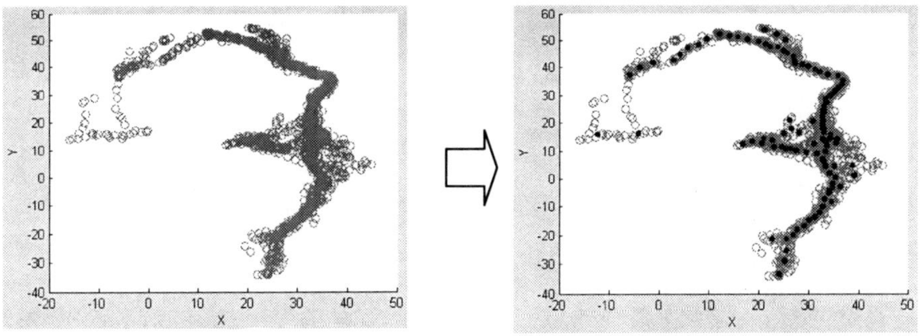

**Fig. 3.3.** Fuzzy C-means Clustering, Third Trial: Overestimate

These examples show the general behavior of above mentioned methods Subtractive Clustering and Fuzzy C-means Clustering. Detailed experimental study follows.

The clustering process stops when the maximum number of iterations is reached, or when the objective function improvement between two consecutive iterations is less than the minimum amount of improvement specified. Thus we do not consider, *Min Improvement* in our experimental study. The default value, *1e-5*(6), is considered small enough to satisfy the stopping criteria. Considering three of the input variables, during the each step of our experimental study, the value of one variable is changed and the value of the rest is defined to be equal to default value given by Fuzzy Logic Toolbox User Guide. Variable values and the resulting objective function values are given in Table 1, 2 and 3.

**Table 1**. Fuzzy C-means Experimental Study I

| *Cluster number* | *Max. Iteration number* | *Min Improvement* | *Exponent* | *Objective Function* |
|---|---|---|---|---|
| *30*(6)3 | *100*(6) | *1e-5*(6) | *2.0*(6) | - |
| -default- | -default- | -default- | 1.00 | NaN |
| -default- | -default- | -default- | 1.01 | NaN |
| -default- | -default- | -default- | 1.02 | 8598.8974 |
| -default- | -default- | -default- | 1.03 | 8614.7766 |
| -default- | -default- | -default- | 1.04 | 9739.2891 |
| -default- | -default- | -default- | 1.05 | 10355.0855 |
| -default- | -default- | -default- | 1.10 | 8471.6932 |
| -default- | -default- | -default- | 1.50 | 7336.3639 |
| -default- | -default- | -default- | 2.00 | 5231.4857 |
| -default- | -default- | -default- | 2.50 | 1751.7093 |
| -default- | -default- | -default- | 3.00 | 489.7839 |
| -default- | -default- | -default- | 5.00 | 0.94517 |
| -default- | -default- | -default- | 10.00 | 5.5798e-008 |

**Table 2**. Fuzzy C-means Experimental Study II

| *Cluster number* | *Max. Iteration number* | *Min Improvement* | *Exponent* | *Objective Function* |
|---|---|---|---|---|
| *30*(6) | *100*(6) | *1e-5*(6) | *2.0*(6) | - |
| 1 | -default- | -default- | -default- | 3.0599e-047 |
| 10 | -default- | -default- | -default- | 24068.9422 |
| 25 | -default- | -default- | -default- | 6600.9008 |
| 50 | -default- | -default- | -default- | 2748.9016 |
| 100 | -default- | -default- | -default- | 999.2239 |
| 500 | -default- | -default- | -default- | 123.9284 |
| 1000 | -default- | -default- | -default- | 34.6358 |
| -default- | -default- | -default- | -default- | 0.0011433 |

**Table 3.** Fuzzy C-means Experimental Study III

| *Cluster number* | *Max. Iteration number* | *Min Improvement* | *Exponent* | *Objective Function* |
|---|---|---|---|---|
| *30*$^{(6)}$ | *100*$^{(6)}$ | *1e-5*$^{(6)}$ | *2.0*$^{(6)}$ | - |
| -default- | 1 | -default- | -default- | 49067.8864 |
| -default- | 5 | -default- | -default- | 27860.4277 |
| -default- | 50 | -default- | -default- | 4768.51 |
| -default- | 100 | -default- | -default- | 5222.105 |
| -default- | 250 | -default- | -default- | 5199.1296 |
| -default- | 500 | -default- | -default- | 5609.0615 |
| -default- | 1000 | -default- | -default- | 5205.0908 |
| -default- | 1907 | -default- | -default- | 5696.8879 |

Table notes, Table 1,2,3:

[1] Except cluster number, these values are given as default (Fuzzy Logic Toolbox For Use with MATLAB, User's Guide Version 2, January 1999). Default value for cluster number is given as 2. However we select 30 as our default value.

During the run of a single experiment, the iterations are based on minimizing an objective function that represents the distance from any given data point to a cluster center weighted by that data point's membership grade. For 1.05 and larger values of *exponent;* as the value increases, the objective value decreases, yielding more centered representative data points (black points, Figure 4). Depending on the aim of the experimenter, the minimization of the objective function among the experimentation runs (not among the iteration for a single run) might be an objective. However, we would like to find the representation of many individual trajectories. Smaller *exponent* values provide us with more dispersed depiction compared to higher values of this variable. Thus, for further study, the value of *exponent* is selected o be equal to '1.05'. However, we are not claiming that our selection of exponent value is the optimal one. Other researchers are free to select any value that fits their aim of research and the cluster structure that they prefer to achieve (denser or more dispersed). This situation is illustrated at Figure 4.

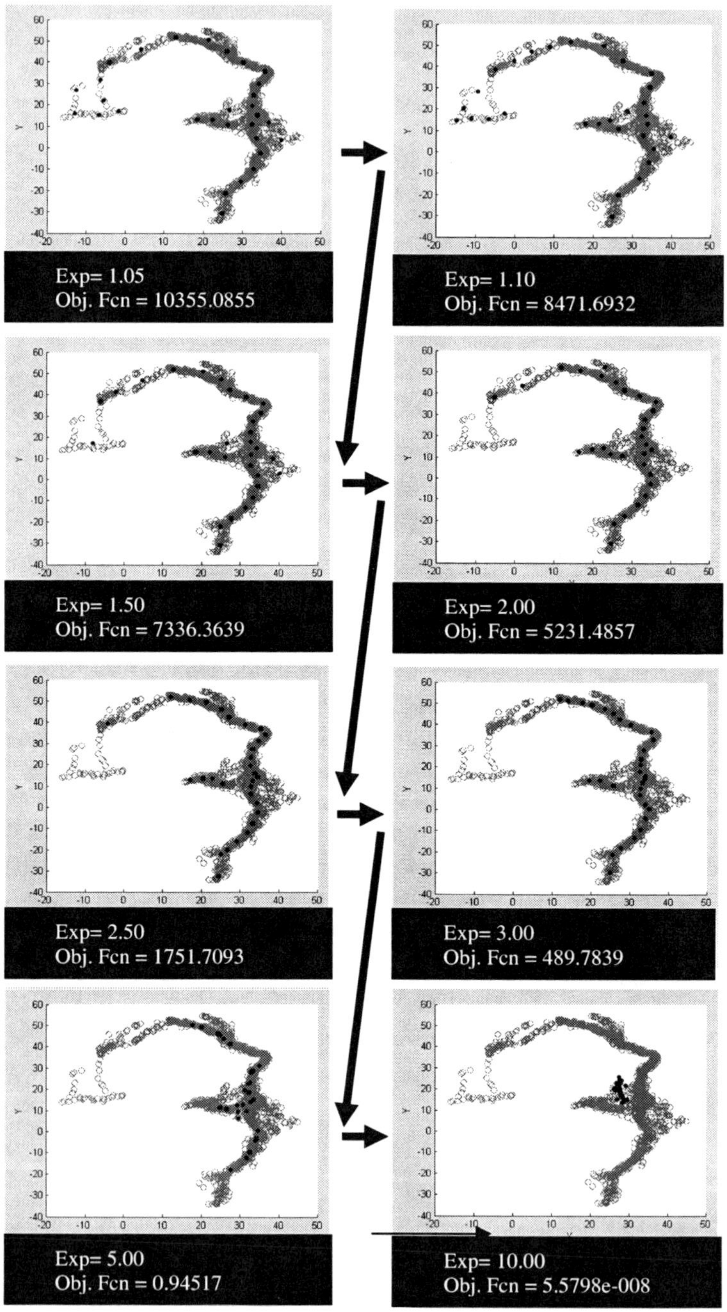


**Fig. 4.** Fuzzy C-means Experimental Study I

As the *Cluster number* value increases, the objective value decreases and this yield more detailed representative data points (black points). For an extreme case experiment, i.e. *Cluster number* is selected to be equal to the total number of points, achieving the best possible objective function value of ~0 but a coarse representation including all data points. Therefore, it is hinted that there is a trade-off between the objective function value and the number of clusters. The selection of the value for the cluster number parameter depends on the aim of the experimenter. Different values of this parameter will result in different granularities. For further study, particularly for the Gaussian Mixture Model and Contour Plot of the Distribution Function, the value of *Cluster number* is selected to be equal to '10'. A smaller number of cluster centers generate more room for a clearer representative explanation for Gaussian Density Model.

As the *Maximum Iteration number* limit increases the objective value decreases until a point which is a certain value of this variable. During this phase of the experimental study it was observed that it takes the algorithm approximately 200-400 iterations. Thus, after ~250 iterations, the experiment produces almost similar objective function values. Therefore, a large number of iteration limits might leave more room for better clustering. However, this consumes more computing time. The conclusion is that there is a trade-off for the selection of iteration limit. For this case we select the value of *Maximum Iteration number* to be equal to '400'.

The k-means algorithm is well known for its efficiency; especially for manipulating and clustering large data sets. However, some major issues are addressed in application of the k-means-type (non-fuzzy or fuzzy) algorithms. As we discussed before, in some clustering algorithms, the number of clusters needs to be determined in advance and the k-means-type algorithms are very sensitive to the initial cluster centers. Making the clustering process not sensitive to the initial cluster centers may be one method to overcome the problem. However, in this case we need to verify whether we are able to get more consistent new clustering results or not. Also we need to be sure if we are able to determine the number of clusters correctly or not.

Analysis of the clustering results relies on checking consistency and our knowledge of the dataset. Due to Talbot et al. (1994) the qualitative evaluation of cluster analysis (cluster validity) is still largely unsolved in the literature. Halkidi et al. (2001) discusses more on clustering validation techniques.

In general the evaluation objective is difficult to prescribe when the analysis purpose is not well defined or is not easily translated into mathematical formulas. As underscored by the authors, since the validity measures tied to properties of the data, sometimes it is hard to provide in general. Oreskes et al. (1994) discusses more on the difficulties of verifying, validating, and confirming numerical models in earth sciences. We believe this kind of approach may be the subject of a further work.

## 6 Our Analysis

This analysis takes into consideration the questions presented in the section 3.1 of this paper.

FCM provides us estimation for the cluster centers and the data points (black points). In this study these cluster centers enable us to represent our characterization of trajectory and manage movement of white storks. We believe this represent our characterization of the path for the whole migration (answer to Question 1). The accumulation of representative data points shows us that not all birds choose the same migration path. Some parts of the migration path are more likely to be chosen by some of the birds than others. As stated before, by using this representative migration trajectory it is also possible to characterize the direction of move by splitting the analysis into sections (answer to Question 2). Our current perception is to consider two directions of the move; migration to south hemisphere and reverse migration (to north hemisphere). We may conclude about the object's movement behavior and possible future trajectory.

The accumulation of representative data points shows us that all birds do not choose same migration path (answer to Question 3). We see some of the bird choose different paths and those birds increase in number for some locations of those paths (answer to Question 4). Namely, we are able to define places that white storks cumulate and increase in number during the move, critical (bottleneck) points or areas (zones) of interest (answer to Question 5). For this, Gaussian density function is placed at each representative data point, and the sum of the density functions is computed over the range of the data. A contour plot of the distribution function provides us with the information concerning the importance of locations subject to interest. Generally speaking, these clustering related techniques and statistical pattern recognition tools assist to interpret our migration data points, and consequently to translate them into an explanatory graph and predict some density estimates. It is also possible to define certain zones as having different degrees of importance. Indeed, in such zones, birds cumulate and increase in number during the migration (Figure 5). Ornithologists, biologists and scientists akin may use this information of critical zones to protect birds from being hunted, or to focus their observatory studies onto these locations (i.e. counting of birds).

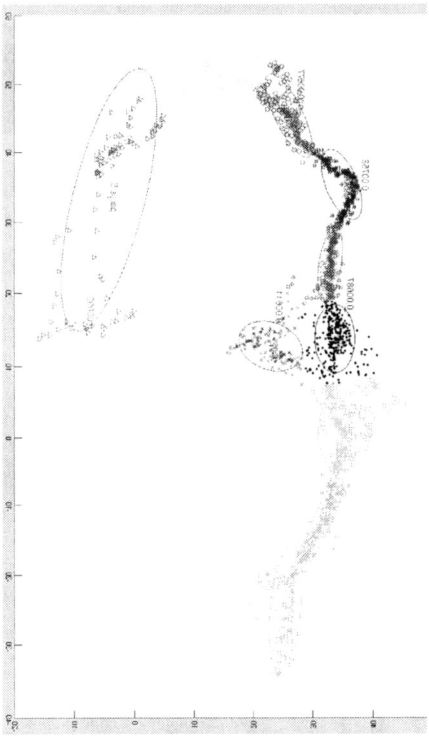

Fig. 5.1. The Gaussian Mixture Model

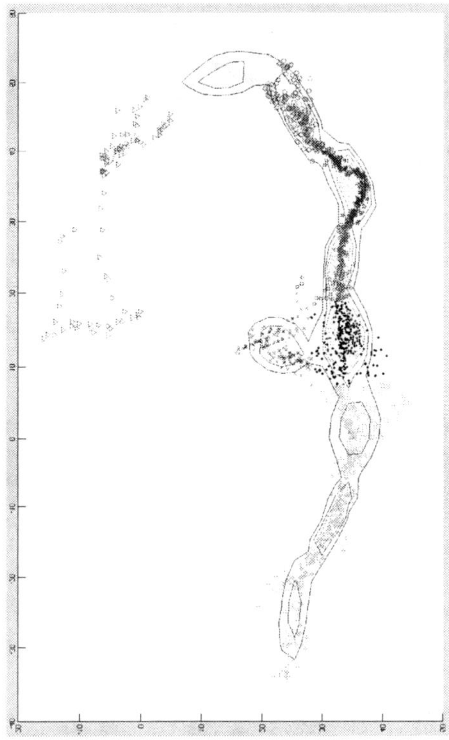

Fig. 5.2. Contour Plot of the Distribution Function

Our current study concentrates on the coordinates (GPS data) gathered for the white stocks. However our analyses are not particular to birds and restricted with the behaviors of animals. As long as the coordinate data is available (i.e. for cars, ships, trains or air planes) we believe that the construct explained in this part of our work can be used for general moving objects and their trajectories. One of the traditional problems with this kind of information (mainly birds' positions that can be spatially located) is the ability to convey implicit information (e.g., covering areas, birds' localization, and route trajectories), therefore omitting important information for the analysis and decision process. Moreover, for moving objects such as cars, trains, ships, airplanes integration of GPS data and recent technological developments may help us to capture the semantics data. Such ability may let the user to integrate the analysis results for entertainment or marketing purposes. For example, if we are observing the trajectory of a traveling salesman, as long as the tracking systems figures out the important places that he visits (i.e. shopping malls or critical traffic zones) the traveling salesman can be informed with the latest advertisements, traffic information that is specific to his particular situation and location. With the advent of Geographical Information Systems (GIS), such implicit information could be included in the result analysis and, thus, related to geographical data such as trough the use of distinct maps according to the same reference system. In fact the use of maps in order to present the information reveals relationships and patterns that otherwise would be difficult to identify. Hence, the modeling of geographical data affects the administration, control and information's dissemination method. Also using reasoning mechanism, we can extract information about stops or trajectory segments as explained before and we can also store the learned experience from the reasoning mechanism.

### *Illustrative Summary: Analysis of the Data Set, Migration Habits and Spatial Regions*

Below (Figure 6) we illustratively, summarize the steps that we used in this study.

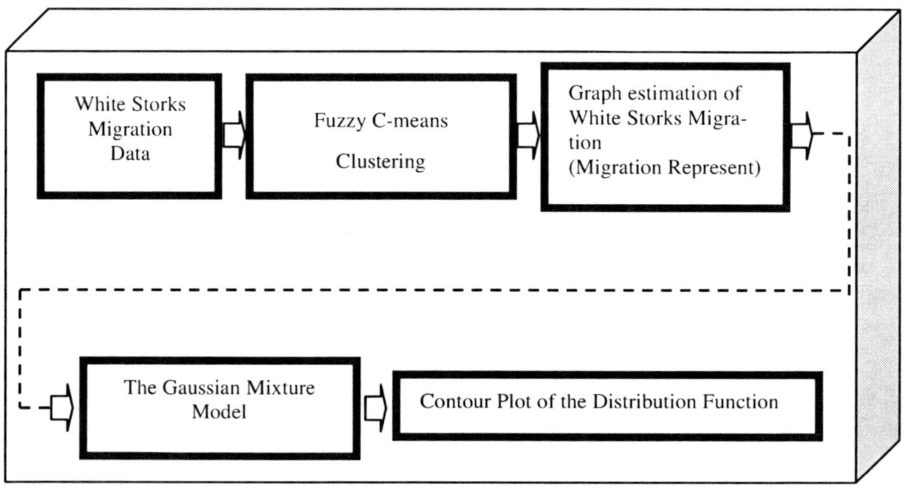


**Fig. 6.** Our 'black-box': Illustrative Summary of this study

For better understanding of our proposed method, we illustrate the whole process using Figures 7.1, 7.2 and 7.3. The main idea behind of our approach is to provide means for clustering spatio-temporal trajectory data in an interactive way. Starting with a spatio-temporal trajectory data (Figure 7.1), which are basic described by time-geography points (i.e. timestamp, longitude and latitude), our method provides a first insight on all data points producing an aggregated trajectory pattern. This pattern is built by several colored clusters and contour plots of interesting regions (e.g.

critical bird zones) as illustrated in Figure 7.2. With this first insight the user gains a whole understanding on the total data coverage and general critical zones. Indeed, this whole coverage in the maximal capture of the data area or more technically, minimal granularity. In our case study, it is the all map covering the north hemisphere and the south hemisphere.

From the analyst (i.e. biologist) point of view, currently an initial understanding is accomplished. This is indeed a first attempt on analysis of the geographic data. However the analyst needs to take further steps. Namely our current result is the output of a single run; our black box approach is explained in Figure 6. The analyst may run several runs to come up different analysis and more elaborated understanding. Indeed each step takes one scale (spatio-temporal trajectory data) into consideration. As the observer's interest gets detailed he needs to focus on more details, namely large scales. This whole observation is an iterative process in which each time a different scale is considered. Thus, at each interaction with different scale, different spatio-temporal trajectory data are taken into account. This will terminate until the observer will yield a satisfactory result, a detailed understating matching with his objective of research.

By using this iterative analysis process, he can focus on different geographic regions of interest even focusing some of them in detail. For example in the later steps on his analysis he can focus on only one of those regions without worrying the rest of the world and the data. After understanding Africa that is important zone for him, he can run the whole analysis approach to further focus on mid-Africa and later more focusing on only few numbers of flocks grouping around a lake or a food source.

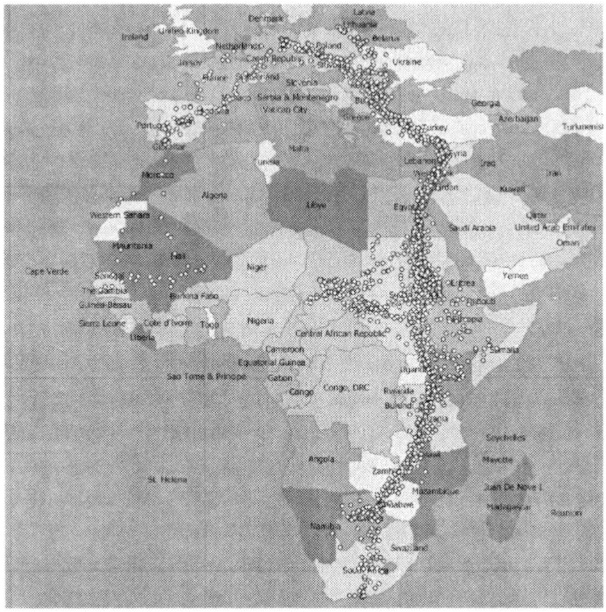

**Fig. 7.1.** Input: White Storks Migration Data (spatio-temporal trajectory data)

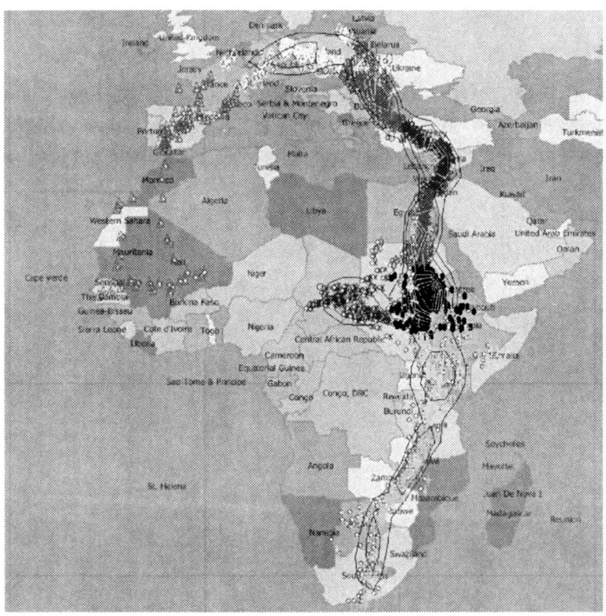

**Fig. 7.2:** – Output: Aggregated trajectory pattern; clusters and contour plots of regions of interest (critical bird zones). Our reasoning for the smallest scale (North and South Hemispheres)

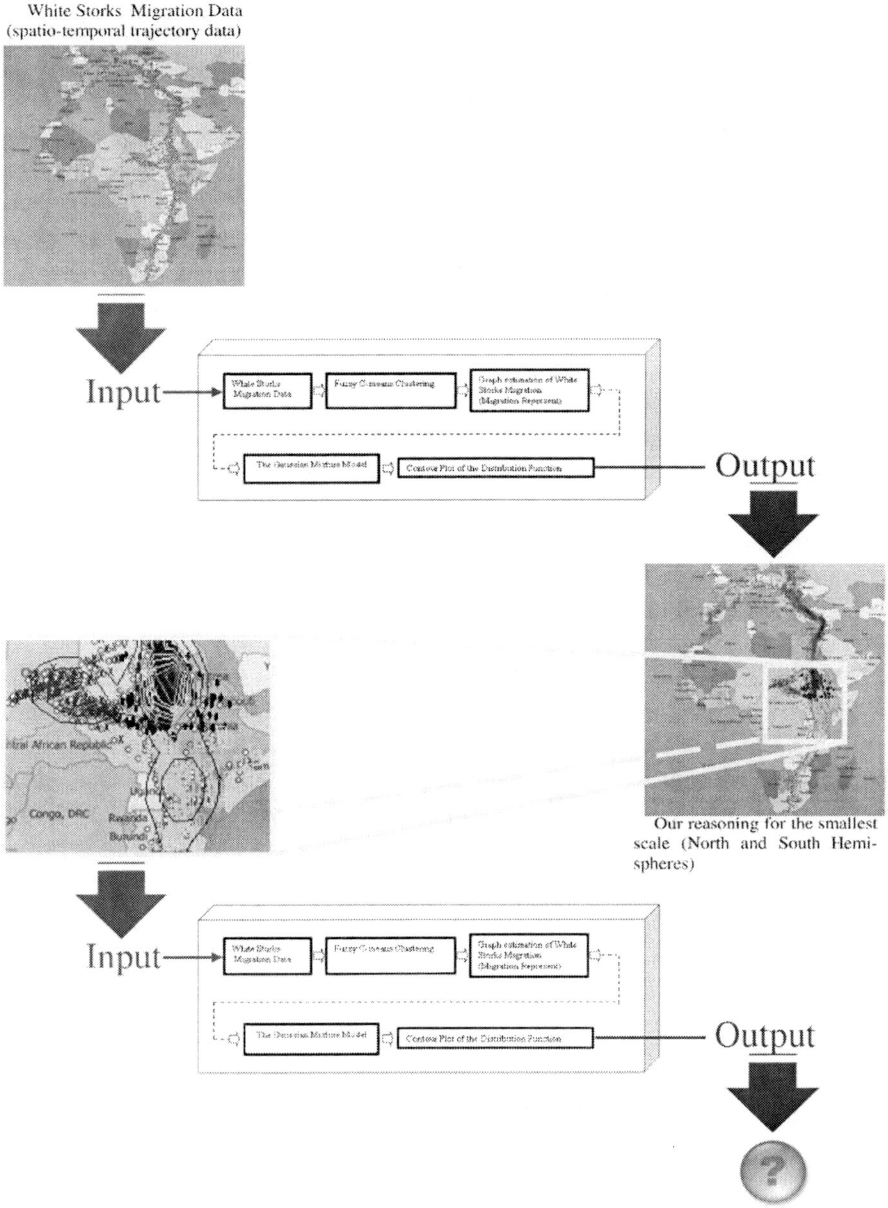


**Fig. 7.3:** – Illustrative explanation: iterative analysis process and consideration of different scales

## 7 Conclusions

This study moves further from a raw data about moving objects to a level of geographic knowledge discovery and knowledge extraction from large spatio-temporal dataset. Using appropriate data mining methods, it is possible to represent and manage regions and trajectories of moving objects (for this study 'white storks').

To this end, we have proposed an interactive method for clustering spatio-temporal trajectory data. This method permits analysts to navigate from a whole observation to a detailed one through scale changes. This iterative analysis process allows focusing on different geographic regions of interest in different levels of detail.

By obtaining our representative migration trajectory and critical zones of interest we are able to find answers to some of our research questions mentioned previously. By using our representative migration trajectory we can conclude about the future migration route of these birds. Also we are able to characterize the direction of the migration by splitting the analysis into sections to differentiate the whole migration into two different habits: i) migration-to-destination ii) reverse-migration and to define places that these birds cumulate and increase in number during the migration (bottleneck points).

We can also conclude the degree of importance for a given point or probability of existence of a bird at a given coordinate within a certain confidence degree. It is possible to determine certain zones having different degrees of importance for the migration. This degree of confidence is highly correlated with the density function that is used to define the characteristics. By using Gaussian Density Estimation we can define critical areas and we can locate these zones on the map. These critical areas and related density functions can be used to define some geographical constraints and may be of the utmost interest to biologists and people interested in urban planning and environmental monitoring (risk management).

Last but not least, the experiments described in this paper showed that the relevance of resulting knowledge is highly dependent on the skills of the experimenter through the tuning of the parameters' values of the algorithms. Clearly, it reinforces the idea that the experimenter's background knowledge must be taken into account during algorithm execution, in order to provide a set of parameters that permits to conduct experiments to a coherent result. In addition, the combination of different methods (e.g. clustering and statistical) may provide different views of the data for the experimenter. Methods that do not provide such flexibility will be rather limited in processing spatio-temporal geographical data.

We can briefly explain our further research directions as follows. In this study we tried to explain our initial approach and we believe the next step to be related to the use of appropriate data warehouse methods for representing, storing and managing trajectories, with varying levels of granularity (accuracy and certainty). Our methodology fully considers 'White Storks Migration Data Set' (All data points, 1998 – 2005). As we explained before it is possible to organize the trajectory data in three ways: classifying according to time (day, week, month and year), classifying according to individual white storks and classifying according to the direction of migration. This classifying constitutes '*time*', '*bird*' and '*direction*' organization into analysis.

We believe in a further study, if we broaden the outlook; if we extend our research interests and if we consider three dimensions than we are more likely to contribute in biological research literature, urban planning and environmental monitoring (risk management).

## Acknowledgements

The authors are definitely indebted to *Christelle Vangenot, Jean-Cédric Chappelier and Othman Tajmouati*. We also extend our thanks to all *GeoPKDD partners* for their comments on a preliminary presentation of this work at a project meeting.

## References

Bezdek, J.C. (1981): Pattern Recognition with Fuzzy Objective Function Algorithms, Plenum Press, New York.

Brakatsoulas, S., Pfoser, D., Tryfona, N. (2004): Modeling, storing, and mining moving object databases. In IDEAS '04, Proceedings of the International Database Engineering and Applications Symposium (IDEAS'04), Washington, DC, USA, IEEE Computer Society, pp. 68–77.

Cao, H., Mamoulis, N., Cheung, D.W. (2006): Discovery of collocation episodes in spatiotemporal data. In ICDM, IEEE Computer Society, pp. 823–827.

Forlizzi, L., Güting, R.H., Nardelli, E., Schneider, M. (2000): A data model and data structures for moving objects databases. In Chen, W., Naughton, J.F., Bernstein, P.A., eds.: SIGMOD Conference, ACM, pp. 319–330.

Giannotti, F., Nanni, M., Pedreschi, D., Pinelli, F., (2007): Trajectory Pattern Mining. KDD'07, August 12–15, ACM Press, San Jose, California, USA.

Gray, R. M. (1984): Vector Quantization. In IEEE ASSP Magazine, pp. 4-29.

Gudmundsson, J., van Kreveld, M., Speckmann, B. (2004): Efficient detection of motion patterns in spatio-temporal data sets. In: GIS '04: Proceedings of the 12th annual ACM international workshop on Geographic information systems, New York, NY, USA, ACM Press, pp. 250–257.

Halkidi, M., Batistakis, Y., and Vazirgiannis, M. (2001): On Clustering Validation Techniques. In: Journal of Intelligent Information Systems, Volume 17, Issue 2-3, Pages: 107 – 145.

Iyengar, V. S. (2004), On detecting space-time clusters. In 'KDD', pp. 587–592.

Laube, P., Imfeld, S., Weibel, R. (2005): Discovering relative motion patterns in groups of moving point objects. International Journal of Geographical Information Science 19(6), pp. 639–668.

MacQueen, J. B. (1967): Some Methods for classification and Analysis of Multivariate Observations. In proceedings of 5-th Berkeley Symposium on Mathematical Statistics and Probability, Berkeley, University of California Press, 1:281-297.

Mouza, C. (2005): Mobility patterns. GeoInformatica 9 (December), pp. 297–319(23).

Oreskes, N., Shrader-Frechette, K., and K. Belitz, (1994): Verification, validation, and confirmation of numerical models in the earth sciences. In: Science, 263, 641–646.

Spaccapietra, S., Parent, C., Damiani, M. L., Macedo J. A. F., Porto, F., Vangenot, C., (2007): A Conceptual View on Trajectories, DKE.

Talbot, L. M., Talbot, B. G., Peterson, R. E., Tolley, H. D., Mecham, H. D., (1999): Application of Fuzzy Grade-of-Membership Clustering to Analysis of Remote Sensing Data. In: Journal of Climate Article: pp. 200–219, Volume 12, Issue 1.

Tsoukatos, I., Gunopulos, D. (2001): Efficient mining of spatiotemporal patterns. In Jensen, C.S., Schneider, M., Seeger, B., Tsotras, V.J., eds.: SSTD. Volume 2121 of Lecture Notes in Computer Science, Springer, pp. 425–442.

Verhein, F., Chawla, S. (2006): Mining spatio-temporal association rules, sources, sinks, stationary regions and thoroughfares in object mobility databases. In Lee, M.L., Tan, K.L., Wuwongse, V., eds.: DASFAA. Volume 3882 of Lecture Notes in Computer Science, Springer, pp. 187–201.

Vorbis I Specification. Xiph.org, Retrieved on 2007-03-09. http://xiph.org/vorbis/doc/Vorbis_I_spec.html.

Wolfson, O., Xu, B., Chamberlain, S., Jiang, L. (1998): Moving objects databases: Issues and solutions. In Rafanelli, M., Jarke, M., eds.: SSDBM, IEEE Computer Society, pp. 111–122.

Žalik, K.R. (2006): Fuzzy C-Means Clustering and Facility Location Problems. In Pasqual del Pobil, A., eds: In ASC 2006: Proceeding (544) Artificial Intelligence and Soft Computing, Palma De Mallorca, Spain.

# Mining Spatio-Temporal Data at Different Levels of Detail

Elena Camossi, Michela Bertolotto, Tahar Kechadi

School of Computer Science & Informatics, University College Dublin, Belfield, Dublin 4, Ireland. {elena.camossi, michela.bertolotto, tahar.kechadi}@ucd.ie

**Abstract.** In this paper we propose a methodology for mining very large spatio-temporal datasets. We propose a two-pass strategy for mining and manipulating spatio-temporal datasets at different levels of detail (i.e., granularities). The approach takes advantage of the multi-granular capability of the underlying spatio-temporal model to reduce the amount of data that can be accessed initially. The approach is implemented and applied to real-world spatio-temporal datasets. We show that the technique can deal easily with very large datasets without losing the accuracy of the extracted patterns, as demonstrated in the experimental results.

**Keywords:** data mining, spatio-temporal data, level of detail, large datasets, DBSCAN

## 1 Introduction

Recently, it has been estimated that 80% of the available datasets have spatial components (Fayyad and Grinstein 2001), and are often related to some temporal aspects. Such a considerable amount of information needs suitable analysis techniques to be applied correctly. In the last few years, several systems providing an integrated approach for the management of spatial and temporal information have been proposed (e.g., Chen and Zaniolo 2000; Güting et al. 2000; Huang and Claramunt 2002).

The application of knowledge management tailored to the exploitation of implicit semantics of spatio-temporal data has emerged as the key technology to address the application of spatio-temporal data mining techniques and algorithms to real-world problems. Spatio-temporal data mining is a user-centric, interactive process, where data mining experts and domain experts work closely together to gain insight on a given problem. Several open issues have been identified ranging from the definition of the mining techniques capable of dealing with spatio-temporal information to the development of effective methods for interpreting and presenting the final results.

In this study, we focus on a specific data mining technique that deals with clustering. Clustering is one of the fundamental techniques in data mining. It groups data objects into clusters based on some similarity or distance measures. These clusters contain information found in the data that describes similar objects and their relationships. The goal is to optimise similarity within a group of objects and dissimilarity between the groups in order to identify interesting structures in the underlying data. While the complexity of spatio-temporal clustering is far higher than its traditional counterpart, the ideas behind it are similar i.e., it focuses either on characteristic features of objects in a spatio-temporal region or on the spatio-temporal characteristics of a set of objects (Ng and Han 1994).

The mining process for spatio-temporal data is complex in terms of both the mining efficiency and the complexity of patterns that can be extracted from spatio-temporal datasets (Roddick and Lees 2001). The reason is that the attributes of the neighbouring patterns (i.e., close in either space or time or both) may have significant influence on a pattern and should also be considered. Therefore, new techniques are required to efficiently and effectively mine these datasets. The main problem for analysing spatio-temporal data is the size of the data. Today's GIS systems are collecting Gigabytes and even Terabytes of data each day. So the major goal for such a strategy is to process these datasets within a reasonable response time and memory space, without affecting the accuracy of the findings.

In this paper we propose a spatio-temporal clustering technique to deal with the data at different levels of detail, i.e., granularities, to improve the algorithm efficiency. Such a technique relies on a hierarchical multi-granular model in which datasets are generalised to generate less detailed representations of reduced size. Thus, the mining can first be applied to the reduced dataset, and then refined only for those objects, which have been filtered through the first step. In other words, the mining can be further deepened on spatial areas or temporal intervals of interest. The corresponding objects are converted at finer spatial and temporal granularity before applying the mining. Our approach handles the data at different levels of detail both from a spatial and a temporal point of view. The conversions of data at different levels of detail are performed by applying the operators available in the underlying multi-granular spatio-temporal model, whose definitions are described in (Camossi et al. 2006).

The paper is organised as follows. In Section 2 we present recent related work. In Section 3 we describe a multi-granular spatio-temporal model that enables the conversion of spatio-temporal data at different levels of detail. In Section 4 we introduce the spatio-temporal data mining system and show how we apply it to spatio-temporal data represented at higher spatial and temporal levels of granularity. In Section 5 we discuss some experimental results. Finally, Section 6 concludes the paper and outlines future research directions.

## 2 Related Work

The proposals for the integrated management of spatio-temporal information can be mainly classified into: temporal extensions of GIS (Claramunt and Thériault 1995; Langran 1992); extensions of relational, object relational (Chen and Zaniolo 2000) and object oriented standards (Griffiths et al. 2004,0 Huang and Claramunt, 2002); algebraic frameworks for moving points and regions (Güting et al. 2000); and independent frameworks (Tryfona and Jensen 1999; Worboys 1994).

Recent systems have addressed the issues related to multi-granularity, multi-resolution and multiple representations of spatial (Balley et al. 2004; Fonseca et al. 2002; Kulik et al. 2005; Vangenot 2001) and spatio-temporal data (Bittner 2002; Camossi et al. 2006; Claramunt and Jiang 2000; Hornsby and Egenhofer 2002), Hurtado and Mendelzon, 2001, Khatri et al. 2002). In particular, Claramunt and Jiang (2000) defined nested hierarchies for modelling space and time from which quantitative information about spatio-temporal relationships are obtained. Khatri et al. (2002) extended a semantic formalism to support the specifications of spatio-temporal data at multiple granularities, relying on the concepts of temporal indeterminacy and spatial imprecision. The resulting model and the granularity systems described are effective for data specification. In (Camossi et al. 2006) a framework enabling the conversion of spatio-temporal values at different spatial and temporal granularities is defined as extension of the ODMG data model (Cattel et al. 1999). In the spatial domain, Fonseca et al. (2002) and Kulik et al. (2005) proposed the use of anthologies to multi-resolution.

The progressive application of data mining techniques for spatio-temporal data to improve efficiency is discussed in (Mennis and Liu 2005; Tsoukatos and Gunopulos 2001). Tsoukatos and Gunopulos (2001) presented an incremental algorithm for discovering frequent spatio-temporal sequences by decomposing the search space in a hierarchical structure, addressing its application to multi-granular spatial data. Mennis and Liu (2005) discussed multi-level association rule mining of spatio-temporal data, i.e., mining of rules at varying levels of a concept hierarchy to fit the best resolution for the rule. Hierarchical data mining is discussed also for spatial (Koperski 1999; Shahabi et al. 2001) and temporal (Abraham and Roddick 1999) data separately. Recently, there has been a growing interest in the application of wavelet transforms in some processes of data mining (Li et al. 2002; Shahabi et al. 2001).

## 3 Multi-Granular Representation of Spatio-Temporal Data

In this section we describe the data model for the representation of data at multiple spatio-temporal granularities used in our mining approach. The model relies on the work presented in (Camossi et al. 2006), where the ODMG type system (Cattel et al. 1999) has been extended to enable the representation and the conversion of spatio-temporal object attributes at different levels of details, for both the spatial and the temporal dimensions. The same set of conversions has been applied in the definition of an object-relational spatio-temporal multigranular model (Bertino et al. 2005). In this paper we follow the object-relational approach, instead of the full object oriented approach, because it is adopted by current commercial DBMS. Furthermore, like most of them (e.g., ORACLE™ 2008, PostgreSQL 2008), the model applies an integrated approach for the representation of geometric aspects of data. In the following, we first present the notion of spatial and temporal granularities supported by the model; then, we describe how multi-granular spatio-temporal data can be represented and converted.

### 3.1 A Spatio-Temporal Multi-Granular Data Model

The data model supports the definition of temporal granularity formalised by Bettini et al. (2000), which is commonly adopted by the temporal databases and reasoning community, and integrates the notion of spatial granularity compliant with the formalization of *stratified map spaces* proposed by Stell and Worboys (1998). Temporal and spatial granularities are specified as mappings from an index set to the power set of the *TIME* and the *SPACE* domains, respectively. *TIME* is totally ordered. The supported

*SPACE* domain is 2-dimensional (i.e., a proper subset of $R^2$). For instance, *days*, *weeks*, *years* are temporal granularities; *meters*, *kilometres*, *feet*, *yards*, *provinces* and *countries* are spatial granularities. Each portion of the temporal and spatial domain corresponding to a granularity mapping is referred to as a (temporal or spatial) *granule*. Spatial granularities can include 2-dimensional granules (e.g., units of area: $m^2$, *acre*, etc.; administrative boundaries classifications: *municipalities*, *countries*, etc.), or in 1-dimensional granules (e.g., measures of length: *km*, *mile*, etc.; map scales: 1:24 000, 1:62 500, etc.). Granules give the validity bounds of spatio-temporal for the definition of spatio-temporal values. For instance, we can say that a value reporting the measure of the daily temperature in Dublin is defined for the first and the second of January 2000, and so on. "01/01/2000", "02/01/2000", and "Dublin" are textual labels that univocally identify two temporal and one spatial granule. Granules of the same granularity cannot overlap. Moreover, non-empty temporal granules must preserve the order given by the index set.

Spatial and temporal granularities are related by the *finer-than* relationship. Such a relationship formalises the intuitive idea that different granularities correspond to different partitions of the domain, and that, given a granule of a granularity *G*, usually a granule of a coarser granularity exists that properly includes it. For example, granularity *days* is finer than *months*, and granularity *months* is finer than *years*. Likewise, *municipalities* is finer than *countries*. If a granularity *G* is finer-than *H*, we also say that *H* is *coarser-than G*. According to the finer-than relationship, spatial and temporal granularities are related to form two directed graphs, usually two lattices.

Beyond the conventional relational and object-relational database values, the database schemas can include spatial, temporal, and spatio-temporal values. 2-dimensional geometric vector features (i.e., points, lines, and polygon) can then be represented. Multi-granular spatial and temporal data are uniformly defined by instances of two parametric types, *spatial* and *temporal*, which are specified according to granularities (spatial and temporal, respectively) and an inner conventional (i.e., without spatio-temporal characteristics) or geometric type.

The model enables the conversion of multigranular spatio-temporal data at different granularities, to improve or reduce the level of detail employed for data representation. Granularity conversions are crucial in order to represent data at the most appropriate level of detail for a specific task, and enable consistent comparisons of data defined in the schema at different granularities, improving the expressive power of spatio-temporal query languages. Granularity conversions enable to apply different conversion semantics.

The conversion of multi-granular geometrical features is obtained through compositions of model-oriented and cartographic map generalisation operators (Muller et al. 1995) that guarantee topological consistency (Bertolotto 1998; Saalfeld 1999), an essential property for data usability, and refinement operators that perform the inverse functions. Such operators can be classified with respect to the semantics of the conversion performed: *contraction* and *thinning* operators reduce the dimension of vector features, whereas *expansion* operators increase their dimension; *merge* operators merge adjacent features of the same dimension into a single one, while *splitting* operators subdivide single features in adjacent features of the same dimension; *abstraction* and *simplification* operators discard isolated features from polygons and remove shape points from a line, respectively, whereas *addition* operators add isolated features to polygons and shape points to lines.

On the other hand, to retrieve for instance the annual trend of a phenomenon with a daily frequency (e.g., the national values of sales in shops located in several countries, the model supports also the conversion of quantitative (i.e., not geometrical) attribute values supported for both temporal and spatial data. These conversions are classified in families according to the semantics of the operation performed (Camossi et al. 2006): *selection* (e.g., projection); *aggregation* (e.g., sum, average); *restriction*, by which, if a granular value assumes value $v$ in a granule $g$, value $v$ also refers to any finer granule $g'$ included in $g$; *splitting*, which subdivides each coarser value among the finer granules included in it either uniformly (i.e., all the finer values are the same), or according to non-uniform distribution.

The given set of granularity conversions can be extended with user-defined granularity conversions that are specified as class methods in a database schema. Granularity conversions have been proved to return legal values of the type system defined, and to preserve the semantics of the spatio-temporal data represented (Camossi et al. 2006).

## 3.2 Multi-Granularity to Improve Mining

In this paper, we take advantage of the multi-granularity support provided by the data model to enhance the effectiveness of the clustering algorithm. In particular, the mining process can benefit of multi-granularity in different ways.

First of all, multi-granularity enables to apply the mining to data represented at different levels of detail, e.g., semantically homogeneous data coming from different sources. In this case, data can be converted into uniform spatial and temporal granularities before applying the mining process. The level of detail is chosen in order to represent the specific dataset. Usually the choice falls on the *greatest lower bound* (*glb*), or the *least upper bound* (*lub*), of the spatial and temporal granularity available. Given two granularities *G* and *H* of the same type (i.e., either spatial or temporal), *glb(G,H)* is the coarsest granularity *K* (not necessarily different from *G* and *H*) among the granularities finer than both *G* and *H*. By contrast, *lub(G,H)* is the finest granularity *J* (not necessarily different from *G* and *H*) among the granularities coarser than both *G* and *H*.

Then, once the level of detail used for the representation is uniform for the whole dataset, granularity conversions are applied before the refinement process. Indeed, spatio-temporal data are pre-processed for reducing the size of the starting dataset, i.e., data are converted to coarser spatial and temporal granularities. This conversion allows us to focus on the relevant dataset, which is, in general, much smaller than the original data, hence, improving response time of the overall mining process. The choice of the level of detail can be iterative, and depends on a trade-off between mining efficiency and maximum detail required by the mining process. Finally, the conversion depends on the generalisation process used for a given dataset. Once the semantics for generalising certain dimensions or attributes of the data has been defined, the conversion is straightforward and mainly for the model defined above. Therefore, the mining process needs only the level of accuracy as an input parameter and the conversion and even the number of levels of detail that need to be explored is done automatically through hierarchical navigation. The way that these levels are explored depends on the type of the algorithm implemented.

After the application of the clustering algorithm, the selected spatio-temporal data of interest are converted into finer granularities, for more detailed representation, once a deepen analysis is required on significant data. In the first case, granularity conversions are applied globally, to the whole dataset to materialise those objects, which will be accessed frequently and in more detail. In the second case, granularity conversions are applied locally, zooming in on specified and restricted pieces of information, whenever the user asks for a more detailed mining of such data, specified with respect to a given spatio-temporal area. In both cases, spatio-temporal data are converted to different granularities without losing information. Indeed, the conversions are performed by applying the granularity conversions supported by the data model that preserve semantics and then usability of the data.

# 4 Proposed System

To address the issues of mining and managing spatio-temporal datasets we have proposed a 2-layer system architecture (Bertolotto et al. 2007; Compieta et al. 2007) including a mining layer and a visualisation layer. The mining layer implements a mining process along with the data preparation and interpretation steps. For instance, the data may need some cleaning and transformation according to possible constraints imposed by some tools, algorithms, or users. The interpretation step consists of visualising the selected models returned during the mining phase to effectively study the application behaviour. The interpretation is carried out in the visualisation layer. More details on the visualisation tools can be found in (Bertolotto et al. 2007; Compieta et al. 2007). In the next section we will focus on the mining strategy implemented in the mining layer.

## 4.1 2-Pass Strategy

To reduce the amount of memory and computational complexity that these data spaces require without affecting the information presented by the data, the first task in our strategy is to find the data points that are most similar according to their static (non spatial and temporal) attributes. This part of the strategy is the key to the whole success of the generalisation process, so that we do not lose any important information that might have an adverse effect on the results. To further reduce the complexity in space of the algorithm, the raw datasets are pre-processed in order to obtain, through granularity conversions defined in Section 3, a coarser representation of their spatio-temporal dimensions. Since the granularity conversion preserves the semantics of data (Camossi et al. 2006), the application of spatio-temporal mining algorithms to coarser representation does not affect the algorithm outputs.

The second task is to cluster these groups of closely related data points in a meaningful way to produce a new (meta-)dataset suitable and acceptable for further analysis (i.e., models, patterns, rules, etc.).

## 4.2 First Pass

The algorithm for this first pass produces clusters of data points that are closely related. The goal here is to produce new data objects, where each object represents one cluster of raw data. Therefore, the main objective is to reduce the size of the initial data without losing any relevant information. Figure 1 shows a high level view of the steps carried out by the algorithm for this first phase of this approach. It is important to note that only the data points that have a very high similarity between them will be grouped together. As a result, the new dataset is much smaller than the original data. It contains more information about individual clusters. This will help the clustering performed during the second pass.

This pass is basically implementing the generalisation and conversion model defined above. The process of exploring the generalised data and its conversion either from top-to-bottom or bottom-to-top is linear. Usually the generalisation process is implemented as a tree structure, which is efficient in exploring relevant branches and the memory space needed to store them. In this phase, we access the higher-level generalised data. The second pass will deal with the detail when necessary.

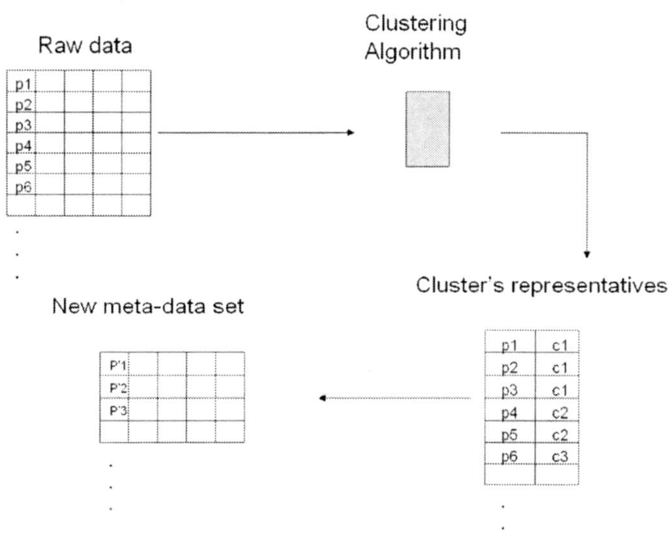


**Fig. 1.** Step-by-step view of the first pass of the strategy

## 4.3 Second Pass

The second pass involves clustering the tightly grouped data points from the first pass to produce a new representation of the data (meta-dataset). This meta-dataset should be reduced by a certain degree of magnitude so that it can now be analysed and mined more easily. In Figure 2, a 2-D example shows how a larger dataset is processed to create a much smaller meta-dataset. There are locations in the space that are highly similar; these are represented within each of the small location groups (small circular shapes). It is important that no location group overlaps with another so that the integrity of the data is not affected. The next step shows data mining on the meta-data using clustering for an example.

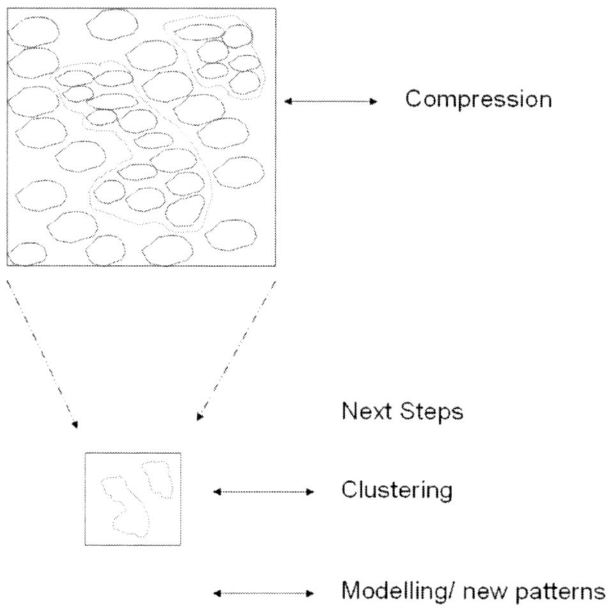


**Fig. 2.** A 2-d example of dataset compression

The clustering technique proposed for this second phase of the strategy will be DBSCAN (Ester et al. 1996). It is a density-based clustering algorithm that produces disjoint clusters, in which the number of clusters is automatically determined by the algorithm. It is relatively resistant to noise (as it detects noisy data and outliers) and can handle clusters of arbitrary shapes and sizes. The main reason for choosing DBSCAN is twofold: 1) to illustrate our methodology and our conversion model, and 2) while DBSCAN is not highly scalable; it is interesting to study its performance on very large datasets using our methodology as from our first phase the amount of tiny clusters representing highly similar data is very large, and we would like to take advantage of finding the regions that are very similar. These regions can then form clusters that will present a new compact representation of the dataset.

The next step is to mine this new representation of the dataset. The space and computational complexities for these algorithms have been reduced greatly from the original data. This strategy and mainly the mining algorithm is also suitable for interactive data mining and visualisation since it is so quick and efficient that it can be incorporated in a visualisation tool of the data. The data can be explored and analysed using this approach interactively and with ease as shown in the next section.

## 5 Preliminary Experimental Results

We have implemented a 2-pass strategy that uses the DBSCAN algorithm for the mining process. The technique has been implemented within a data-mining engine and includes also a visualisation layer for interactive data interpretation. The experiments conducted so far were obtained from the Hurricane Isabel dataset (National Hurricane Center 2003), which is a proper instance for geographical spatio-temporal dataset. Figure 3 lists different variables contained in the dataset and for more details about these variables we refer the reader to (National Hurricane Center 2003).

All variables are real-valued (4 bytes) and were observed along 48 time steps (hourly-sampled), in a space having 500 x 500 x 100 = 25x106 total points. So, each variable in each time step is stored in a different file, resulting in 624 files of 100MB each. This raw data can be represented by the following parameters; the number of time steps (Nts), the number of data points (N), and the number of static parameters (Nsp). Nts = 48 time steps, N = 25x106 data points, and Nsp = 13. This fine fragmentation allows for great flexibility in choosing different subset of data for each mining task.

**Variable Descriptions**

| Variable | Descripton | Min/Max | Units |
|---|---|---|---|
| QCLOUD | Cloud Water | 0.00000/0.00332 | kg/kg |
| QGRAUP | Graupel | 0.00000/0.01638 | kg/kg |
| QICE | Cloud Ice | 0.00000/0.00099 | kg/kg |
| QRAIN | Rain | 0.00000/0.01132 | kg/kg |
| QSNOW | Snow | 0.00000/0.00135 | kg/kg |
| QVAPOR | Water Vapor | 0.00000/0.02368 | kg/kg |
| CLOUD | Total cloud (QICE+QCLOUD) | 0.00000/0.00332 | kg/kg |
| PRECIP | Total Precipitation (QGRAUP+QRAIN+QSNOW) | 0.00000/0.01672 | kg/kg |
| P | Pressure; weight of the atmosphere above a grid point | -5471.85791/3225.42578 | Pascals |
| TC | Temperature | -83.00402/31.51576 | Degrees Celsius |
| U | X wind component; west-east wind component in model coordinate, postive means winds blow from west to east | -79.47297/85.17703 | |
| V | Y wind component; south-north wind component in model coordinate, postive means winds blow from south to north | -76.03391/82.95293 | |
| W | Z wind component; vertical wind component in model coordinate, positive means upward motion | -9.06026/28.61434 | |

**Fig. 3:** Description of the dataset layers.

Figure 4 shows one of the clusters we extracted, whose shape resembles the shape of the hurricane or one of its features. In Figure 4, DBSCAN algorithm outputs a spherical type cluster that represents the shape of the Hurricane's eye for different values of pressure. The eye is clearly visible in the low, where the cluster represents high values for pressure in pink with the hole in the middle representing very low values for pressure. These clusters provide some clues about direction or strength of the hurricane. We can track and represent in real time the movement of the hurricane eye over time by clustering different time steps of the dataset.

The application of the mining algorithm on the reduced dataset produces results visually comparable to those obtained with the fully detailed one, and with improved efficiency in response time and memory occupation. The results we obtained on the test-bed application (i.e., the Hurricane Isabel) are very promising, and this technique suites very well interactive environments. Our technique is designed according to the multi-level granularity model explained above. In this paper, we presented our results by using a 2-pass (i.e., 2-level) algorithm. The first level generalises the dataset to reduce its size and complexity. The second pass refines only the data of interest. However, these improved results may depend on the specific dataset. For instance, if the final patterns cover different objects, which were not identified to be neighbours in the first pass, the cost will be higher as one has to explore different spatio-temporal regions to refine the final results. This can be solved by adapting our algorithm to support multiple levels of clustering relying on the model defined above. We are currently implementing a version for multi-level clustering using decision-tree approach.

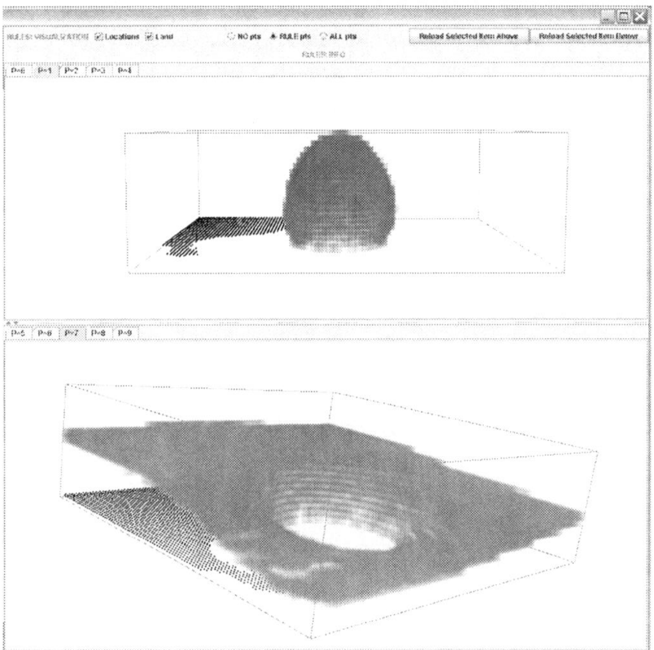

**Fig. 4:** The eye of the Hurricane in isolation. This is represented by one cluster

## 6 Conclusion

The approach proposed in this paper is different from the approaches presented in the literature (Abraham and Roddick 1999; Koperski 1999; Mennis and Liu 2005; Tsoukatos and Gunopulos 2001) with respect to the specific data mining problem addressed, and mainly the use of multi-granularity concept to both be able to design scalable technique for data mining and analysis and speed up the process of the mining and its accuracy. The work presented in (Tsoukatos and Gunopulos 2001) focuses on mining frequent patterns, while (Abraham and Roddick 1999; Koperski

1999; Mennis and Liu 2005) address the mining of association rules, meta-rules and classification. (Tsoukatos and Gunopulos 2001) uses spatial granularities defined according to boundary regions, and the operator supported to perform granularity conversions is region merge. Likewise, only spatial concept hierarchy are supported in (Abraham and Roddick 1999; Koperski 1999; Mennis and Liu 2005). In all these projects, spatial granularities are employed for rules representations. Instead, we focus on clustering, and the multi-granularity concept is used to reduce the size of the datasets, mainly at the beginning. Furthermore, we apply multi-granularity for both the spatial and the temporal domains, supporting a wide range of granularity conversions, specifically designed to preserve data usability. We will extend this approach to other clustering techniques and also we will study their effectiveness in real-world environments. Moreover, we have planned further experimentations considering different spatio-temporal datasets to the test of the efficiency of the approach.

## Acknowledgement

The work of Elena Camossi is supported by the Irish Research Council for Science, Engineering and Technology.

## References

Abraham T., Roddick J.F. (1999) Incremental Meta-Mining from Large Temporal Datasets. Advances in Database Technologies, In Proc. of the *1st Int'l Workshop on Data Warehousing and Data Mining*, Springer-Verlag Berlin. LNCS 1552:41-54.

Balley S., Parent C., Spaccapietra S. (2004) Modelling Geographic Data with Multiple Representations. *International Journal of Geographical Information Science*, Taylor & Francis. 18(4):327-352.

Bertino E., Cuadra D., Martìnez P. (2005) An Object-Relational Approach to the Representation of Multi-granular Spatio-Temporal Data. In Proc. of the *17th Int'l Conf. on Advanced Information Systems Engineering*, Springer-Verlag Berlin. LNCS 3520:119-134.

Bertolotto M. (1998) Geometric Modeling of Spatial Entities at Multiple Levels of Resolution. Ph.D. Thesis, Università degli Studi di Genova, Italy.

Bertolotto M., Di Martino S., Ferrucci F., Kechadi T. (2007) A Visualisation System for Collaborative Spatio-Temporal Data Mining. *International Journal of Geographical Information Science*, Taylor & Francis. 21(7): 895-906.

Bettini C., Jajodia S., Wang X. (2000) Time Granularities in Databases, Data Mining, and Temporal Reasoning, Springer-Verlag Berlin.

Bittner T. (2002) Reasoning about qualitative spatio-temporal relations at multiple levels of granularity. In Proc. of the *15th European Conf. on Artificial Intelligence*, IOS Press. 317-321.

Camossi E., Bertolotto M., Bertino E. (2006) A multigranular Object-oriented Framework Supporting Spatio-temporal Granularity Conversions. *International Journal of Geographical Information Science*. Taylor & Francis. 20(5): 511-534.

Cattel R., Barry D., Berler M., Eastman J., Jordan D., Russel C., Schadow O., Stanienda T., Velez F (1999). The Object Database Standard: ODMG 3.0. Morgan-Kaufmann.

Claramunt C., Thériault M. (1995) Managing Time in GIS: an event oriented approach. In Proc. of the *Int'l Workshop on Temporal Databases: Recent Advances in Temporal Databases*, Springer-Verlag. 23-42.

Claramunt C., Jiang B. (2000) Hierarchical Reasoning in Time and Space. In Proc. of the *9th Int'l Symposium on Spatial Data Handling*. 41-51.

Compieta P., Di Martino S., Bertolotto M., Ferrucci F., Kechadi T. (2007) Exploratory spatio-temporal data mining and visualization. *Journal of Visual Languages and Computing*, Elsevier. 18(3):255-279.

Chen C.X, Zaniolo C. (2000) $SQL^{ST}$: A Spatio-Temporal Data Model and Query Language. In Proc. of *19th Int'l Conf. on Conceptual Modeling / the Entity Relational Approach*. Springer-Verlag Berlin. LNCS 1920:96-111.

Ester M., Kriegel H.-P., Sander J., Xu X. (1996) A Density-Based Algorithm for Discovering Clusters in Large Spatial Databases with Noise. In *Proc.* of the *2nd Int'l Conf. on Knowledge Discovery and Data Mining*. 226-231.

Fayyad U.M., Grinstein G.G. (2001) Introduction. Information Visualization in Data Mining and Knowledge Discovery, Los Altos, CA: Morgan Kaufmann. 1-17.

Fonseca F., Egenhofer M.J, Davis C., Câmara G. (2002) Semantic Granularity in Ontology Driven Geographic Information Systems. *Annals of Mathematics and Artificial Intelligence, Special Issue on Spatial and Temporal Granularity*. 36(1-2).

Griffiths T., Fernandes A.A.A., Paton N.W., Barr R. (2004). The Tripod spatio-historical data model. *Data Knowledge and Engineering*, Elsevier. 49(1): 23-65.

Güting R.H., Bhölen M.H., Erwig M., Jensen C.S., Lorentzos N.A., Shneider M., Vazirgiannis M. (2000) A Foundation for Representing and Querying Moving Objects. *ACM Transaction On Database Systems*, 25:1-42.

Hornsby K., Egenhofer M.J. (2002) Modeling Moving Objects over Multiple Granularities. *Annals of Mathematics and Artificial Intelligence*. Special Issue on Spatial and Temporal Granularity. Kluwer Academic Press. 36(1-2):177-194.

Hurtado C.A., Mendelzon A.O. (2001) Reasoning about summarizability in Heterogeneous Multidimensional Schemas. In Proc. of the *8th Int'l Conf. on Database Theory*. 375-389.

Huang B., Claramunt C. (2002) STOQL: An ODMG-based Spatio-Temporal Object Model and Query Language. In Proc. of the *10th Int'l Symposium on Spatial Data Handling*, Springer-Verlag Berlin. 225-237.

Khatri V., Ram S., Snodgrass R.T., O'Brien G. (2002) Supporting User Defined Granularities and Indeterminacy in a Spatio-temporal Conceptual Model. *Annals of Mathematics and Artificial Intelligence*. Special Issue on Spatial and Temporal Granularity, 36(1):195-232.

Koperski K.(1999) A Progressive Refinement Approach to Spatial Data Mining. Ph.D. Thesis, Simon Fraser University, Canada.

Kulik L., Duckham M., Egenhofer M.J. (2005) Ontology driven Map Generalization. *Journal of Visual Language and Computing*, 16(3):245-267.

Langran G. (1992) Time in Geographic Information Systems. Taylor & Francis.

Li T., Li Q., Zhu S., Ogihara M. (2002) A Survey on Wavelet Applications in Data Mining. *ACM SIGKDD Explorations Newsletter*. 4(2):49-68.

Mennis J., Liu J.W. (2005) Mining Association Rules in Spatio-Temporal Data: An Analysis of Urban Socioeconomic and Land Cover Change. *Transactions in GIS*, Blackwell Publishing. 9(1):5–17.

Muller J-C., Lagrange J.P., Weibel R. (eds.) (1995) GIS and Generalization: methodology and practice. Taylor and Francis.

National Hurricane Center (2003), *Tropical Cyclone Report: Hurricane Isabel*, http://www.tpc.ncep.noaa.gov/2003isabel.shtml.

Ng R.T., Han J. (1994) Efficient and Effective Clustering Methods for Spatial Data Mining. In Proc. of the *20th Int'l Conf. on Very Large Data Bases*. 144-155.

ORACLE™ (2008), Oracle Corp. http://www.oracle.com. Last date accessed: 01/2008.

PostgreSQL (2008), PostgreSQL Inc. http://www.postgresql.org. Last date accessed: 01/2008.

Roddick J.F., Lees B.G. (2001) Paradigms for Spatial and Spatio-Temporal Data Mining. *Geographic Data Mining and Knowledge Discovery*. Taylor and Francis. 33-50.

Saalfeld A. (1999) Topologically consistent line simplification with the Douglas-Peucker algorithm. *Cartography and Geographic Information Science*. 26(1):7-18.

Shahabi C., Chung S., Safar M., Hajj G. (2001) 2D TSA-tree: A Wavelet-Based Approach to Improve the Efficiency of MultiLevel Spatial Data Mining. In Proc. of the *13th Int'l Conf. on Scientific and Statistical Database Management*. 59-68.

Stell J.G., Worboys M. (1998) Stratified Map Spaces: A Fomal Basis for Multi-Resolution Spatial Databases. In Proc. of the *8th Int'l Symposium on Spatial Data Handling*. 180-189.

Tsoukatos I., Gunopulos D. (2001) Efficient Mining of Spatiotemporal Patterns. In Proc. of the *7th Int'l Symposium on Spatial and Temporal Databases*. LNCS 2121:425-442.

Tryfona N., Jensen C.S. (1999) Conceptual Modeling for Spatiotemporal Applications. *Geoinformatica*, Springer Netherlands. 3(3):245-268.

Vangenot C. (2001) Supporting Decision-Making with Alternative Data Representations. *Journal of Geographic Information and Decision Anaysis*. 5(2):66-82.

Worboys M. (1994) A Unified Model for Spatial and Temporal Information. *The Computer Journal*, Oxford University Press. 37(1):26-34.

# Automated Boundary Creation: Atomic Small Areas in Ireland

A Stewart Fotheringham, Peter F Foley, Martin Charlton

National Centre for Geocomputation, National University of Ireland Maynooth

**Abstract.** This paper describes the creation of a set of small-areas for the reporting of census data in the Republic of Ireland. The current areas used for reporting the results of the quinquennial population censuses are known as Electoral Divisions; they are large compared with similar reporting areas in Northern Ireland, they have widely varying populations and considerable internal social heterogeneity which makes them unsuitable for a wide variety of planning tasks. We describe an automated method of creating a suitable census geography which uses existing digital map and gazetteer data. We describe its structure and operation, validation and its application to nationwide. The areas have a prescribed minimum size, are designed to be consistently small, nest into the existing ED geography, cover the whole country, are constrained by natural boundaries, use streets as their unifying feature, and are reasonably homogenous.

**Keywords:** census, electoral divisions, gazetteer, spatial patterns

## 1 Introduction

In many countries census data is reported for a set of zones whose extent covers the entire country. Such zoning systems are usually hierarchical in nature with the upper tiers of the hierarchy being administrative or governmental units. In the United Kingdom the areal units used to collect census information (Enumeration Districts) are different from those used to report it (Output Areas) although this was not the case until the 2001 Census – there are 116895 Enumeration Districts in England and Wales with 175434 Output Areas. The Output Areas have a minimum size of 40 households and 100 residents, although the recommended size is 125 households. Northern Ireland is part of the United Kingdom, with 5022 Output areas and 2591 Enumeration Districts with an average size of 260 households or 650 residents (ONS 2007).

The situation in the Republic of Ireland is slightly different. The areas used for the collection of the quinquennial census are known as Enumeration Areas, but the areas used for reporting the data are known as Electoral Divisions (EDs). EDs nest into adminstrative counties. There are 3440 EDs with an average size of 1145 residents or 376 households. EDs very greatly in size; in the published Small Area Population Statistics the largest has 24400 residents and the smallest 55 (or 7859 and 17 households). One ED is split between two counties; 32 EDs have low populations (<55) and their data is amalgamated with an adjacent ED. The number of EDs in the SAPS is 3409. The average household size across the Republic is 3.04 residents per household, although there is wide variation across the EDs from 1.81 to 12.60 residents per household. Not only do the EDs vary wildly in size, there is considerable internal social heterogeneity. The ED boundaries have been stable for many years, and the recent boom in the Irish economy (the 'Celtic Tiger') has generated a rapid demand for new housing. The result is that increasingly EDs are an unsuitable spatial unit for attempting to understand local social changes, particularly in urban areas, and especially for targetted spatial delivery of policies designed to alleviate social problems.

A related issue is in recent efforts to consider social patterns in both Northern Ireland and the Republic. The spatial granularity of the census reporting units in the two countries inhibits reliable joint analysis of census and other information due to the different scales involved. There are notable border effects when regression modelling is attempted with ED data in the south and Output Area or Ward level data in the north.

The question arose as to whether a set of atomic small areas (SAs) could be constructed which would become basic collection and reporting units. Such areas would require a consistent definition, would be large enough to avoid any confidentiality issues, be small enough to allow social micrography to be uncovered, and be reasonably similar in size. It would be desirable if they be created from existing data sources, and should be capable of long term maintenance. They should also form a complete partition of the Republic, and they should nest into the existing ED boundaries.

## 2 Design Issues

Designing a new grography is not a task to be approached lightly or wantonly, and initial thoughts were whether the approach used in the United Kingdom might be employed to create a set of SAs. The underyling spatial building block for the OAs is the unit postcode. The postal and administrative geographies in England and Wales are misaligned. For the 1981 and 1991 Censuses many postcodes straddled Enumeration District boundarie. In 1991 OPCS created a lookup table (OPCS/GROS 1992) which provided population counties for every intersection of postcode and Enumeration District as well as the 'majority' ED for each split postcode: the postcode is a convenient surrogate for a full address and the majority linkage provided a method of assigning postcoded records (whether for households to individuals) to EDs on an 'all-or-nothing' basis. The unit postcodes are also misaligned with larger units in the administrative hierarchy.

The solution for the UK exercise was to create 'postcode polygons' (Martin 1998) which contain those postal delivery points in a single postcode which do not cross a parish or ward boundary. Thiessen polygons were created around individual delivery points (using the ADDRESS-POINT gazetteer from the Ordnance Survey) which were then merged within each postcode/ward intersection. These basic spatial units are then aggregated into larger units with the desired characteristics (100 residents and 40 households) and social homogeneity (Martin 1998). The algorithm which was used to aggregate the postcode polygons is modified version of the Automatic Zoning Procedure (Openshaw 1977). A further design modification allowed for the minimisation of the perimeter$^2$/area ratio of the resulting polygons, although in the final allocation, the shape criterion has been based on minimising the dispersion of postcodes in each polygon.

This approach creates a problem for the Republic of Ireland as it is one of the few developed countries in the world which does not have a postcode system: this rules out the postcode as a building block. Like the UK, there is a gazetteer of postal delivery points which are geocoded using two projection systems (Irish National Grid and Irish Transverse Mercator) which is known as GeoDirectory (Fahey and Finch 2007). Initial thoughts prompted by a suggestion that a street-based allocation might be possible considered whether groups of delivery points could be created which would match the initial design criteria above.

# 3 Stage I and II Pilot

Initial design and testing was carried out on two EDs in Northern Kildare, Maynooth and Leixlip. In 2002 Maynooth had a population of 10387 with 3199 households (average size 3.25) and Leixlip had 15154 residents in 4430 households (average size 3.42). Both are commuter towns in North East Kildare about 20 miles west of Dublin, both have urban and rural parts, and Leixlip is home to Intel's microprocessor fabrication facility, a major local employer, and Maynooth is the location of the National University of Ireland Maynooth.

After discussions with the Central Statistics Office, it was decided that the minimum SA size would be fixed at 65 households. There would be no cap on the SA size – this would be determined by the algorithm depending on individual cases. However, as individual geocoded Census records are not available, the residential delivery point would be used as the surrogate for a household: the minimum SA size is 65 residential delivery points.

## 3.1 Basic Strategy

The initial approach was to join road centrelines into skeletons where the segments forming the centrelines were tagged with the number of residential delivery points nearest to that segment. For each skeleton a list was made of segments which could join it; one segment was chosen from this list and joined to the skeleton. Joining terminated once the property count for the skeleton reached 65. The initial segment was chosen at random from the list of unallocated segments in the ED. The criterion for joining was to chose either (a) the next unallocated segment with the most delivery points, (b) the segment with the fewest delivery points, (c) the longest segment or (d) the shortest segment. These choices correspond to using as attributes the delivery point count or the segment length and with either a 'greedy' or 'abstemious' option.

A FORTRAN program was written to carry out the allocation allowing for the various options. Three outcomes from this process arose. First, a completed skeleton is created with more than 65 households. Second, it is possible for the remaining unallocated segments to have insufficient delivery points to reach the acceptance threshold: these were referred to as 'orphans'. Third, in some cases segments do not join with any other segments: these were called 'singletons'. Orphans were dealt with by redistributing their segments among the already-created skeletons according to the creation criterion (delivery point/segment length and greedy/abstemious). Singletons then had to be merged with neighbouring areas.

The outcome of the program is a list of segments and skeleton codes. The segment which is closest to each delivery point (both residential and commercial) is already known, so the skeleton codes can be transferred to the delivery point locations. Proto-small areas are then formed by creating Thiessen polygons for all the delivery points inside an ED, constrained by the ED boundary, and merging the internal boundaries between those delivery point polygons which have the same skeleton code. Thiessen polygons created for singleton skeletons are merged with the neighbour that shares the longest boundary.

Some GIS operations are required to extract the data from various data sources, integrate it, dump it into a suitable format for the FORTRAN allocation program, and then assemble the pieces back together to form small areas. Consideration was given to the choice of software platform – the final decision was made to use ESRI's ArcINFO product, and code the GIS operations into a set of linked macros using the Arc Macro Language (AML).

Examination of the alternative approaches, together with mapping geocoded household data made available from the Central Statistics Office, suggested that using the residential delivery point counts as the criterion attribute and the greedy allocation strategy produced the most satisfactory set of boundaries.

## 3.2 Addressing

Most properties in urban areas in Ireland have what we might term a 'well formed address'; that is, an address which uniquely identifies the property. This might be some combination of a number or house name, street name, locality name, and district name. However, this is not always the case, particularly in rural areas. In some small villages the road running through the village does not have a name, and the houses along it have neither numbers nor names; the address for each delivery point is just the name of the settlement. It is due to the local knowledge of the postman that letters are delivered to the right households, although the arrival of a new postman to an area can cause problems. The problem of non-uniqueness of address is greatest in the most rural areas; an examination of the addresses in GeoDirectory suggests that some 66% of addresses in the county of Roscommon are not unique. Whilst GeoDirectory does indicate whether an address is unique or not, the BUILDINGS table does indicate whether a property is on a named throughfare. Most unnamed thoroughfares are in rural areas.

### 3.3 Incorporating Natural Boundaries

Some enhancements were sought. While the SA boundaries nest into their parent ED boundaries, they do not take into account 'natural' boundaries such as watercourses or railway lines. The question of modifying the algorithm to take into account these boundaries was then explored.

The Thiessen polygon algorithm in ArcINFO takes a set of points and returns a set of polygons which have Thiessen properties which can be clipped using the ED boundary. The locations of the polygon boundaries cannot be influenced – there is no way of modifiying the algorithm.

A raster equivalent is to use the costallocation function in ESRI's GRID module together with the locations of the delivery points to produce an allocation grid in which any cell is closest to the cell representing its residential delivery point and no other. Creating the grid of residential delivery points is easy – the pointgrid function is used. The other input required is a cost grid – each cell contains the cost of traversing one unit of distance. If every cell in the cost grid contains the same value, the allocation grid which is output will correspond to a rasterised version of the Thiessen polygon vector coverage of the type created in the pilot. The key to modifying the allocation is to introduce varying traverse costs to represent the 'importance' of the various natural boundaries. Initally all cells in the cost grid were set to 1, and then those which were crossed by watercourse, a railway line, or a main road were set to 10000. The resulting output grid is then vectorised to obtain polygons.

Clearly the raster size is important here – larger cells require less processing than smaller cells, but smaller cells are closer to the original boundaries when the output allocation grid is vectorised. Raster sizes of 1m and 2m were used. This raised problems which will be discussed further below.

The watercourse, railway and road centrelines were rasterised from Ordnance Survey Ireland’s large scale data using the linegrid function. This caused problems – for instance, the Royal Canal which flows through Maynooth is represented by separate lines for its north and south banks. Each track on the railway that runs along the canal is represented by a separate line. In essence, this level of data provided too much detail for the polygon formation, and a decision was made to use the representations from OSi's 1:50000 vector data where watercourses have, counterintuitively, a richer feature coding, but are represented by single centrelines.

## 3.4 Segments without Delivery Points

During testing it became clear that there were some segments in the road centreline data that lacked delivery points. In general these were outside the urban areas, and usually along the longer segments. This resulted in some of the rural skeletons growing in rather unexpected ways. This was a problem whether the delivery point count or segment length was used as the choice criterion.

A solution to this problem was to treat the 'urban' small-area and 'rural' small area formation as separate tasks and then merge the 'urban' and 'rural' proto small areas. The question arises of deciding what parts on an ED are urban and which are rural.

There are smaller officially-defined geographical areas in Ireland than EDs which are known as Townlands. There are some 50000 of these and their boundaries reflect a pre-medieval division of the landscape. Like EDs townlands vary widely in size, shape, and population. In most cases they nest into EDs, but their boundaries are not always aligned. Intersecting the townland and delivery point coverages allows us to count the numbers of delivery points on both named and un-name throughfares in each townland. Examination of the results for the pilot areas suggested that if any delivery point in a townland was on a named thoroughfare, then the townload could be regarded as urban, otherwise it was treated as rural.

## 3.5 Rural Allocation

The initial processing of the data for an ED was to decide which were the rural townlands. The townland residential delivery point counts and a townland adjacency matrix were extracted from the ArcINFO coverages in ASCII form and passed to a FORTRAN program. It was decided that a brute force approach would be taken. A townland is selected at random and its un-allocated neighbours examined. The one with the largest property count was joined. The unallocated neighbours of these fused areas are examined and the one with the largest property count is added until the delivery point count reaches the threshold of 65. This continues until all townlands have been processed. The mean and variance of the delivery counts is computed. The process is iterated from different random starting points until the solution with the lowest mean and variance is obtained (this is usually in fewer than 10000 iterations).

Analogous allocation problems occur with this method as with the skeleton building process. Orphan and singleton allocations appear. Orphans are dealt with by reallocating their component townlands among any neighbours. Singletons are flagged to be treated using the skeleton building method.

## 3.6 Rural/Urban Merging

With the development of the separate procedure for handling the rural parts of each ED the final set of operations require merging of the separate sets of proto small areas.

## 3.7 Random Numbers

In the pilot phase the system random number generator was used – the software runs on a Sun workstation under Solaris 5.9. It is clear that applying this process to the whole of Ireland will require enormous streams of reliable random numbers. The programs were redesigned to use to Mersenne Twister (Matsumoto and Nishimura 1998). This has been shown be reliable in operational situations using spatial data (van Niel and Laffan 2003). The seeds are set using the system clock so every run will use a different stream of random numbers.

## 3.8 Testing and Validation

Producing a prototype algorithm is just a starting point. An algorithm which appears to work on two EDs out of over 3400 might not be expected to work in a wider variety of contexts. Some EDs are entirely 'urban', some EDs are entirely 'rural'. Some EDs have interesting geometry. Nine additional Eds were selected to test the algorithm for stability. The test EDs are listed below in Table 1.

**Table 1.** Test ED characteristics

| ED Type | ED Name(s) |
|---|---|
| Rapidly expanding commuter town | Maynooth (Kildare) |
| | Leixlip (Kildare) |
| Old inner city area | Merchants Quay A (Dublin) |
| New urban development | Ashtown A (Dublin) |
| Large county town | Longford Urban No. 1 |
| Small rural town | Abbeyleix (Laois) |
| Rural | Ardamine (Wexford) |
| Rural with holiday homes | Moy (Clare) |
| Island community | Inishmore (Clare) |
| Unusual geometry: Doughnut | Kilkenny Rural |
| Ground truthing | Botanic A (Dublin) |

Whilst the initial algorithm had been developed using Maynooth and Leixlip, these areas were included in Stage II to assess the impact of the refinements. Both Eds are in areas of very rapid urban expansion, and Maynooth is an ED with many student houses. Merchant's Quay is in the Liberties area of Dublin – an old inner city area which has been gentrifying but has a good mix of old corporation housing, new apartments, and some low value older owner occupied houses. Ashtown A by contrast is undergoing rapid development on green/brownfield sites although there are some older estate developments; it lies on the edge of Dublin. Abbeyleix in a classic medium-size rural ED with a town in the centre. Inishmore is a challenge: it contains the three Aran Islands with a combined population of about 1300. Kilkenny Rural is one of several polygons which surrounds completely one or more EDs – these doughnut polygons were to prove an interesting challenge. Botanic A was the subject of initial manual exploration: it lies in north-central Dublin.

The results of the application of the algorithm to the test areas is in Table 2:

**Table 2.** Summary of Small Areas for Test EDs

| ED | Small Areas | Mean Size | Max Size | Min Size | Total H/holds |
|---|---|---|---|---|---|
| Leixlip | 35 | 137 | 250 | 70 | 4825 |
| Longford | 13 | 126 | 215 | 74 | 1629 |
| Maynooth | 33 | 118 | 226 | 66 | 3887 |
| Moy | 4 | 107 | 133 | 80 | 428 |
| Merchant's Quay | 5 | 200 | 308 | 103 | 1002 |
| Inishmore | 6 | 109 | 198 | 65 | 656 |
| Abbeyleix | 8 | 134 | 217 | 80 | 1071 |
| Kilkenny Rural | 40 | 132 | 391 | 68 | 5311 |
| Ardamine | 13 | 137 | 283 | 80 | 1793 |
| Ashtown A | 17 | 133 | 266 | 66 | 2261 |
| Botanic A | 11 | 124 | 174 | 68 | 1359 |

The table shows some summary statistics for the EDs chosen as the test areas. As well as the ED name, the table reports the number of Small Areas created within the ED, the average number of residential delivery points in each Small Area in the ED, the minimum number of delivery points, the maximum number of delivery points and the total number of residential delivery points in the ED.

It is clear that there's quite a strong and re-assuring relationship between the number of small areas which the algorithm produces and the number of households in an ED. The average number of households is roughly on a par with that average for the Output areas in Northern Ireland. The minimum threshold of 65 is not breached. As expected there are anomalies.

The statistics for Merchant's Quay are skewed by the number of apartment blocks along a small number of street segments. However, the initial impressions of this testing stage were that the algorithm was robust and was able to produce sensible results.

## 3.9 From Prototype to Production

The transition from a two stage pilot to the full roll out took 18 months. Applying the algorithm to 11 EDs was a relatively simple process, and the initial estimates were that, given the average processing time on one of the authors' laptop, that running the whole county on a Sun workstation would be relatively straightforward. This proved not to be the case.

The prototype of the algorithm consisted of a series of loosely linked macros and two fortran programs. The data from GeoDirectory has been pre-processed and the extracts from the buildings table made for each ED. The macros were run from a simple GUI where the user clicked on an ED name and a macro was called which ran macros to extract the road centre-lines, natural boundaries, townland boundaries, dumped the data, ran the external programs, and then merged the results back together creating a coverage/shapefile called small-areas, and some two dozen intermediate coverages which were used in processing.

It was decided that the EDs in an entire county would be processed in a single run. The results would be stored in a subdirectory named after the county, and underneath would be separate subdirectories for each ED. Within the ED subdirectory would be the extracted data, intermediate coverages and grids, and the final coverage named small-areas, so that when problems arose, we could track backwards through the formation process for any ED in the county or country.

The data for the whole country was conveyed to the NCG on a 500Gb disk drive – 212Gb of the drive were the large scale orthophotographs which would be used during the verification stage of the process. Handling such large amount of data requires some thought. Copying the data took several hours, and the orthos were reduced in physical size by conversion from TIFF to JPEG format – this took over 8 hours.

The macros were rewritten to make them amenable to batch style processing with the name of the county as a parameter. After some early setbacks it became clear that we would need to retain all the printed output from the ArcINFO commands. These are stored in 'watchfiles' which can grow unexpectedly large.

# 4 Production Algorithm

The implementation is through a set of linked macros and four fortran programs. The macros total some 1300 lines of ArcINFO AML, and the fortran programs are just short of 3000 lines of code. There is extensive pre-inter- and post-processing of the outputs from the various fortran programs in the GIS. Some of this arises because of the requirements of the fortran, other manipulations are required because of the architecture of the small-areas creation method, and a final group of manipulations might be described as housekeeping – these are due to the data representation in ArcINFO.

The allocation for the urban and rural parts of the EDs has been described above. It is useful to consider how the various component operations are brought together, and what parameters are needed to control the system.

For any ED there are 90 separate GIS operations required, as well as the running of four external programs, to create the final set of small-areas. The editing phase is then followed by a final merge of the EDs in each county to create a county set of small-areas.

There is a small set of utility routines, in particular, a clipping macro. The clipping macro takes an ED boundary and returns points, lines, or polygons which lie within the ED boundary, and also removes any internal polygons which are not part of the ED – this assists in the processing of doughnut polygons. As with many 'simple' operations provided as part of GIS software, the desired result is often the outcome of a series of linked operations. The clipping routine is an example of a generic routine, so it was coded separately rather than the operations being coded 'in-line' in the main processing macro.

## 4.1 Data Extraction

There are several components required for the creation of the small areas within an ED. The townland and ED boundaries are misaligned – they do not form a neat spatial hierarchy. At the county edges, there are slight misalignments in the source data which create thousands of sliver polygons – these must be removed before further processing continues. After the application of the clipping routine, slivers of less than $1m^2$ are removed but the processing is organised so that the external edge of the clipped townland coverage remains coterminous with the ED boundary. This requires 8 separate GIS operations, including an edit session to remove the resulting pseudo-nodes (in the ArcINFO terminology nodes with a valency of 2 are referred to as pseudo-nodes).

The road centreline segments are extracted from the national road centrelines data – this is provided as part of the 1:50000 scale data, but the centrelines are based on 1:1000 and 1:2500 source. The building centroids are extracted from the GeoDirectory coverage.

An important parameter is the spatial tolerance. In the ArcINFO model this is known as the fuzzy tolerance and determines when nodes will be snapped and whether coordinates are moved. As positional accuracy is important, a spatial tolerance of 0.001m is used. This ensures that the final boundaries can be adjusted to be coterminous with the supplied ED boundaries.

A parallel operation is the extraction of a list of orthophotographs which cover this ED. The orthos are used in plotting the results for individual EDs. There are 25500 orthos covering the extent of the Republic, so for any ED only a handful are required as backdrops. As part of the initial data preparation for the project we created an index coverage which contains the name and extent of every ortho. This is intersected with the ED boundary to obtain a list of the orthos which cover the ED.

## 4.2 Rural Townland Processing

Initial experiments using the EDs in Kildare as a test bed revealed that the original criterion from Stage II for the identification of 'rural' townlands was too crude. After some investigation it was decided that two thresholds were important: the proportion of delivery points which were not on named throughfares and the total number of delivery points in each townland. Townlands with more than 27.5% of delivery points on unnamed thoroughfares and with fewer than 232 residential delivery points are deemed to be rural, as is any townland with no delivery points. These thresholds were determined after analysis of Townland characteristics in County Kildare. Kildare contains a wide range of settlement types from very small villages to large commuter towns.

The centres of doughnut EDs have to removed – they are not considered as part of the outside polygon in the ArcINFO model so have to be separately flagged at each stage in the processing. The polygon attribute table (which contains the ID of each rural townland) and the arc attribute table (which contains the ID of the polygons on each side of any boundary) are dumped into ASCII files for input into the townland allocation program. The output is a list of proto-small areas codes and townland IDs. These are merged back with the original rural townland coverage, and internal boundaries between adjacent townlands in the same small area are dissolved to create the set of rural proto small-areas.

## 4.3 Urban Townland Processing

Urban townlands and rural townlands flagged for processing as urban are extracted from the townlands coverage, and used to clip the road centreline coverage and the buildings coverage. The next stage is data pre-processing for input into the urban centreline allocation program.

The throughfare codes in GeoDirectory do not match any segment coding in the road centreline data. This precludes a simple tally of residential and commercial delivery points over road segments. A proximity analysis is carred out to determine the ID of the closest road segment to each residential delivery point, and the number of delivery points is then tallied over the segment IDs. These counts are merged back with the road centreline segment coverage.

A problem arises in a few cases where the proximity analysis misallocates building locations to an incorrect road segment – this usually occurs at road junctions. A re-allocator program was written in C++ to correct the misallocation – correction is possible in about 50% of cases.

Records from the arc attribute table for the roads coverage are dumped to an ASCII file – this contains, for each segment, the IDs of the start and end nodes of the segment, the ID of the segment itself, its length, and the number of residential delivery points. The skeleton creation program takes these as input and produces an ASCII file with the segment IDs and their associated skeleton IDs which is merged back with the road centrelines coverage.

The proximity analysis for the buildings also gives a segment ID to each delivery point, so the skeleton IDs are also merged with the buildings coverage – we now know in which skeleton each building lies – this includes commercial buildings as well as residential buildings.

## 4.4 Raster Processing

The next stage is to form the urban proto-small areas. As these are based on constrained Thiessen polygons, this stage is carried out in a raster environment. There is a rich collection of raster processing functions, based on Tomlin's Map Algebra (Tomlin 1990), available in the GRID module which is part of ArcINFO.

A fundamental aspect of raster processing is deciding on the size of the raster to be used. Large cells can be processed quickly, but suffer from low spatial resolution when they are re-vectorised. Small cells take longer to process, halving the side length quadruples the number of cells, but creates vectors which it was thought can be more easily merged back into the townland based results. We shall discuss this problem further below. After some experimentation, 1m was chosen as the raster size.

The urban townlands are extracted from the boundary and rasterised at 1m relative to the lower left corner of the bounding box of the ED. This will be used as both a window and a masking grid. The urban parts of the rail, motorway, primary road, watercourse, and path coverages are clipped (using the clipping routine described earlier) and rasterised – cells through which any of these constraints pass are given a passage cost of 10000 and all others are given a passage cost of unity. The resulting grid is the cost grid. The buildings coverage is finally rasterised, with the skeleton code as the attribute.

The building skeleton code grid and the cost grid are used as arguments to the costallocation function. The output from costallocation is a grid in which each cell contains the skeleton code of the building to which it is closest. Vectors are then recovered from this grid using the gridpoly function. This gives us the vector boundaries of the feature constrained Thiessen polygons. The polygon IDs are adjusted by the addition of 30000 to differentiate them from the rural proto small-area polygons when the two sets of polygons are merged.

The result of vectorising the raster data is that lines which are not vertical or horizontal are created from a set of 'steps'. Aesthetically these are unpleasing when seen close-up, so they are smoothed to remvoe the worst effects. The spline operation is used in ArcEDIT to accomplish this; the requisite parameter is the grain tolerance, which is set at 2.5m.

## 4.5 Merging the Urban and Rural Proto Small Areas

The final stage in the algorithm turns out to be as complex as any of the previous steps. There are several special cases which require careful handling, in particular the preservation of the doughnut hole, uninhabited islands, and sliver polygons remaining after the vectorisation which have not been sufficently smoothed.

The proto small area coverages from the urban and rural processing are merged. Remaining slivers from the re-vectorisation are removed if they within a 2.5m buffer of the ED boundary, and the ID of the doughnut is adjusted to preserve its 'external' status. A final tally of residential delivery points is made for the proto small-area coverage which is then merged with the proto small-area coverage polygon attribute table.

Although the result of the spline smoothing in the vectorisation of the urban small area rasters is designed to produced asethetically pleasing results, the boundaries may not be conterminous with the original spatial constraint (rail, road, watercourse, path) locations, nor with townland boundaries. The ArcEDIT module allows a snapping operation to take place, such that any feature which is closer than, in this case, 1.415m of a snapfeature can be re-aligned with that snapfeature. Small area boundaries are snapped to roads, watercourses, railway lines, paths, urban townland boundaries, and finally to the ED boundary.

At this stage there may be a small residue of polygons with residential delivery point counts which are below the threshold of 65. These are merged with the adjacent polygon with which they share the longest border. Care has to be taken here to preserve the external and any doughnut boundaries. There may still be islands with a delivery point count lower than 65 – these are flagged. There are also a few complete EDs with a delivery point count which fails the threshold; these are flagged as well, as they will consist of only one small area.

The final stage is to assign small-area codes. Each ED has a 5 or 6 digit code of the form CCEEE or CCCEEE where the CC or CCC is a county code, and the EEE is a sequence number of the alphabetic order of the ED within the county. These codes are multiplied by 1000 and the sequence number of the small-area is added. Although the largest ED has 92 small areas, it may be that further building expansion within this ED results in more than 100 small areas – the coding system needs the redundancy to allow this.

## 4.6 Aesthetic Improvement

A final stage, which unfortunately is somewhat time consuming is to remove artefacts which arise out of the application of a completely automatic algorithm. These can be quite bizarre – a river and a railway running parallel to one another can produce something which looks like a heron's beak sticking from the side of a polygon. Some polygons have a bowtie shape, often in rural areas where a townland with zero delivery points has been added to one small area rather than another. These artefacts must be identified and removed by hand. A final external program computes a variety of shape statistics (Folk 1968, Moellering and Rayner 1979). Analysis of the distributions of the shape statistics reveals some helpful thresholds for identifying the various eccentrically shaped polygons which are candidates for manual adjustment. At the time of writing this process is being undertaken.

## 4.7 Time Requirements

A final consideration is the time required for the production. Processing is being undertaken on a Sun Blade 2500 workstation with an UltraSPARC chip – the CPU clock speed is 1.28GHz. The data are stored on a Dell PowerEdge 2950 server with 1TB of disk store. Kildare, which has 92 EDs takes around 8 hours to process, Dublin takes nearly 50 hours, Kerry, with fewer EDs but a more complex coastline takes nearly 60 hours.

# 5 Discussion

The preliminary results are shown in the table below and represent work in progress – there is much work to be completed before the final set of small Areas can be released.

**Table 3.** Preliminary results

| County | Eds | Pop'n | H'holds | SAs | Pop/SA | HH/SA |
|---|---|---|---|---|---|---|
| Carlow | 54 | 46014 | 17195 | 194 | 237 | 89 |
| Dublin | 322 | 1122821 | 420429 | 4092 | 274 | 103 |
| Kildare | 89 | 163944 | 60957 | 635 | 258 | 96 |
| Kilkenny | 113 | 80339 | 29651 | 341 | 236 | 87 |
| Laois | 98 | 58774 | 22591 | 249 | 236 | 91 |
| Longford | 55 | 31068 | 12111 | 170 | 183 | 71 |
| Louth | 42 | 101821 | 38703 | 423 | 241 | 91 |
| Meath | 92 | 134005 | 53938 | 604 | 222 | 89 |
| Offaly | 87 | 63663 | 23769 | 264 | 241 | 90 |
| Westmeath | 106 | 71858 | 27064 | 322 | 223 | 84 |
| Wexford | 124 | 116596 | 45566 | 621 | 188 | 73 |
| Wicklow | 82 | 114676 | 42870 | 467 | 246 | 92 |
| Clare | 155 | 103277 | 38210 | 485 | 213 | 79 |
| Cork | 398 | 447829 | 167234 | 1900 | 236 | 88 |
| Kerry | 166 | 132527 | 48110 | 674 | 197 | 71 |
| Limerick | 173 | 175304 | 64225 | 749 | 234 | 86 |
| Tipperary | 175 | 140131 | 52367 | 645 | 217 | 81 |
| Waterford | 130 | 101546 | 38580 | 448 | 227 | 86 |
| Galway | 238 | 209077 | 78661 | 1011 | 207 | 78 |
| Leitrim | 78 | 25799 | 10646 | 164 | 157 | 65 |
| Mayo | 154 | 117446 | 43431 | 636 | 185 | 68 |
| Roscommon | 112 | 53774 | 20734 | 289 | 186 | 72 |
| Sligo | 82 | 58200 | 21480 | 302 | 193 | 71 |

| | | | | | | |
|---|---|---|---|---|---|---|
| Cavan | 93 | 56546 | 21929 | 297 | 190 | 74 |
| Donegal | 149 | 137575 | 50415 | 719 | 191 | 70 |
| Monaghan | 70 | 52593 | 18655 | 255 | 206 | 73 |
| RoI | 3437 | 3917203 | 1469521 | 16956 | 231 | 87 |

While these are preliminary results, the average sizes of 231 residents and 87 households compare well with the equivalent areas in Northern Ireland which have an average of 326 residents and 125 households. The apparently anomalous result for Leitrim (65 per small area) is because 9 of the EDs already fail the household count threshold. Figure 1 shows the existing Electoral Division boundaries in County Kildare which extend about 70km north-south and 50km east-west. The small-areas created using the algorithm described in this paper are shown in figure 2. The detailed subdivision in the urban areas is quite clear.

**Fig. 1.** Electoral Divisions in County Kildare

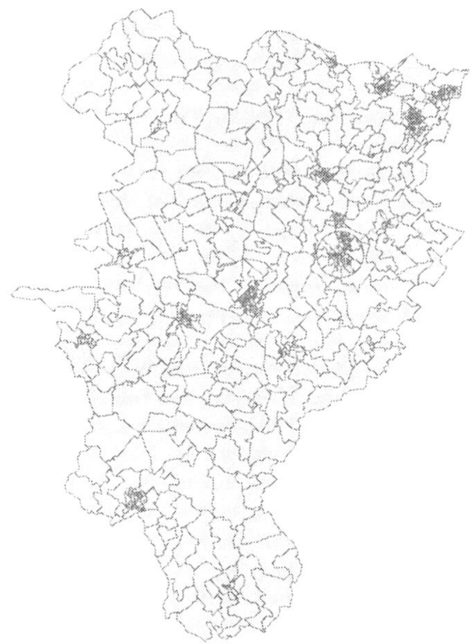

**Fig. 2.** Small Areas in County Kildare

A bespoke algorithm and methodology has been developed to generate small areas automatically with Electoral Divisions in Ireland. The method produces robust and sensible results. The small areas meet the design criteria and have a number of attractive features

- In urban areas the small areas are based on communities
- In rural areas the small areas are based on historic spatial units
- In urban areas streets are cohesive rather than dividing features
- The small areas boundaries take into account natural features
- The small areas are large enough for data to be reported without breaking any confidentiality thresholds. Where an ED fails to reach this threshold it will contain only one small area, and current official practice is to merge this and an adjacent ED for data reporting.
- The small areas are spatially similar to the Output Areas used in Northern Ireland which will greatly facilitate the creation of all-island datasets and encourage all-island analysis.

## References

Fahey D, Finch F (2007) GeoDirectory Technical Guide. An Post/Ordnance Survey Ireland, Dublin

Folk RL (1968) Petrology of Sedimentary Rocks. Hemphill's, Austin TX

Martin D (1998) Census output areas: from concept to prototype. Population Trends 94:19-24

Matsumoto M, Nishimura T (1998) Mersenne Twister: A 623-dimensionally equidistributed uniform pseudorandom number generator. ACM Transactions on Modeling and Computer Simulation 8(1):3-30

Moellering H, Rayner JN, (1979) Measurement of shape in geography and cartography, Numerical Cartography Laboratory Report No SOC77-11318. Ohio State University

van Niel K, Laffan S (2003) Gambling with randomness: the use of pseudo-random number generators in GIS. International Journal of Geographical Information Science 17(1):49-68

ONS (2007) Census Geography, URL:http://www.statistics.gov.uk/geography/census_geog.asp

OPCS/GROS (1992) ED/Postcode directory: Prospectus, 1991 Census User Guide 26. Office of Population Censuses and Surveys, Titchfield

Openshaw S (1977) A geographical solution to scale and aggregation problems in region building, partitioning and spatial modeling. Transactions of the Institute of British Geographers, NS 2:425-446

Tomlin CD (1990) Geographic Information Systems and Cartographic Modeling. Prentice-Hall, Englewood Cliffs, NJ.

# Climate-Change Adaptations in Land-Use Planning; A Scenario-Based Approach

Eric Koomen[1], Willem Loonen[2], Maarten Hilferink[3]

[1]Vrije Universiteit Amsterdam, Faculty of Economics and Business administration, de Boelelaan 1105, 1081 HV Amsterdam, the Netherlands, ekoomen@feweb.vu.nl and Geodan next, President Kennedylaan 1, 1079 MB Amsterdam, the Netherlands;
[2]Netherlands Environmental Assessment Agency (MNP), PO Box 303, 3720 AH Bilthoven, the Netherlands, willem.loonen@mnp.nl;
[3]Object Vision BV, c/o CIMO-Vrije Universiteit Amsterdam, De Boelelaan 1085, 1081 HV Amsterdam, the Netherlands
mhilferink@objectvision.nl

**Abstract.** Socio-economic and climatic changes are expected to alter the current land-use patterns in the Netherlands. In order to study these uncertain developments and propose adaptation and mitigation strategies to cope with the possible changes in the physical and societal environment a set of future scenarios is developed. These scenarios integrate possible socio-economic and climatic changes and are used in the *Land Use Scanner* model to simulate future land-use patterns. Based on these simulations sector-specific adaptation and mitigation measures are developed in related research projects as will be described in this paper.

**Keywords:** climate change, land use modelling, spatial patterns, spatial planning

## 1 Introduction

It is widely believed that climate change and increased climatic variability will impact land use through affecting different economic sectors such as agriculture, housing, nature and ecosystems, and by changing the water resources system (Commissie Waterbeheer 21e eeuw 2000; IPCC 2001; Verbeek 2003). Climate change directly affects, for example, local agricultural and hydrological conditions and consequently influences the economic development potential. Climate change thus modifies the demand and supply for space, as well as the suitability of space for certain uses (Beinat and Nijkamp 1998). These processes can be assessed through land-use simulation models that integrate sector-specific demands (for housing, agriculture, etc.) and land suitability for certain uses and provide an indication of the likely use at a specific location in the future under different climate conditions.

Climate change modifies the mechanisms of the demand-supply interplay as well as the boundary conditions and scenarios within which it unfolds. The main processes through which climate change and socio-economic developments may affect the interaction between the demand and supply of land are:

- The physical modification of the suitability of certain areas for some uses of the land;
- The modification of productivity and production processes within sectors such as agriculture, forestry, and nature;
- Changes to the primary functioning of economy and society leading to a different set of policies that influence for instance economic development (growth) or the type of development (e.g. free market versus government);
- The extra demand for space as a result of adaptation strategies within various sectors.

In order to accommodate these impacts, pro-active adaptation measures within the area of spatial planning are prerequisite to cope with climate change and will offer new opportunities for rearranging land use (Parry 2000a; Parry 2000b). However, such rearrangements will pose challenges and conflicts between the national and regional policy levels, and between sectors. For instance, when problems concerning water storage and flooding are tackled with spatial rather than technical measures, the capital-intensive agricultural or urban functions of these buffering areas will be highly restricted (Borsboom-van Beurden et al. 2005).

The research programme *Climate changes Spatial Planning* aims to develop an adequate and timely set of policies for mitigation and adaptation to cope with the impacts of climate change in the Netherlands. The research programme is centred on four main research themes:

- Climate scenarios: climate scenarios and climate data management for decision support in spatial planning;
- Mitigation: decreasing greenhouse gas emissions in relation to land use and spatial planning;

- Adaptation: dealing with the effects of climate change in spatial planning;
- Integration: methods for research exchange and integration.

Obviously, climate change is not the only factor driving land-use change. Socio-economic developments are another major driving force. In fact, these developments interact with climatic changes (Dale 1997; Watson et al. 2006). For example, economic and population growth cause increased emission of greenhouse gasses, which influence the global climate. As a result, changes in annual regional rainfall patterns could impact agricultural production or cause the tourist industry to migrate to other regions. Prolonged droughts and other extreme weather are other examples of climatic changes that impact the economy.

Integration of climate-change and socio-economic factors is, in our opinion, thus needed in any long-term study on future land-use configurations and related spatial planning measures. However, the scenarios used in most land-use allocation models, are usually neutral to climate change, as only socio-economic factors are taken into account. This assumption appears inappropriate in relation to the expected substantial climatic changes. Therefore, we present here an integrated set of future scenarios, based on socio-economic scenarios and climate models. These scenarios are used as input in various sector-specific models and are subsequently fed into the *Land Use Scanner* model for an initial simulation of land-use change. In a following phase these results will be analysed on their possible adverse impacts on different sectors. Based on this analysis sector-specific adaptation and mitigation measures are drafted that can eventually be fed back into the *Land Use Scanner* to come to an integrated view on possible land-use changes that results from expected societal and climatic developments.

This paper starts by introducing the renewed *Land Use Scanner* version that contains a 100 metres grid and offers a discrete description of land-use limited to only one type of use per cell. This model differs considerably from the previous version that contained a 500 metres grid with a continuous description of the fraction that all land-use types took up in a cell. We then go on to describe the socio-economic and climatic characteristics and land-use simulations of the proposed future scenarios. The paper concludes with a discussion on the applicability of the presented results in developing land-use related adaptation measures for climate change.

## 2 Simulating Land-Use Change

In this long-term scenario-study we use the *Land Use Scanner*. This GIS-based model produces simulations of future land use that are based on the integration of sector specific inputs from dedicated models (Dekkers and Koomen 2007; Hilferink and Rietveld 1999). The model is based on demand-supply interaction for land, with sectors competing for allocation within suitability and policy constraints. Land-use simulations are generally scenario driven, with series of coherent assumptions regarding variables such as economic growth or level of government intervention, determining the way the land demand-supply unfolds (Borsboom-van Beurden et al. 2007; Koomen et al. 2005). The renewed model-configuration used for this project applies a 100-meter grid offering a very detailed view on possible spatial patterns in the future. It distinguishes 17 land-use types, out of which the model allocates 11. The remaining six types, mainly related to infrastructure and water, have a pre-defined location that is not influenced by model-simulation. Their location is either a continuation of current land use or consists of pre-defined, approved plans, as is the case with, for example, long-planned railway links. For a more detailed description of the most recent model version and its calibration and validation the reader is referred to other publications (Loonen and Koomen 2007).

Unlike many other land-use models the objective of the *Land Use Scanner* is not to forecast the dimension of land-use change but rather to integrate and allocate future land-use demand from different sector-specific models or experts. Figure 1 presents the basic structure of the *Land Use Scanner* model. External regional projections of land-use change, which are usually referred to as demand or claims, are used as input for the model. These are land-use type specific and can be derived from, for example, sector-specific models of specialised institutes. The predicted land-use changes are considered as an additional claim for the different land-use types as compared with the present area in use for each land-use type. The total of the additional claim and the present area for each land-use function is allocated to individual grid-cells based on the suitability of the cell. This definition of local suitability may incorporate a large number of spatial datasets referring to the following aspects that are discussed below: *current land use*, *physical properties*, *operative policies* and *market forces* generally expressed in distance relations to nearby land-use functions.

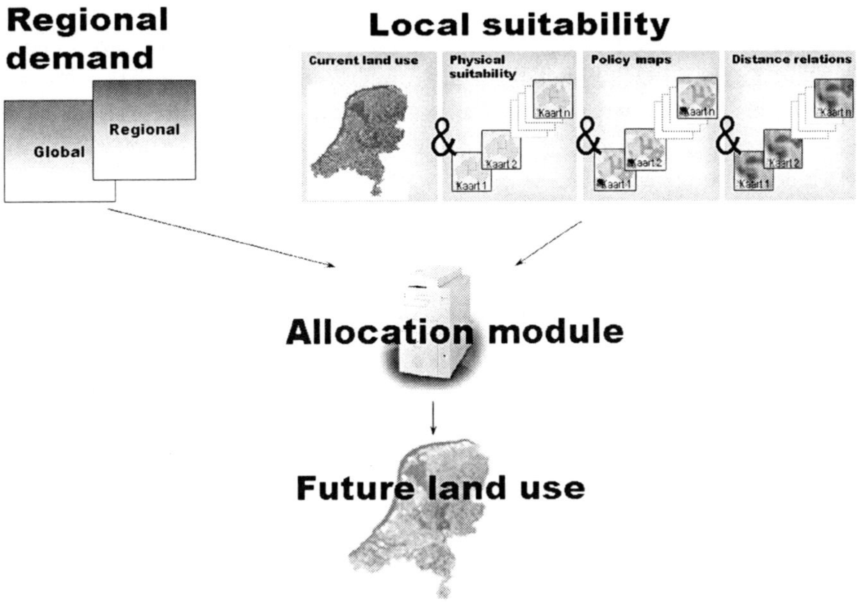


**Fig. 1:** Basic layout of the *Land Use Scanner*

*Current land use*, of course, offers the starting point in the simulation of future land use. It is thus an important ingredient in the specification of both the regional claim and the local suitability. Current land-use patterns are, however, not necessarily preserved in model simulations. This offers the advantage of having a large degree of freedom in generating future simulations according to scenario specifications, but calls for attention when current land-use patterns are likely to be preserved.

The *physical properties* of the land (e.g. soil type and groundwater level) are especially important for the suitability specification of agricultural land-use types as they directly influence possible yields. They are generally considered less important for urban functions, as the Netherlands have a long tradition of manipulating their natural conditions.

*Operative policies*, on the other hand, help steer Dutch land-use developments in many ways and are important components in the definition of suitability. The national nature development zones and the municipal urbanisation plans are examples of spatial policies that stimulate the allocation of certain types of land use. Various zoning laws related to, for example, water management and the preservation of landscape values offer restrictions on urban development.

The *market forces* that steer, for example, residential and commercial development are generally expressed in distance relations. Especially the proximity to railway stations, highway exits and airports are considered important factors that reflect the locational preferences of the actors that are active in urban development. Other factors that reflect such preferences are, for example, the number of urban facilities or the attractiveness of the surrounding landscape.

The selection of the appropriate factors for each of these components and their relative weighing is a crucial step in the definition of the suitability maps and determines, to a large extent, the simulation outcomes. The relative weights of the factors that describe the market forces and operative policies are normally assigned in such a way that they reflect the scenario storylines. The following sections describe the new allocation algorithm and the simulation method applied in this application.

## 2.1 New Discrete Allocation Model

The *Land Use Scanner*'s new discrete allocation model allocates equal units of land (cells) to those land-use types that have the highest suitability, taking into account the regional land-use claim. This discrete allocation problem is solved through a form linear programming. The solution of which is considered optimal when the sum of all suitability values corresponding to the allocated land use is maximal.

This allocation is subject to the following constraints:

- the amount of land allocated to a cell cannot be negative;
- in total only 1 hectare can be allocated to a cell;
- the total amount of land allocated to a specific land-use type in a region should be between the minimum and maximum claim for that region.

Mathematically we can formulate the allocation problem as:

$$\max_{X} \sum_{cj} S_{cj} X_{cj} \tag{1}$$

subject to:

$X_{cj} \geq 0$ for each $c$ and $j$

$\sum_{j} X_{cj} = 1$ for each $c$

$L_{jr} \leq \sum_{c} X_{cj} \leq H_{jr}$ for each $j$ and $r$ for which claims are specified in which:

$X_{cj}$ is the amount of land allocated to cell $c$ to be used for land-use type $j$

$S_{cj}$ is the suitability of cell $c$ for land-use type $j$

$L_{jr}$ is the minimum claim for land-use type $j$ in region $r$

$H_{jr}$ is the maximum claim for land-use type $j$ in region $r$

The regions for which the claims are specified may partially overlap, but for each land-use type $j$, a grid cell $c$ can only be related to one pair of minimum and maximum claims. Since all of these constraints relate $Xcj$ to one minimum claim, one maximum claim (which cannot be both binding) and one grid cell with a capacity of 1 hectare it follows that if all minimum and maximum claims are integers and feasible solutions exist, the set of optimal solutions is not empty and cornered by basic solutions in which each $Xcj$ is either 0 or 1 hectare.

The problem at hand is comparable to the well-known Hitchcock transportation problem that is common in transport-cost minimisation and, more specifically, the semi-assignment problem (Schrijver 2003; Volgenant 1996). The objective of the former problem is to find the optimal distribution in terms of minimised distribution costs of units of different homogenous goods from a set of origins to a set of destinations under constraints of a limited supply of goods, a fixed demand, and fixed transportation costs per unit for each origin-destination pair. The semi-assignment problem has the additional characteristic that all origin capacities are integer and the demand in each destination is one unit. Both are special cases of linear programming problems. The discrete allocation algorithm has two additional characteristics that are not incorporated in the classical semi-assignment problem formulation: (1) we can specify several, (partially) overlapping regions for the claims (although the regions of claims for the same land-use type must be disjoint); and (2) it is possible to apply distinct minimum and maximum claims.

Our problem, with its very large number of variables, calls for a specific, efficient algorithm. To improve the efficiency we apply a scaling procedure and, furthermore, use a threshold value. Scaling means that we use growing samples of cells in an iterative optimisation process that has proven to be fast (Tokuyama and Nakano 1995). For each sample an optimisation is performed. After each optimisation, the sample is enlarged and the shadow prices in the optimisation process are updated in such way that the (downscaled) regional constraints remain respected. To limit the number of alternatives under consideration we use a threshold value: only allocation choices that are potentially optimal are placed in the priority queues for each competing claim. An important advantage of the applied algorithm is that we are able to find an exact solution with a desktop PC (Pentium IV-2.8 GHz, 1 GB internal memory) within several minutes, provided that feasible solutions exist.

The constraints that are applied in the new discrete allocation model are equal to the demand and supply balancing factors applied in the original logit-based version of the model (Dekkers and Koomen 2007; Hilferink and Rietveld 1999). In fact, all the extensions to the original model related to the fixed location of certain land-use types, the use of regional claims, the incorporation of minimum/maximum claims and the monetary scaling of the suitability maps also apply for the discrete model. Similar to the original model, the applied optimisation algorithm aims to find shadow prices for the regional demand constraints that increase or decrease the suitability values, such that the allocation based on the adjusted suitability values corresponds to the regional claims. The main difference of the discrete model is that each cell only has one land-use type allocated, meaning that for each land-use type the share of occupation is zero or one. From a theoretical perspective the models are, however, equivalent when the scaling parameter that defines the importance of the suitability values would become infinitely large. In the latter case the continuous model would also strictly follow the suitability definition in the allocation and produce homogenous cells. This procedure is, however, theoretical and cannot be applied in the calculations due to computational limitations. An extensive calibration study of the two available allocation algorithms showed that both models provided very similar results given equal land-use claims and suitability definitions.

The new allocation model is similar to other well-known modelling frameworks, such as *MOLAND* (Engelen et al. 2004; 2007) and *CLUE-s* (Verburg et al. 2002; Verburg and Overmars 2007), in its flexibility, wide range of applications and use of high resolution, homogenous, grid cells that contain only one type of land use. The latter characteristic makes it easier to use the simulation results in subsequent impact assessments and visualisation efforts as is demonstrated in, for example, Van der Hoeven et al. (2008) and Rodríguez-Lloret et al. (2008). The current model differs from the *MOLAND* and *CLUE-s* frameworks in the sense that it is not dynamic but comparatively static, although recent applications now use one or more intermediate time steps to arrive at the final year. The *Land Use Scanner* is, furthermore, different in its modelling approach that has its roots in economics. This is exemplified by the monetary scaling of the suitability values reflecting their interpretation as initial bid prices and the use of shadow prices to solve the allocation problem. The model does not rely on neighbourhood interactions as does the cellular automata based *MOLAND* model, but it can incorporate neighbourhood characteristics in its suitability definition. The main difference with the hybrid *CLUE-s* model is that it does not use a conversion matrix that defines whether certain transitions are allowed. The model can thus easily be used to generate simulations of future land-use that deviate considerably from the current situation. This is useful in assessing the impact of novel policy options or visualising long-term scenarios that include relatively extreme societal changes.

## 2.2 Simulation Methodology

Land-use simulation starts by creating a 2015 land-use map from a 2000 base map. In this step current, explicit land-use plans, mainly taken from the new map of the Netherlands survey (NIROV 2005) are included in the simulation to represent autonomous developments. Based on this situation the simulations for 2040 are made for the different scenarios according to the specific assumptions and sector-related developments discussed in the following sections. The general scenario descriptions have been made spatially explicit with the help of several sector-specific models and a number of additional assumptions. These calculations have been performed by various specialised institutes: CPB and ABF have provided the expected amount of residential development (ABF 2006; CPB et al. 2006b), CPB has delivered the claim for industrial and commercial land use and office space (CPB 2002; CPB et al. 2006b) and LEI the projections for agricultural land-use changes (Helming, 2005). A concise description of the basic characteristics of the underlying regional models and a short discussion on the related quality issues is provided elsewhere (Dekkers and Koomen 2006). The claims were subsequently inserted in the *Land Use Scanner* model together with a spatially explicit translation of the scenario-assumptions into suitability maps.

The presented land-use simulations integrate expert-knowledge from various research institutes and disciplines and thus represent the best-educated guess regarding the possible spatial patterns. It should be noted however that the simulations are based on many assumptions. They can by no means be seen as exact predictions and should therefore not be treated like that.

# 3 Integrating Scenarios of Socio-Economic and Climate Change

This section presents the socio-economic and climatic dimension of the proposed scenarios. These scenarios are used in the LANDS project and related projects in the *Climate changes Spatial Planning* programme. We refer to these scenarios as the G and W scenarios, following the names of the original climate scenarios that are central to whole the research programme. To these scenarios we have added the socio-economic components of the scenario study *Prosperity, well being and quality of the living environment* (CPB et al. 2006b) according to the relations we have established in the previous chapter. A further addition consists of the regional projections of anticipated land-use change and the related land-use simulations. An extensive report on the presented scenarios is provided by Riedijk et al. (2007).

## 3.1 G scenarios

The G scenarios have the following general characteristics:

- Moderate population growth until 2010 and a slight decline thereafter;
- Modest economic growth;
- Trade blocks and taxes for protection of the environment;
- Emphasis on national environmental policy;
- Increased public environmental awareness;
- Extension of rail and motorway infrastructure.

This section first summarises the macro-economic changes and their regional impact (3.1.1). We then describe the more detailed land-use projections (3.1.2) and finally provide the climatic projections for the G and G+ scenarios (3.1.3).

### *3.1.1 Macro-Economic Changes and Regional Distribution*

The expected macro-economic changes in the G scenarios can be grouped into four themes. The main characteristics of each theme are provided below and in Table 1.

As compared to the 1971-2001 period, *population growth* is expected to slow down in the future, but considerable differences exist between the scenarios. In the G scenarios population increases until 2010, while it later decreases due to strict immigration policy and low birth rates, leading to approximately the same population size in 2040 as in 2000. In spite of this more or less stable population, the number of households will increase slightly due to ageing and increased prosperity.

**Table 1.** Overview of socio-economic, demographic and land-use change indicators for the G and W scenarios in 2040 (CPB et al. 2006b)

| Indicator | G scenarios | W scenarios |
|---|---|---|
| Population, in millions | 15.8 | 19.7 |
| Households, in millions | 7.0 | 10.1 |
| Labour participation (as % of professional population) | 68% | 74% |
| Ageing population (share 65+) | 25% | 23% |
| GDP per capita (2001=100) | 133 | 221 |
| Energy usage per capita, compared to 2002 | -5% | +30% |
| Car use in km, compared to 2002 | +5% | +40% |
| Goods traffic, ton per km, compared to 2002 | -5% | +120% |
| Total waste production | +11% | +100% |
| Land demand for waste disposal in other countries | +4% | +55% |
| Housing demand, compared to 2002: | | |
| - new single family dwellings, in millions | +0.3 | +1.9 |
| - new multiple family dwellings, in millions | +0.1 | +1.2 |
| Commercial area demand, compared to 2002: | | |
| - business parks | -3% | +43% |
| - offices | +1% | +34% |
| - informal locations | +7% | +46% |
| - sea harbour | -9% | +30% |
| Rural area development, compared to 2002: | | |
| - area agriculture | -10% | -15% |
| - area cultivation under glass | -45% | +60% |
| - dairy farming | -15% | +25% |
| - pigs | -55% | -5% |
| Demand for nature and recreation, compared to 2002: | | |
| - demand nature (x1000 ha) | +123 | +115 |
| - demand recreational green (x1000 ha) | +10 | +49 |
| - demand sports fields (x1000 ha) | +2 | +20 |
| Additional area for water retention, compared to 2002, | | |
| - near major rivers (x1000 ha) | +2 | +5 |
| - near urban extensions (x1000 ha) | +2 | +3 |

This scenario is characterised by a modest increase in *economic growth*, as is expressed in the growth of the gross domestic product (GDP) per capita by 1 % per year until 2020, and 0.5% between 2020 and 2040. This increase is partly due to the increasing labour productivity. Because of ageing, the potential labour force decreases severely. Labour supply decreases each year by nearly 0.5% due to a stagnating population growth and increasing ageing pressure. However, this is partly compensated by increased labour participation of women and elderly.

The physical environment is not only influenced by the magnitude of economic growth, but also by the performance of different *economic sec-*

*tors*. The sector-specific prospects influence to a large extent the actual spatial developments and emission rates. In line with recent trends, the sectors agriculture, industry, energy and building will decrease marginally, whereas commercial services, health care and public services will grow slightly in this scenario.

Both scenarios indicate that the *amount of land* used for sectors such as residences, industry, nature and recreation increases at the expense of agriculture. Figure 2 depicts these national changes and especially shows that the increase in residential land is more moderate in the G scenarios. However, urban sprawl is expected to continue due to the continuing demand for rural residences and green, spacious urban housing. The extent of this sprawl is limited however, due to population growth coming to a halt in all regions.

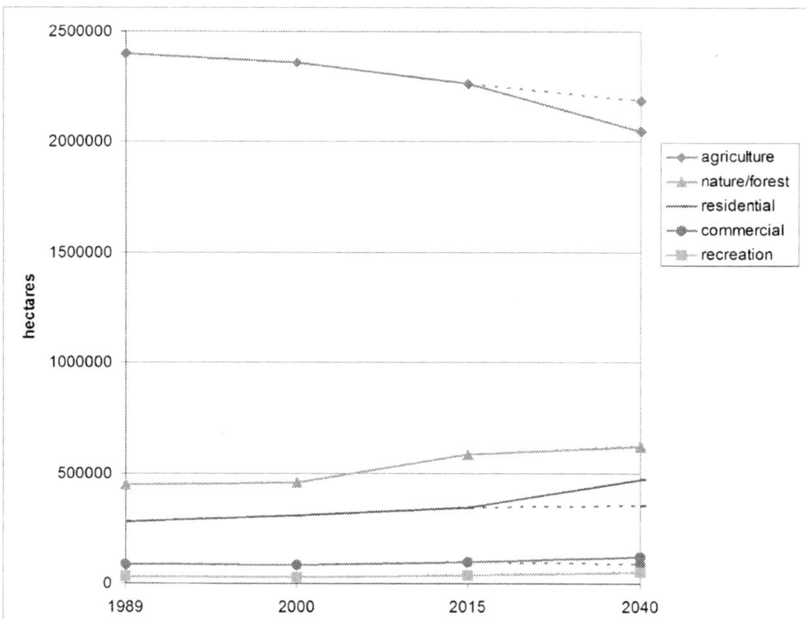


**Fig. 2.** Land use in the Netherlands 1990-2040. The land use in 2040 is shown for both the G (dotted line) and W (solid line) scenarios

The *regional development* in employment is strongly related to the provision of consumer services. Employment in this type of services (banks, care institutions, retail etc.) increases mostly in the Intermediate Zone. A further decrease is expected in agriculture and industry, mostly in the Randstad. Unemployment rates are slightly higher in the Periphery, leading to a marginal migration towards the Intermediate Zone and Randstad in the long term. Furthermore, the willingness to commute will increase. Most commuter traffic takes place within the so-called COROP regions. Based on the aforementioned assumptions related to macro-economic changes and their regional impact a number of developments are anticipated in specific socio-economic sectors (CPB et al. 2006b). These are included in the *Land Use Scanner* as land-use specific regional claims and suitability definitions. The most important anticipated developments are also listed in Table 1.

### *3.1.2 Land-Use Simulations*

The resulting simulation (Figure 3 at left) shows a modest increase in residential areas, despite the fairly limited population growth. This growth can be largely ascribed to the minor increase in households and residential preferences for a rural living environment. Urban growth is most notable in the central and western part of the Netherlands. Arable farming diminishes strongly in this scenario. Greenhouse horticulture disappears in many areas; especially from its current stronghold south of The Hague. From the map it can be concluded that existing nature areas are enlarged in a number of cases. New nature areas are developed along the rivers Waal, Rhine, Meuse and IJssel. Clusters of outdoor recreation arise in attractive landscapes, in particular in the northern and western part of the Netherlands.

### *3.1.3 Climate Dimension*

The G scenarios are characterised by a 1 degree Celsius temperature rise between 1990 and 2050. Due to a change in atmospheric circulation the winters (December, January and February) will be mild and wet because of increased westerly winds in the G+ scenario. Summers (June, July and August) will be warmer and dryer because of increased easterly winds in this scenario. Table 2 provides more information on the climate dimension.

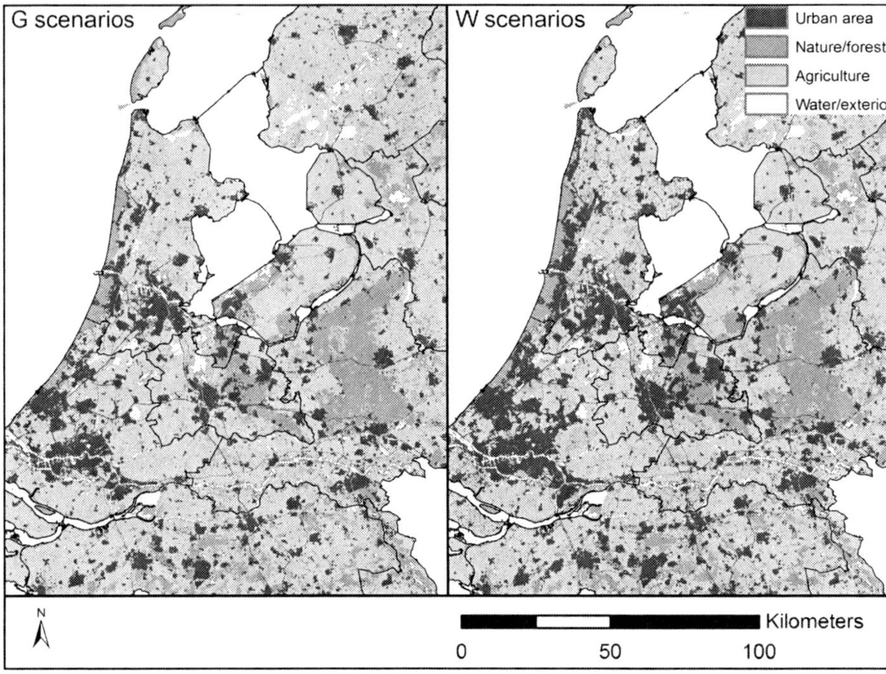


**Fig. 3.** Simulated land use for the G and W scenarios in 2040

**Table 2.** Dutch climate change scenarios for 2050 relative to 1990 (Van den Hurk et al. 2006)

| | G | G+ | W | W+ |
|---|---|---|---|---|
| Absolute sea level rise (cm) | 15-25 | 15-25 | 20-35 | 20-35 |
| *Winter* | | | | |
| Mean temperature | +0.9°C | +1.1°C | +1.8°C | +2.3°C |
| Yearly coldest day | +1.0°C | +1.5°C | +2.1°C | +2.9°C |
| Mean precipitation | +4% | +7% | +7% | +1.4% |
| Wet day frequency | 0% | +1% | 0% | +2% |
| 10 yr return level daily precipitation sum | +4% | +6% | +8% | +12% |
| Yearly maximum daily mean wind speed | 0% | +2% | -1% | +4% |
| *Summer* | | | | |
| Mean temperature | +0.9°C | +1.4°C | +1.7°C | +2.8°C |
| Yearly coldest day | +1.0°C | +1.9°C | +2.1°C | +3.8°C |
| Mean precipitation | +3% | -10% | 6% | -19% |
| Wet day frequency | -2% | -10% | -3% | -19% |
| 10 yr return level daily precipitation sum | +13% | +5% | +27% | +10% |
| Potential evaporation | +3% | +8% | +7% | +15% |

## 3.2 W scenarios

The W scenarios have the following general characteristics:

- High population growth;
- High economic growth;
- Expansion of the EU to the east;
- Global free trade without political integration;
- No initiatives on international environmental agreements;
- Extension of rail and motorway infrastructure

This section first summarises the macro-economic changes and their regional distribution (3.2.1). We then describe the local land-use projections (3.2.2) and finally provide the climatic projections for the W and W+ scenarios (3.2.3).

### *3.2.1 Macro-Economic Changes and Regional Distribution*

The anticipated macro-economic developments for the W scenarios are provided below and in Table 1.

In this scenario, *population growth* continues because of a high birth rate and an open immigration policy that brings many new people to the country. The total population in 2040 is estimated at 20 million people. The number of households increases even more as a result of further individualisation and ageing.

*Economic growth* will lead to a doubling of the Gross Domestic Product (GDP) per capita as compared to 2001. The working population will decrease due to ageing, but this will be counterbalanced by the increased labour participation of women and elderly (50+) people.

A further shift towards a service economy is the main underlying change in the structure of the *economic sectors* in this scenario. Nevertheless, all economic sectors show an increase in the volume of production in this particular scenario. Also agricultural production increases significantly. The total employment in this sector, however, decreases. The same development is seen in the industrial sector. The demand for health care increases significantly, as well as employment in this sector.

A strong increase in the *amount of land* for urban functions is expected. Urban sprawl is anticipated to continue due to the need for rural living. A detailed account of the projected regional land-use changes is provided in Appendix 2.

The trends in *regional development* are comparable to the G scenarios, but their magnitude increases. The well-educated immigrants that are expected to come to work in the country will spread across the country. They will, relatively, reside more in the Intermediate Zone where employment is growing strongly. This domestic migration process will initially focus on the Intermediate Zone but later also spread to the Periphery. Regional employment development is influenced by an increase in consumer services like banks, care institutions and retail. Employment in this sector increases most in the Intermediate Zone. Particularly in the Randstad employment in agriculture and industry will decreases strongly. Just as in the G scenarios, unemployment rates are slightly higher in the Periphery, leading to a marginal, overall migration towards the Intermediate Zone and Randstad in the long term. Furthermore, the willingness to commute will increase, with most commuter traffic taking place within COROP regions.

### 3.2.2 Land-Use Simulations

The most striking land-use change in the W scenarios is the strong increase in urban land use (Figure 3 at right). Residential land use expands substantially around the larger cities in the Randstad, as well as around many smaller villages in the rural areas. Commercial land use also increases strongly. This increase takes place in the Randstad and bordering parts of the Intermediate Zone. The urbanisation causes a serious deterioration of the quality and openness of the landscape, also in the designated national landscapes.

The appearance of the rural areas will also change. Arable farming will, to a large extent, disappear and be replaced by grassland for dairy farming. Capital-intensive forms of farming will also demand more land. Greenhouse horticulture expands around the main ports of Rotterdam and Schiphol. New nature will mainly be developed along the major rivers, where strong constraints are imposed on the expansion of urban functions and capital-intensive forms of farming. Recreation also claims more space, especially in the attractive small-scale landscapes of, for example, the Achterhoek. A limited description of the land-use simulations for the W scenarios has also been published in the 6th National Environmental Outlook (MNP 2006).

### 3.2.3 Climate Dimension

The W scenarios are characterised by a 2 degrees Celsius temperature rise between 1990 and 2050. Due to a change in atmospheric circulation the winters will be mild and wet because of increased westerly winds in the W+ scenario. Summers will be warmer and dryer because of increased easterly winds in this scenario. More details are provided in Table 2.

## 4 Discussion

The current paper describes a combination of existing scenarios of climate and socio-economic change as a starting point for further research into the adaptation and mitigation measures that may be needed to face the future in the Netherlands. The selected scenarios are fed into the *Land Use Scanner* model that simulates future land-use patterns. The G scenarios with a moderate temperature rise are combined with a socio-economic scenario that combines an orientation on national sovereignty with clear public responsibilities. The related land-use simulation for 2040 depicts a limited amount of urban growth, predominantly in the central and western part of the country. The W scenarios with a higher temperature rise are combined with socio-economic conditions that favour international cooperation and private responsibilities. These simulations show a much stronger urban growth, especially of rural types of residences in attractive landscapes.

The study presented in this paper is carried out within the larger *Climate changes Spatial Planning* research programme. As such the results are closely related to a number of projects within this programme. These are foremost two climate scenarios projects and the *Socio-economic scenarios for climate change assessments* Integration project. The former projects have provided the climate scenarios described in this paper and continue to elaborate on specific climatologic variables. The latter project discusses possible combinations of socio-economic and climate scenarios in a similar fashion as the current paper. The project aims to provide researchers with a set of components to analyse possible adaptation and mitigation measures. The reader is referred to their upcoming publications for further information.

For the LANDS project, that initiated this study, the presented results are a first step towards the creation of an integrated outlook on a climate-proof future for the Netherlands. To achieve this ambitious objective the project relies on the integration of the results from many other projects in the *Climate changes Spatial Planning* research programme that analyse and propose possible adaptation and mitigation measures for different societal sectors. The relation of the LANDS project and its current scenario study with these other projects is depicted in Figure 4. The figure shows that the integrated climate and socio-economic scenarios, described in this paper, are not only intended to feed into the *Land Use Scanner* model for the simulation of future land-use patterns. The scenarios also provide the starting point for the different sector-specific studies. The possible impact on water management and nature is studied as part of the LANDS project, as well as in several other climate-related projects focussing on flood risk, adaptation measures in the Rhine basin, the spatial distribution of vegetation, optimizing the nature conservation potential. Other projects dealing with possible climate-adaptation measures for, amongst others, agriculture, transport, specific local or regional circumstances and fen meadows can also use the scenario assumptions and simulated land-use patterns as a reference point for their analyses. Subsequently, the results of these projects will be fed into the *Land Use Scanner* to simulate adjusted land-use patterns that take the possible impact of climate change into account. To quantify the differences between the different simulations, indicators are being developed that characterise the outcomes at the local, regional and national level. Specific attention will be paid to evaluation measures that identify potential conflicts or synergies between different land-use types.

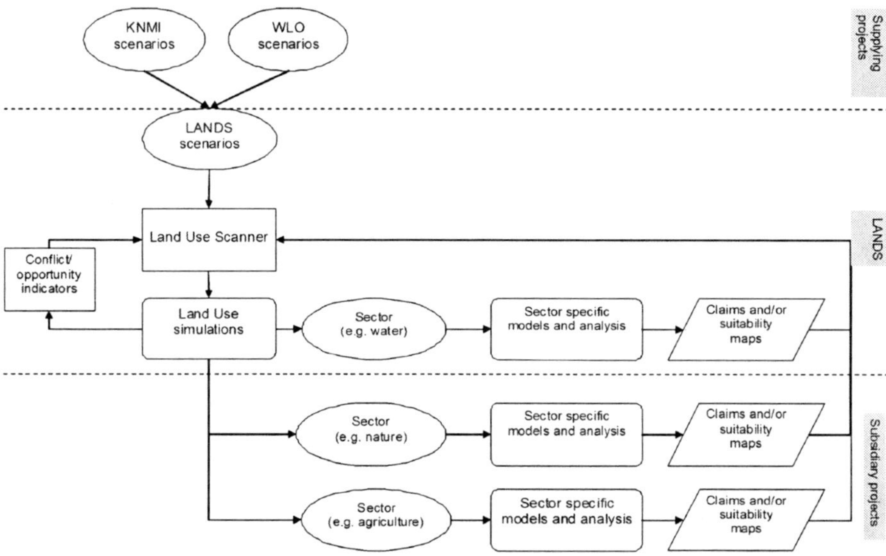


**Fig. 4.** The LANDS project scheme

To exemplify this approach we present an assessment of the potential water management problems that may arise from new urban development. Figure 5 shows the simulated urban extensions on a detailed elevation map of the country. It marks the new urban areas, lying well below sea-level, that are likely to face serious problems in maintaining a dry surface with the anticipated increases in sea-level and precipitation. The figure also indicates those areas in the immediate vicinity of the major rivers that are expected to receive higher peak discharges. These may experience an increased risk on flooding and can be appointed as additional water retention areas. Both types of areas have in common that the timely implementation of building restrictions may prevent costly reallocation programs at a later stage.

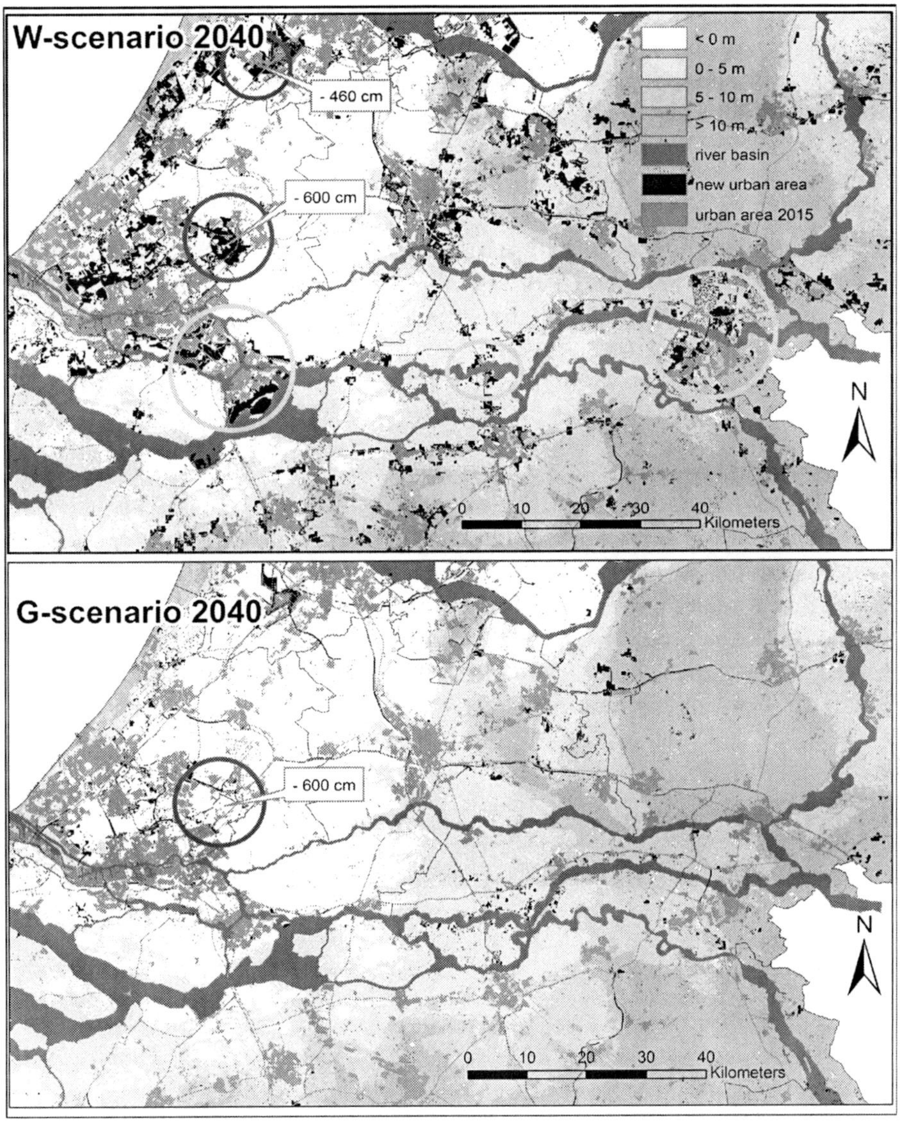


**Fig. 5.** Simulated urban extensions in 2040; the circles indicate areas facing possible water management problems near the major rivers (light circles) and in low-lying areas (dark circles)

In order to integrate the outcomes of the different, sector-oriented projects into coherent maps of future land use, as is one of the ultimate goals of the *Climate changes Spatial Planning* research programme, it is essential that they all start from the same assumptions regarding the possible future developments. To be able to actually use the output of the individual projects it is needed that they provide quantitative information related to the regional land-use demand (claim) and location preferences (suitability) for their specific sector. This information can then be included in the *Land Use Scanner* model to simulate new land-use projections.

The preliminary attempts to integrate results from the flood risk and agriculture project indicate the importance of actually starting from the same assumptions (see Van der Hoeven et al. 2008). The diverging character of their respective research results makes clear that substantial post-processing and intensive cooperation are essential to successfully integrate the various results.

## Acknowledgements

The authors are grateful to The Dutch National research program *Climate changes spatial planning* for financing the presented research. They are, furthermore, thankful to numerous individuals at the Vrije Universiteit Amsterdam and the Netherlands Environmental Assessment Agency (MNP), KNMI and other research institutes that contributed to the scenario definition, land-use simulations, and preceding reports.

## References

ABF (2006) Achtergrondrapport bevolking. ABF research, Delft.

Beinat, E. and Nijkamp, P. (1998) Multicriteria analysis for land use management. Kluwer, Dordrecht.

Borsboom-van Beurden, J.A.M., Bakema, A. and Tijbosch, H. (2007) A land-use modelling system for environmental impact assessment; Recent applications of the LUMOS toolbox. In: Koomen, E., Stillwell, J., Bakema, A. and Scholten, H.J. (eds.). Simulating land-use change Kluwer Academic Publishers, New York.

Borsboom-van Beurden, J.A.M., Boersma, W.T., Bouwman, A.A., Crommentuijn, L.E.M., Dekkers, J.E.C. and Koomen, E. (2005) Ruimtelijke Beelden; Visualisatie van een veranderd Nederland in 2030. RIVM report 550016003, Milieu- en Natuurplanbureau, Bilthoven.

Commissie Waterbeheer 21e eeuw (2000) Waterbeleid voor de 21ste eeuw; geef water de ruimte en de aandacht die het verdient.

CPB (2002) De BLM: opzet en recente aanpassingen. Centraal Planbureau, Den Haag.

CPB, MNP and RPB (2006a) Welvaart en Leefomgeving. Achtergronddocument. Centraal Planbureau, Milieu- en Natuurplanbureau en Ruimtelijk Planbureau, Den Haag.

CPB, MNP and RPB (2006b) Welvaart en Leefomgeving. Een scenariostudie voor Nederland in 2040. Centraal Planbureau, Milieu- en Natuurplanbureau en Ruimtelijk Planbureau, Den Haag.

Dale, V.H. (1997) The relationship between land-use change and climate change. Ecological Applications 7(3): 753-769.

Dekkers, J.E.C. and Koomen, E. (2006) De rol van sectorale inputmodellen in ruimtegebruiksimulatie; Onderzoek naar de modellenketen voor de LUMOS toolbox. SPINlab research memorandum SL-05. Vrije Universiteit Amsterdam, Amsterdam.

Dekkers, J.E.C. and Koomen, E. (2007) Land-use simulation for water management: application of the Land Use Scanner model in two large-scale scenario-studies. Chapter 20. In: Koomen, E., Stillwell, J., Bakema, A. and Scholten, H.J. (eds.). Modelling land-use change; progress and applications Springer, Dordrecht, pp. 355-373.

Engelen, G., Lavalle, C., Barredo, J.I., Van Der Meulen, M. and White, R. (2007) The MOLAND modelling framework for urban and regional land-use dynamics. Chapter 17 in: Koomen, E., Stillwell, J., Bakema, A. and Scholten, H.J. (eds.) Modelling land-use change; progress and applications, Springer, Dordrecht, pp. 297-319.

Engelen, G., White, R., Uljee, I., Hagen, A., van Loon, J., van der Meulen, M. and Hurkens, J. (2004) The Moland Model for Urban and Regional Growth, Research Institute for Knowledge Systems, Maastricht, The Netherlands.

Helming, J. (2005) A model of Dutch agriculture based on Positive Mathematical Programming with regional and environmental applications, PhD thesis. Wageningen University, Wageningen.

Hilferink, M. and Rietveld, P. (1999) Land Use Scanner: An integrated GIS based model for long term projections of land use in urban and rural areas. Journal of Geographical Systems 1(2): 155-177.

IPCC (2001) Impacts, Adaptation & Vulnerability. Contribution of Working Group II to the Third Assessment Report of the Intergrovernmental Panel on Climate Change (IPCC). Cambridge University Press, Cambridge, UK.

Koomen, E., Kuhlman, T., Groen, J. and Bouwman, A.A. (2005) Simulating the future of agricultural land use in the Netherlands. Tijdschrift voor Economische en Sociale Geografie 96(2): 218-224.

Loonen, W. and Koomen, E. (2007) Calibration and validation of the Land Use Scanner allocation algorithms. MNP report, Milieu- en Natuurplanbureau, Bilthoven.

MNP (2006) Nationale Milieuverkenning 6: 2006-2040. MNP-rapportnr. 500085001, Milieu- en Natuurplanbureau, Bilthoven.

NIROV (2005) Nieuwe Kaart, Nieuwe Ruimte: Plannen voor Nederland in 2015. Nirov, Den Haag.

Parry, M.L. (2000a) Assessment of Potential Effects and Adaptations for Climate Change in Europe: The Europe ACACIA Project. Jackson Environmental Institute, University of East Anglia, Norwich, UK.

Parry, M.L. (2000b) Scenarios for climate impact and adaptation assessment. Global Environmental Change 12: 149-153.

Riedijk, A., R. van Wilgenburg and E. Koomen (2007) Integrated scenarios of socio-economic and climate change; a framework for the 'Climate changes spatial planning' program, Vrije Universiteit Amsterdam.

Rodríguez-Lloret, J., N. Omtzigt, E. Koomen and F.S. de Blois (2008) 3D visualisations in simulations of future land use: exploring the possibilities of new, standard visualisation tools, Journal of digital Earth 1(1) (forthcoming).

Schrijver, A. (2003) Combinatorial optimization - polyhedra and efficiency. Springer-Verlag, Berlin.

Tokuyama, T. and Nakano, J. (1995) Efficient algorithms for the Hitchcock transportation problem. SIAM Journal on Computing 24(3): 563-578.

V&W and VROM (2006) Nota Mobiliteit. Ministerie van Verkeer en Waterstaat/Ministerie van Volkshuisvesting, Ruimtelijke Ordening en Milieubeheer, Den Haag.

Van den Hurk, B., Klein Tank, A., Lenderink, G., van Ulden, A., van Oldenborgh, G.J., Katsman, C., van den Brink, H., Bessembinder, J., Hazeleger, W. and Drijfhout, S. (2006) KNMI Climate Change Scenarios 2006 for the Netherlands. KNMI Scientific Report WR 2006-01, KNMI, De Bilt.

Van der Hoeven, N., Aerts, J., Van der Klis, H. and Koomen, E. (2008) An Integrated Discussion Support System (DSS) for New Dutch Flood Risk Management Strategies. In: Geertman, S. and Stillwell, J. Planning Support Systems: Best Practices and New Methods, Springer, Berlin (forthcoming).

Verbeek, K. (2003) De toestand van het klimaat 2003. KNMI, De Bilt.

Verburg, P.H. and Overmars, K.P. (2007) Dynamic simulation of land-use change trajectories with the CLUE-s model. Chapter 18 in: Koomen, E., Stillwell, J., Bakema, A. and Scholten, H.J. (eds.) Modelling land-use change; progress and applications, Springer, Dordrecht, pp. 321-335.

Verburg, P.H., Soepboer, W., Limpiada, R., Espaldon, M.V.O., Sharifa, M. and Veldkamp, A. (2002) Land use change modelling at the regional scale: the CLUE-S model, Environmental Management, 30: 391–405.

Volgenant, A. (1996) Linear and semi-assignment problems: a core oriented approach. Computers and Operations Research 23(10): 917-932.

Watson, R.T., Noble, I.R., Bolin, B., Ravindranath, N.H.V.D.J. and Dokken, D.J.E. (2006) Land Use, Land-Use Change, and Forestry. A Special Report of the Intergovernmental Panel on Climatic Change. Cambridge University Press, Cambridge.

# Quantifying and Analysing Neighbourhood Characteristics Supporting Urban Land-Use Modelling

Henning Sten Hansen

Aalborg University, Department of Development and Planning
National Environmental Research Institute, Roskilde
Fibigerstræde 11, DK-9200 Aalborg East,
Phone: +45 46 30 18 07, Fax : +45 46 30 12 12
HSH@LAND.AAU.DK

**Abstract.** Land-use modelling and spatial scenarios have gained increased attention as a means to meet the challenge of reducing uncertainty in the spatial planning and decision-making. Several organisations have developed software for land-use modelling. Many of the recent modelling efforts incorporate cellular automata (CA) to accomplish spatially explicit land-use change modelling. Spatial interaction between neighbour land-uses is an important component in urban cellular automata. Nevertheless, this component is calibrated through trial-and-error estimation. The aim of the current research project has been to quantify and analyse land-use neighbourhood characteristics and impart useful information for cell based land-use modelling. The results of our research is a major step forward, because we have estimated rules for neighbourhood interaction from really observed land-use changes at a yearly basis. This higher temporal granularity gives a more realistic foundation for estimating neighbourhood interaction rules to be applied in for example land-use cellular automata.

**Keywords:** cellular automata, land use modelling, spatial patterns, spatial planning

## 1 Introduction

Cities are complex systems that arise through mutual interactions between many factors. As a result, the mechanism of urban growth and its interaction with social, economic and environmental systems are still poorly understood. Policy makers and planners often face tremendous difficulties in decision making with a lack of vision into the future of urban growth.

Land-use modelling and spatial scenarios have gained increased attention as a means to meet the challenge of reducing uncertainty in the decision-making. Several organisations have developed software for land-use modelling. Traditionally, Europe and North America have been the frames for developing systems and software for land-use modelling, but also India and China are putting resources into the development and use of land-use modelling.

Many of the recent modelling efforts incorporate cellular automata (CA) to accomplish spatially explicit land-use change modelling. In contrast to top-down large-scale urban models, models based on cellular automata simulate land-use change processes using local neighbourhood interactions, from which complex urban patterns emerge (Batty 1998). The neighbourhood rules specify how the combined effects of spatial externalities, in the sense of mutual attraction and repulsion of land-use types, work out over distance. Spatial externalities are the effects of one land-use on another. One example could be the noise and smoke from an industrial plant causing inconvenience and even health problems in a nearby residential area. The rules, as applied in practice, are subject to serious limitations, the main one being their lack of theoretical foundation and empirical validation. This has lead to the conclusion that cellular automata based land-use models are largely driven by technology and in a less degree on theories and empirical evidences (O'Sullivan et al. 2000). Unfortunately, we cannot just ignore criticism like this. To be able to justify the application of land-use models for policy support, planners and policy makers must trust the theoretical foundation behind all land-use modelling. Otherwise, there will be no future for land-use scenarios. Luckily, we have seen some recent efforts dealing with the issue of improving the way to define the rules for spatial interaction in land-use modelling (Verburg et al. 2004; Hagoort et al. 2008), and the current paper is just one step further.

The aim of the current research has been to quantify and analyse land-use neighbourhood characteristics and impart useful information for cell based land-use modelling. The paper is divided into 5 parts. After the introduction follows a discussion of spatial interaction in land-use dynamics and modelling. Then in the third section we present the methods and data used in the current research. The fourth section presents the results of the efforts and discusses the potential implications for urban land-use modelling. The paper ends with some conclusions and an outline for subsequent work.

## 2 Spatial Interaction in Land-Use Dynamics

Models of land-use change can address two separate questions: a) where are land-use changes likely to take place – i.e. the location of change); b) and at what rates are changes likely to progress – i.e. the quantity of change (Veldkamp et al. 2001). A prerequisite to the development of realistic land-use simulation models is the identification of the most important drivers of change, and how to represent these drivers in a model. The theoretical understanding of urban land-use patterns started nearly one hundred years ago with Burgess' studies of Chicago leading to the so-called Concentric Zone Model. This model was later on revised by Hoyt, who emphasised a sectoral structure of urban land-use, and Harris and Ullman with the so-called multiple nuclei model. Although these models gave some theoretical insight in the urban land-use structure, the patterns in real world cities are much more complicated and with huge amounts of varieties.

In cellular automata models complicated and realistic patterns and processes emerge from simple rules. The simplicity of the cellular automata approach offers many advantages for urban simulation (Torrens et al. 2000). Cellular models of urban development that are based on classical cellular automata have four main components. Firstly, the simulation process usually operates on a uniform lattice of cells in a 2-dimensional space. Secondly, each cell in the lattice can adopt only *one* state of a set of possible states defined by the system being modelled. Thirdly, the configuration of the neighbourhood of a cell defines the current state of the cell. In classical cellular automata, the neighbourhood is usually the four or eight nearest neighbours. Fourthly, there is a set of transition rules that govern the types of changes in cell states in relation to the neighbourhood configuration. Although urban CA models are not cellular automata strictly speaking, they are generally considered to belong to the family of CA based models. Clearly, such a limited range of action spaces is too limiting for urban applications, and most CA based urban models have modified the neighbourhood parameters. Additionally, distance decay effects have been introduced, often as weights applied to neighbourhoods in transition calculations. Also, neighbourhoods have been extended to comprise larger spaces.

Thus an important component in cellular automata based urban land-use simulation models is the spatial interaction between neighbouring land-use types. The neighbourhood interaction is often addressed based on the notion that urban development can be regarded as a self-organising system in which natural constraints and land-use policies mitigate the way in which local decision-making processes produce macroscopic urban form. The so-called First Law of Geography – 'Everything is related to everything else, but near things are more related than distant things' - is fundamental for every discussion of neighbourhood interaction in the spatial domain. Land-use patterns generally exhibit spatial autocorrelation. Residential areas are often clustered having a positive spatial autocorrelation, whereas other land-uses prefer to be located at some distance from each other – e.g. an airport and residential areas. In Denmark like many other countries in Northern Europe, the spatial policies during more than 30 years have prevented uncontrolled urban sprawl. We can therefore expect that most of the urban expansion will take place along the border of existing cities.

Different processes at various scales can explain the importance of neighbourhood interaction. At large scale, simple mechanisms for economic interaction between locations were provided by Walter Christaller's Central Place Theory (Christaller 1933), that describes the uniform pattern of towns and cities in space as a function of the distance that consumers in the surrounding region travel to the nearest facilities. Spatial interaction at lower scales between the location of facilities, residential areas and industries has been given more attention by Krugman (Krugmanr 1999), who explains the spatial interaction by a number of factors that either causes concentration of urban functions (centripetal forces: economies of scale, localised knowledge spill-over, thick labour markets) and others that lead to a spatial spread of urban functions (centrifugal forces: congestion, land rents, factor immobility etc.).

The neighbourhood effect represents the attraction (pull) and repulsion (push) effects of the various land uses and land covers within the neighbourhood. For each urban land-use function a set of rules determines if is attracted or repelled by the other land-uses, and the strength of the interaction is dependent on the distance. Basically six different forms of the distance functions can be identified (see fig. 1). Additionally, we could add a seventh indifferent distance decay function, but in practice this is neglected.

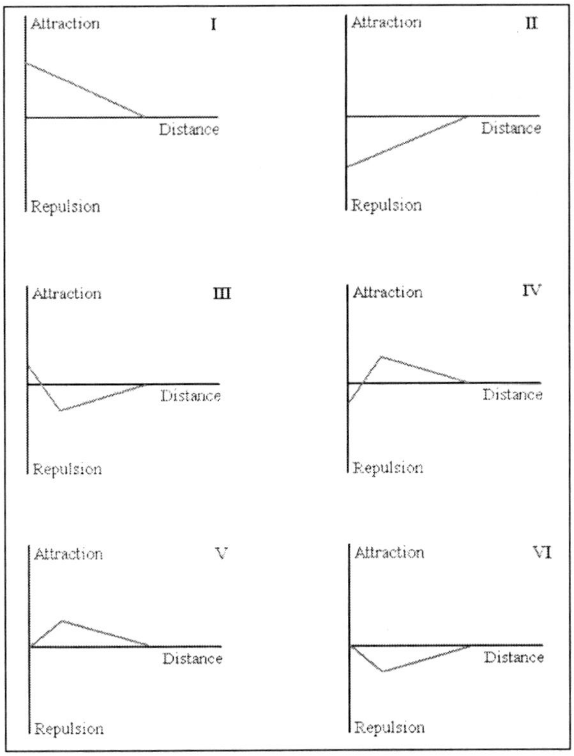


**Fig. 1.** Typology of distance decay functions after Hagoort (Hagoort et al. 2008).

## 3 Methods and Data

As stated above we have used an empirical spatial metric approach to quantify and analyse neighbourhood interaction in urban land-use dynamics. Process-based relationships may be easier to discern at fine scales, where land-use changes can be explicitly linked to the activity of agents in the landscape. These relationships are less apparent when the emergent land-use-change patterns are observed over coarser scales where other processes, such as environmental or macro-economic factors, become dominant (Kok et al. 2001). Study areas with large geographic extent typically have a coarse spatial resolution, due to data and processing costs. The drawback is that some patterns visible at higher resolutions are not appreciable. A minor geographic extent will permit a finer spatial resolution, but in this situation the study area is taken out of the larger context, which it belongs to. Thus decisions concerning extent and spatial resolution have to be balanced against each other.

The Region of Northern Jutland was identified as study area to confirm the approach, as this area possesses a wide range of land-uses. Furthermore, we have had easy and free access to abundant dataset of land-use. Currently, the cell space is a rectangular 2-dimensional lattice of square cells each representing an area of 1 ha – i.e. the edges of each cell are 100 meter. The grid contains 1370 rows and 1905 columns – altogether more than 2.6 million cells. The 100-meter cell size is a reasonable choice aiming at homogeneity concerning land-use and simultaneously reducing the number of holes in a continuous urban area.

## 3.1 Estimating Neighbourhood Characteristics

The many land-use types in principle lead to an incalculable number of neighbourhood relationships and estimating the distance decay functions is not a straightforward process. However, this huge challenge must be taken up, and basically, there exist three different approaches to estimate the neighbourhood rules: A) calibrate the neighbourhood rules using mathematical and statistical methods; B) estimate the neighbourhood rules based on experience and knowledge; C) derive the neighbourhood rules from empirical analysis of observed land-use changes. Below follows a short description of each approach:

A) Calibration - some authors Straatman et al. (2004) have developed advanced numerical calibration methods, but the huge number of coefficients to be estimated means that the obtained solution is not necessarily unique – rather quite the reverse. Besides, this approach has too little emphasis on the theories and processes behind urban development.

B) Knowledge based - opposed to this pure statistical method Hagoort et al. (2008) have tried to estimate neighbourhood rules based on semi-structured expert interviews. More than 30 experts in land-use and spatial planning were involved, and the derived neighbour rules based on experience and knowledge might probably give better results than the numerical calibration method.

C) Empirical - using spatial metrics Verburg et al. (2004) have made an empirical analysis of real changes in the Dutch land-use and derived neighbourhood characteristics to assist the definition of neighbourhood rules for spatial interaction. The biggest advantage of this approach is that it is based on observed land-use changes, while the biggest disadvantage is the requirement of land-use data with a high temporal resolution. Nevertheless, we believe that the spatial metrics approach will give the most reliable results, because it is empirical and non-biased. Therefore this methodology has been used in current research, and below we give a more detailed description of the spatial metrics applied.

### 3.2 Spatial Metrics for Land-Use Changes

The analysis of spatial structures and patterns are central to understand land-use dynamics. Under the name of landscape metrics, several spatial metrics are already commonly used to quantify the shape and pattern of vegetation in natural landscapes. Many landscape metrics were developed in the late 1980s and were partly based on information theory by Shannon & Weaver (Shannon et al. 1964) and fractal geometry by Mandelbrot (Mandelbrot 1983). Landscape metrics are used to quantify the spatial homogeneity / heterogeneity for a specific landscape property of interest, for example landscape fragmentation.

Verburg et al. (2004) has defined a landscape metric – the so-called *mean enrichment factor* - which is very appropriate for quantifying and analysing neighbourhood characteristics. The enrichment factor is a measure that characterises the over- or under-representation of different land-use types in the neighbourhood of a specific grid cell. To measure this over- or under-representation, it compares the amount of occurrences of a particular land-use type in the vicinity of a specific location as relative to the volume of occurrences of that land-use type in the study area in total.

When the proportion of a land-use type in a neighbourhood equals the national average, the neighbourhood possesses an enrichment factor of 1 for that land-use type. If the neighbourhood of a specific location (cell) consists of 20% summer cottages, whereas the proportion of summer cottages in the study area as a whole in total is 5%, we can characterise the neighbourhood by an enrichment factor of 4 for summer cottages. Contrary an under representation of a certain land-use type in the neighbourhood will result in an enrichment factor between 0 and 1.

The formulas for the enrichment factor are specified in Verburg et al. (2004) as follows:

$$F_{i,k,d} = \frac{n_{k,i,d} / n_{d,i}}{N_k / N} \qquad (1)$$

Where $F_{i,k,d}$ characterises the enrichment of neighbourhood $d$ of location $i$ with land-use type $k$. The shape of the neighbourhood and distance of the neighbourhood from the central grid cell $i$ are identified by $d$. The average neighbourhood characteristic for a particular land-use type $l$ ($F'_{l,k,d}$) is calculated by taking the average of the enrichment factors for all grid cells belonging to a certain land-use type $l$, following:

$$F'_{l,k,d} = \frac{1}{N}\sum_{i\in L} F_{i,k,d} \tag{2}$$

where $L$ is the set of all locations with land-use type $l$, and $N_l$ is the total number of cells belonging to this set.

Based on these formulas Verburg et al. calculated enrichment factors for land-use changes in the Netherlands. The values for enrichment factors where calculated for land-use changes, which have taken place between 1989 and 1996. They found that the new developments are near to already existing occurrences of the same land-use type. Furthermore, they found that new residential areas are located in the vicinity of existing residential, industrial and recreational areas, indicating that urban expansion is much more important than new isolated urban development. The main weakness of the results obtained by Verburg et al. is that the changed land-uses cover a time period of seven years. Particularly in rapidly developing areas this will give less trustworthy results, because the near neighbours of the most recent urbanised cells will be e.g. arable land – and not an urban land-use. However, the required data seems not to be available.

## 3.3 Data Layers

To quantify and analyse the neighbourhood interaction between various land-use classes require reliable high quality land-use data for several *consecutive* years. Only in this way you can make a real analysis of the neighbourhood preferences of new land-uses. The basic source for land-use information in the current research is the vector versions of CORINE land-cover for the years 1990 and 2000, but the temporal resolution with a time interval of ten years is clearly insufficient. Furthermore, the level of thematic detail in CORINE land-cover does not satisfy our requirements for the built-up areas. Therefore we introduced two auxiliary data sets. First – and most important – we used the Danish Building and Housing Register, which contains detailed information about each building in Denmark, and this register has been in operation for about 30 years. The Danish local authorities maintain the Building & Housing Register as a part of the general administration of real estates and buildings.

The Building & Housing Register (Daugbjerg et al. 2000) uses the following 3 levels of registration: a) Property level with type of ownership, sewage disposal system etc.; b) Building level with use, year of construction, material of outer walls etc.; c) Unit level with area of the unit, number of rooms, kitchen facilities etc (see fig. 2). Only the building level is considered within the current research – a particular focus is given to building use and year of construction, making possible the assignment of a temporal dimension to the data. The temporal granularity in this case is one year, although the Building & Dwelling database is updated continuously. The spatial reference is provided through the address assigned to all buildings in the register. Furthermore all buildings have a reference to the Danish National grid with cell sizes of 100m, 250m and 1km.

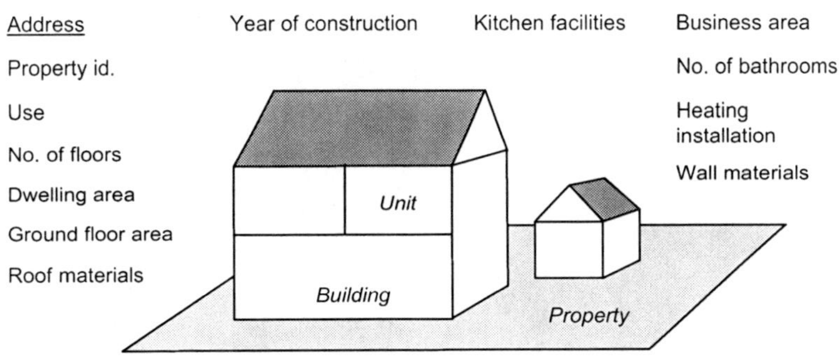


**Fig. 2.**The Building and Housing Register.

The register allocates all buildings to one of 25 use categories (Daugbjerg et al. 2000). Currently we aggregate the 25 categories into six – residential, industry, service, farmhouses, summer cottages, and (other) recreation. Using the Danish national 100-meter square lattice we summarised the built-up area for each use category within each grid cell and assigned the use having the biggest area to the cell – the so-called majority rule. We are fully aware of the risk of ignoring the minority land-use, which theoretically can represent 49% of the area. However, for 1990 69% of the urban land-use cells were pure – i.e. only one land-use type within the cell. Furthermore, in 83% of the cells the dominant land-use accounted for more than 80%. The remaining 17% of the cells are impure in some degree, and might influence the estimation of the neighbourhood rules. Figure 3 illustrates this problem. A further criterion is that at the number of urban cells (including the central cell) within a Moore neighbourhood must be greater than 1 - i.e. at least 2 - otherwise it is not considered built-up. Second, we used detailed nature type registrations to improve the spatial resolution of these ecologically sensitive areas. Several nature types were aggregated into three categories: semi-nature, wetlands and lakes. Finally, the three data layers were merged into a new land-use grid. Thus ten new land-use grids for each year from 1990 to 2000 were produced.

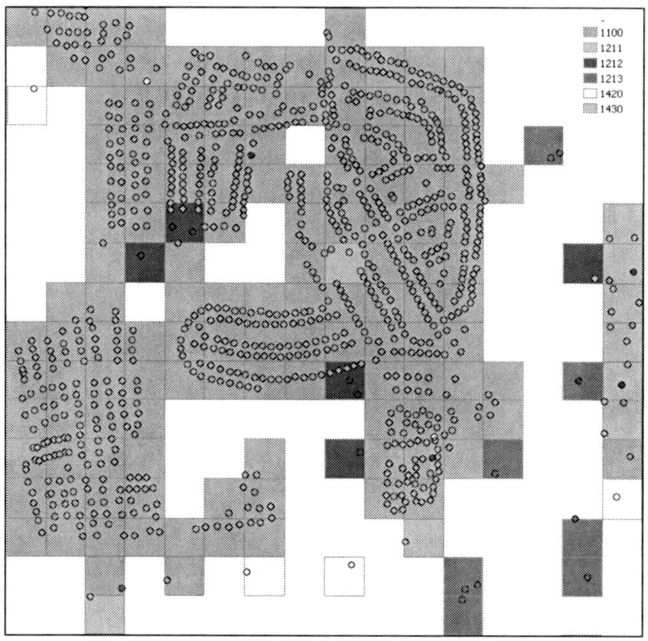


**Fig. 3.** Urban land-use. The points represent the use of each building, whereas the square cells the dominant land-use as s result of the majority rule.

## 4 Results and Discussion

The land-use data set applied in the calculations contains 19 land-use classes as illustrated in table 1. Altogether this gives rise to 19 X 19, which equals to 361 neighbourhood rules. Although the neighbourhood analysis is available through the LUCIA modelling framework (Hansen 2007a, 2007b), it is a time consuming to procedure, and we do only have information about changes among the urban land-uses. Therefore it makes sense to limit the number of land-uses incorporated in the analysis to six so-called active urban land-use classes, whose land-use codes are written with bold in table 1. Altogether, this gives rise to 6 X 19 = 114 neighbourhood rules.

**Table 1.** Land-use types involved in the analysis.

| Land-use code | Description |
|---|---|
| 1100 | Residential areas |
| 1211 | Industrial areas |
| 1212 | Service activities: incl. shopping, administration and public service |
| 1213 | Farm buildings |
| 1230 | Port areas: infrastructure – including quays, dock-yards and marinas |
| 1240 | Airports: runways, buildings and associated land |
| 1300 | Mines, dumps and constructions sites |
| 1410 | Urban green areas – including parks and cemeteries |
| 1420 | Sport and leisure facilities |
| 1430 | Summer cottages |
| 2100 | Arable land: cultivated areas under a rotation system |
| 2300 | Pastures |
| 2400 | Heterogeneous agricultural areas |
| 3100 | Forests |
| 3200 | Shrub and / or herbaceous vegetation |
| 3310 | Beaches, dunes and sand plains |
| 4000 | Wetlands |
| 5100 | Inland lakes |
| 5200 | Marine waters |

Figure 4 illustrates very clearly that new urban development are adjacent to existing urban land-use. This result is clearly in line with the First Law of Geography as well as the analysis carried out in the Netherlands by Verburg et al. (2004).

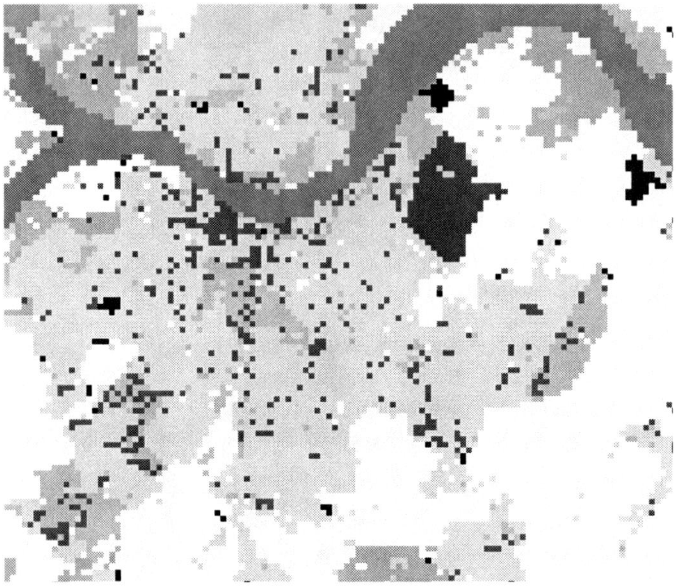

**Fig. 4.** Land-use changes (Aalborg 1990 – 2000). The black cells indicate changes.

The neighbourhood characteristics represented with the enrichment factor are calculated for the whole study area covering the Northern part of Denmark. We have applied a circular neighbourhood with a ten cells radius. The enrichment factor is estimated by calculating the average characteristics of the neighbourhood in 1990 for all cells that have changed into a new land-use type in 1991. Next, new enrichment factors are calculated for all cells changed into new land-use type in 1992 but this time based on a land-use map for 1991. This procedure is repeated for every year until year 2000. We call this method type A. For comparison purposes, we have calculated the average neighbourhood characteristics applying the original method applied by Verburg et al. (2004) - referred to as type B, where the enrichment factor is estimated by calculating the average characteristics of the neighbourhood in 1990 for all cells that have changed into a new land-use type in from 1990 to 2000. Similar to Verburg et al. (2004) we have applied the logarithm function to the enrichment values in order to obtain an equal scale for land-use types, which occur more than average in the neighbourhood (enrichment factor > 1) and land-use types occurring less than average in the neighbourhood (enrichment factor < 1).

Below we present some examples of the calculated neighbourhood characteristics. Figure 5 shows the logarithm of the enrichment factor as a function of distance from the central cell for residential land-use. The black lines represent neighbourhood characteristics calculated using method A, whereas the red lines represent neighbourhood characteristics for the type B method.

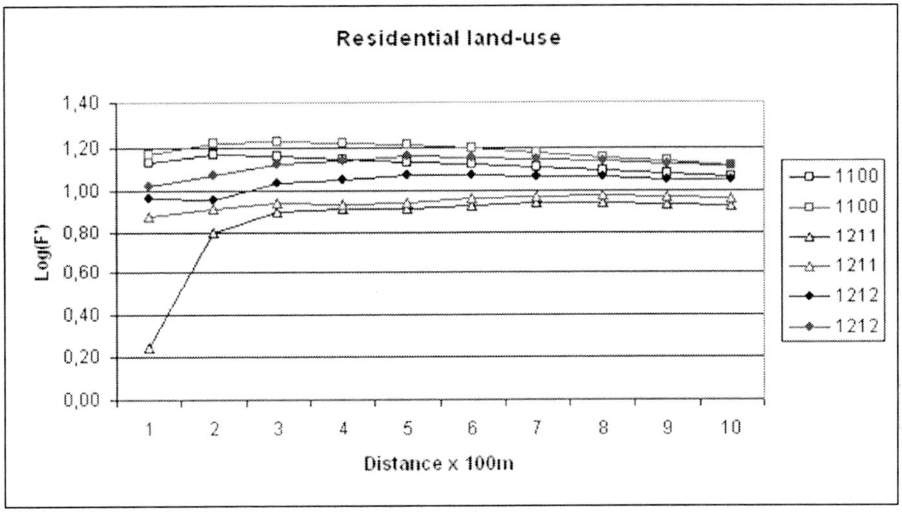


**Fig. 5:** Neighbourhood characteristics (LOG10 of the enrichment factor) as a function of distance for residential land-use.

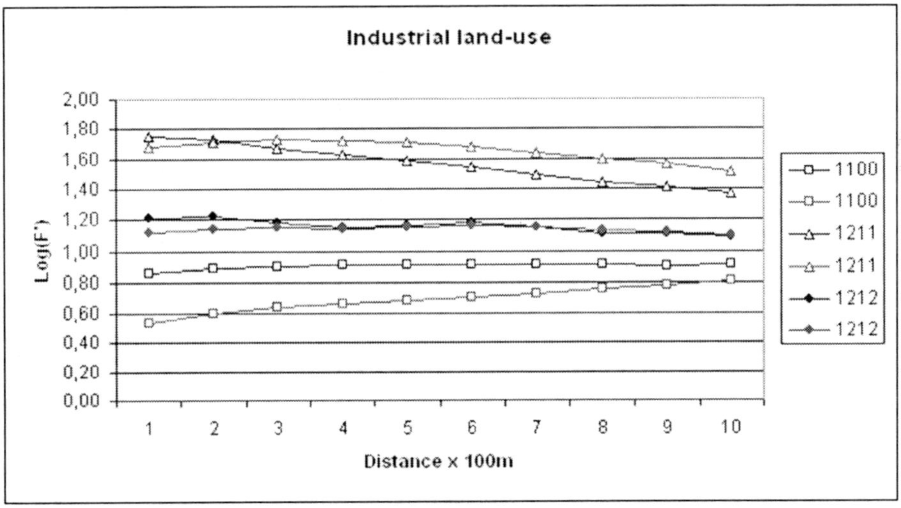


**Fig. 6:** Neighbourhood characteristics (LOG10 of the enrichment factor) as a function of distance for industrial land-use.

The graph on figure 5 shows that new residential development has taken place near existing residential (code 1100) or service areas (code 1212), and the derived neighbourhood characteristics seem to give similar results for the two methods used. On the other hand, new residential areas seem to be repelled by industrial land-use according to type A calculations – at least at shorter distances. However, there is a clear disagreement between the two methods, as method B indicates, that new residential development are not repelled by industry at all – rather the reverse! The result obtained by method A is much more in line with existing knowledge about urban structure, and we believe that this example illustrates the advantage of using land-use changes for individual years to derive neighbourhood characteristics.

Figure 6 shows the neighbourhood characteristics for new industrial land-use. There is a strong tendency that new industrial development takes place near existing industrial land-use, and this a very much in accordance with existing theories for location of new industries. The distance decay function can be considered as type I in the typology of figure 1. New industrial development seems to be rather indifferent regarding distance to existing service land-use, whereas industry seems to be repelled by nearby residential areas – perhaps to avoid conflicts with neighbours due to smoke and noise from the industrial activities. The distance decay function can be considered as belonging to type II in the typology. The two methods – A and B – give the same form of the distance decay function, but the values represented by the curves are very different.

As part of our research we have calculated 114 neighbour characteristics for each of the two methods, and in the above examples we have focused on two of the most important urban land-uses. However, a rapid inspection of the calculated neighbourhood characteristics will give the modeller valuable insight regarding the difficult question: which neighbourhood rules are really needed in our models?

It is often claimed that one single set of neighbourhood rules is used in land-use modelling although urban pattern and processes may differ considerably between regions. This problem is not addressed in the current research project, although we recognise the importance of not disregarding this issue in practical modelling efforts. In principle you should estimate a new set of neighbourhood rules for each case study, but this principle must be weighed against the fact that the area of interest must have a larger spatial extent to include sufficient empirical evidence of land-use changes.

Related to this we must emphasise, that the Region of Northern Jutland generally have a low population increase compared to for example Zealand – and particularly the Copenhagen Metropolitan area. On the other side Northern Jutland have had a huge expansion of summer cottages, which is a major concern in Denmark.

Nevertheless, the proposed method to estimate neighbourhood characteristics will give strong support for land-use modelling by improving the empirical background. We are fully aware, that you should not blindly follow the derived neighbourhood characteristics, but analyse every single distance decay function and try to explain its behaviour based on your knowledge and experience.

## 5 Concluding Remarks

Land-use modelling and spatial scenarios have gained increased attention as a means to meet the challenge of reducing uncertainty in the decision-making, and new models are created every year. Cellular automata based models are very popular, due to their flexibility and ease of implementation. Spatial interaction between neighbouring land-uses is a fundamental component in cellular automata based land-use models, but nevertheless this issue have been treated rather superficially. Too much focus has been put on the technology side of land-use modelling and too little attention has been given to urban theory.

The aim of the current research project has been to quantify and analyse land-use neighbourhood characteristics and impart useful information for cell based land-use modelling. We have had access to very detailed urban land-use data – derived from voluminous information about every Danish building. The results of the completed research is a major step forward, because we have estimated rules for neighbourhood interaction from really observed land-use changes at a yearly basis. We can conclude that the developed methodology where we apply a high temporal granularity, gives more realistic neighbourhood rules than can be obtained using for example Corine land cover for the years 1990 and 2000.

The results presented are mostly of interest for land-use change modelling – and primarily CA based models like MOLAND (Barredo et al. 2003; Engelen et al. 2002), CLUE-S (Verburget al. 2002), and our own LUCIA (Hansen 2007a, 2007b), where neighbourhood interaction plays a crucial role. However, the real advantage of using the described method remains to be demonstrated. Therefore, we are currently carrying out simulations of urban development, were we can compare simulation results with real land-use changes.

Within our current research activities we have started to do the same kind of analysis for the Copenhagen metropolitan area, and we have just provided similar detailed data from the South-western Finland, and we are currently trying to get access to similar data from Sweden in order to test the methodology in an international context. Thus we hope that we can identify the generic nature of neighbourhood interaction rules, although we recognise that they may differ between countries and even between regions.

## References


Barredo J.I., Kasanko, M., McCormick, N. and Lavalle, C. Modelling dynamic spatial processes: Simulation of urban future scenarios through cellular automata. *Landscape and Urban Planning*, vol. 64, pp. 145-160. (2003)

Batty, M.: Urban evolution on the desktop: simulation with the use of extended cellular automata. *Environment and Planning A*, vol. 30, pp. 1943-1967, (1998)

Christaller, W.: *Central Places of Southern Germany* (Edition 1966). Prentice Hall, London, (1933)

Daugbjerg, P. and Hansen, K.V. Property Data. The Danish National Survey and Cadastre. Copenhagen, 2000. (in Danish) (2000)

Engelen, G., White, R. and Uljee, I. (2002). The MURBANDY and MOLAND models for Dublin. Final report, RIKS, 2002.

Hagoort, M., Geertman & Ottens, H. Spatial externalities, neighbourhood rules and CA land-use modelling. *Annals of Regional Science*. Special Issue. (2008)

Hansen, H.S.: An Adaptive Land-use Simulation Model for Integrated Coastal Zone Planning, *Lecture Notes in Geoinformation and Cartography, The European Information Society*, pp. 35 – 53. (2007a)

Hansen, H. S.: LUCIA – a tool for land use change impact analysis. In (Eds. Bjørke, J.T. & Tveite, H) Proceedings ScanGIS'2007. *The 11$^{th}$ Scandinavian Research Conference on Geographical Information Science*. 5 - 7 September 2007, Ås, Norway. pp. 157 – 168. (2007b)

Kok, K. and A. Veldkamp. "Evaluating impact of spatial scales on land use pattern analysis in Central America." *Agriculture, Ecosystems and Environment*, vol.85, pp. 205-221. (2001)

Krugman, P.: The role of geography in development. *International Regional Science Review* vol. 22, pp. 142-161, (1999)

Mandelbrot, B.B.: *The fractal geometry of nature*. New York, NY: W.H. Freeman and Company, (1983).

O'Sullivan, D. & Torrens, P.M.: Cellular models of urban systems. *CASA Working Paper 22*. University College London, Centre for Advanced Spatial Analysis. (2000)

Shannon, C., & Weaver, W.: *The mathematical theory of communication*. Urbana: Univ. Illinois Press, (1964).

Straatman, B., White, R. & Engelen, G: Towards an Automatic Calibration Procedure for Constrained Cellular Automata, *Computers, Environment and Urban Systems*, vol. 28, pp.149-170, (2004)

Torrens P.M. How cellular models of urban systems work: 1. Theory. Centre for Advanced Spatial Analysis, University College London, Paper 28, (2000)

Veldkamp, A. and Lambin, E.F. Predicting land-use change. Editorial. Agriculture Ecosystems and Environment, vol. 85, pp. 1 – 6. (2001)

Verburg PH, de Nijs TCM, van Ritsema Eck J, Visser H, de Jong K.: A method to analyse neighbourhood characteristics of land use patterns. *Computers Environment Urban Systems*, vol. 28, pp. 667–690, (2004)

Verburg, P.H., Soepboer, W., Veldkamp, A., Limpiada, R., Espaldon, V. and Mastura, S. Modelling the spatial dynamics of regional land use: The CLUE-S model. *Environmental Management*, vol. 30, pp. 391 – 405. (2002)

# Interactive Multi-Perspective Views of Virtual 3D Landscape and City Models

Haik Lorenz[1], Matthias Trapp[1], Jürgen Döllner[1], Markus Jobst[2]

[1]Hasso-Plattner-Institute, University of Potsdam, Prof.-Dr.-Helmert-Strasse 2-3, 14482 Potsdam, Germany,
[haik.lorenz, matthias.trapp, doellner]@hpi.uni-potsdam.de
[2]Vienna University of Technology, Erzherzog-Johannplatz 1, A-1040 Vienna, Austria, markus@jobstmedia.at

**Abstract.** Based on principles of panorama maps we present an interactive visualization technique that generates multi-perspective views of complex spatial environments such as virtual 3D landscape and city models. Panorama maps seamlessly combine easily readable maps in the foreground with 3D views in the background – both within a single image. Such nonlinear, non-standard 3D projections enable novel focus & context views of complex virtual spatial environments. The presented technique relies on global space deformation to model multi-perspective views while using a standard linear projection for rendering which enables single-pass processing by graphics hardware. It automatically configures the deformation in a view-dependent way to maintain the multi-perspective view in an interactive environment. The technique supports different distortion schemata beyond classical panorama maps and can seamlessly combine different visualization styles of focus and context areas. We exemplify our approach in an interactive 3D tourist information system.

**Keywords:** multi-perspective views, focus & context visualization, global space deformation, 3D city models, virtual 3D landscape models, geovisualization

## 1 Introduction and Motivation

Virtual spatial environments based on 3D landscape and city models are common tools for an increasing number of commercial and scientific applications and are applied as interactive space and context for planning, simulation, and visualization tasks. One key requirement represents the efficient rendering of large amounts of data based on level-of-detail techniques and multiresolution models. Another key requirement is the effective presentation of the environment and its contents, e.g., by providing detail views for important areas while giving a coarse overview of their spatial context.

While a single-perspective view depicts a scene from a single viewpoint, "a multi-perspective rendering combines what is seen from several viewpoints into a single image." (Yu and McMillan 2004) Mathematically, multi-perspective views rely on non-linear 3D projections or, equivalently, non-planar reference shapes, used to map 3D world space on 2D image space. In this way occlusions become resolvable, scales at which objects are depicted are adjustable, and spatial context information can be included in a single image.

With these techniques, multi-perspective views can visually emphasize or clarify an area of interest while retaining or extending its surrounding area, achieving an effective information transfer (Keahey 1998). Furthermore, they utilize the available screen real estate to a high degree. Their characteristics make multi-perspective views a tool for focus & context visualization. Well-known examples include fisheye maps, which emphasize important information by magnification, or spherical maps, which add context information by non-uniformly integrating a full 360° view.

## 1.1 Multi-Perspective Views for Maps

Multi-perspective views have been developed particularly in landscape depiction and Cartography. Chinese landscape painters used multi-perspective views in the 11th century already (Vallance and Calder 2001).

**Fig. 1a.** A panorama map painted by H.C. Berann (used with permission)

**Fig. 1b.** Multi-perspective view of a virtual 3D city model inspired by (a)

**Fig. 1c.** Multi-perspective focus & context visualization for walk-throughs

**Fig. 1:** A historic panorama map and examples of interactive multi-perspective views of 3D city models

**Fig. 2.** Painting of Venice, Italy (about 1550) (Whitfield 2005). It exhibits a panoramic effect and includes labels

Another example, a 360° panorama view of the London skyline consisting of six separate paintings, was created in the late 18th century. The incorporation of cartographic information yields panorama maps. Fig. 2 shows an early example of Venice, Italy (about 1550). H.C. Berann, an Austrian artist and panorama maker, pioneered one particular kind of panorama map. Beginning in the early 1930's he created a deformation and painting style (Fig. 1(a)), known as "Berann panorama", which became the de-facto standard for tourist maps in recreational areas. This style seamlessly combines a highly detailed image of the area of interest with a depiction of the horizon including major landmarks. The area of interest is shown in the foreground from a high viewpoint, whereas the horizon is shown in the background from a low perspective. The environment is depicted with "natural realism" (Patterson 2000) and key information such as trails or slopes is superimposed in an abstracted, illustrated fashion. As a result of the high viewpoint the foreground shows key information top-down, i.e., free from obstructions and clearly visible. At the same time, the map user can easily orient the map using the horizon, which is visible due to the changed perspective, as reference without the need for a compass. For these reasons, panorama maps are useful specifically to unskilled map readers.

In general, the creation of panorama maps is time consuming and requires a skilled artist. It includes proper viewpoint selection, partial landscape generalization, identification of landmarks, their integration into the map with recognizable shapes, and a smooth transition between the foreground and background perspective (Patterson 2000). Even with the support of digital tools and digital 3D geodata, panorama creation still remains a tedious manual process (Premoze 2002). Despite their effectiveness, panorama maps are rarely created, and the creation techniques can hardly be transferred to interactive systems where the user manipulates the viewpoint.

## 1.2 Multi-Perspective Views for Spatial 3D Environments

Multi-perspective views can be used to visualize 3D landscape models, e.g., mountainous regions with the mountain peaks providing a distinctly recognizable background for orientation purposes. Similarly, they can visualize 3D city models, using the skyline of the city as background. In today's applications, interactive visualization is required to support the user in exploring and analyzing the virtual 3D environment. With respect to the usability of such applications, the navigation and orientation aids represent key issues because users frequently "get lost in space" without guidance (Buchholz et al. 2005). Here, the inclusion of a fixed horizon or skyline similar to a Berann panorama offers an additional orientation cue in the sense of a focus & context visualization.

To obtain an automatic, real-time enabled solution, we need to focus on the projection as major tool for orientation and neglect artistic aspects such as landmark depiction and selective generalization.

Computer graphics knows three approaches to achieve a panorama effect: multi-perspective images, deformations, and reflections on non-planar surfaces (Vallance and Calder 2001). Multi-perspective images either use non-linear, non-uniform projections or combine multiple images from different viewpoints to create the final rendering. Deformations distort the landscape before rendering the final image using a standard projection, which implies recomputation of all geometric data for every image. Finally, reflections on non-planar surfaces use standard projections showing an intermediate object that in turn reflects the landscape.

## 1.3 Interactive Multi-Perspective Views

Techniques implementing multi-perspective views can be classified as multi-pass or single-pass. Multi-pass techniques create several intermediate images that are blended in a final compositing step. Each intermediate image requires separate data processing, which is rather expensive when it comes to complex spatial 3D environments. Specifically, out-of-core algorithms can incur additional penalties because rendering of intermediate images often significantly reduces caching efficiency. Additionally, image quality suffers due to resampling in the compositing step. Single-pass techniques do not exhibit these disadvantages, yet they require customization of the rendering process available only in software rendering (e.g., ray tracing) until recently.

With the advent of a programmable rendering pipeline on GPUs the implementation of interactive single-pass multi-perspective view techniques becomes feasible. We demonstrate a technique that implements a dynamic global deformation and shifts this task to the GPU. This approach exploits best the optimization of current graphics hardware for standard projections both in terms of image quality and speed. We apply our technique to an interactive application that visualizes complex virtual 3D city models in the context of a tourist information system. Our global deformation is not only used to mimic Berann panoramas but also for a novel viewing technique that enables looking ahead the current route in a pedestrian's view.

An important aspect of this contribution is the analysis of view parameters. In contrast to an artist choosing viewpoints for map creation, users of interactive applications are inherently free to move. We analyze how to define the multi-perspective view and how to dynamically adjust our deformation accordingly during the user's navigation. In addition, we discuss the implications for common 3D navigation techniques.

The paper is structured as follows. Section 2 discusses related work. Section 3 explains techniques for interactive multi-perspective views and their use for focus & context visualization. Section 4 describes the implementation. Section 5 discusses the test application and its performance. Section 6 concludes the paper and outlines future work.

## 2 Related Work

The work of H.C. Berann includes maps, panoramas and fine art (Berann 2007). His way of creating panorama maps and techniques are described in (Patterson 2000). (Premoze 2002) introduces a first approach for implementing these techniques except for multi-perspective views by means of 3D computer graphics. Additionally, instructions for manual creation of panorama maps using various tools are available online.

Besides the artistic and visual quality of a Berann panorama, panoramic depictions use a concept known as focus & context in the field of visualization. In general, such visualization not only contains the actual subject but also its embedding context with the goal of supporting the user's interpretation process. Traditionally, focus & context has been regarded as distortion-based view of 2D or 3D information where emphasis is achieved through varying magnification and screen real estate allocation. See (Leung and Apperley 1994) for a survey of different approaches and (Carpendale and Montagnese 2001) for a general definition. (Vallance and Calder 2001) presents ideas on the use of multi-perspective views for focus & context and their different creation techniques. Recently, this concept has been extended to include other methods for emphasis, such as generalization, rendering style, blur, or transparency (Hauser 2003). (Keahey 1998) generalizes focus & context to providing separate information dimensions.

Multi-perspective views have been analyzed mainly in the context of ray tracing, which allows for easy manipulation of the camera model. (Yu and McMillan 2004) defines general linear cameras as affine combinations of 3 sample rays. (Löffelmann and Gröller 1996) proposes a camera model based on arbitrary surfaces to define viewing rays and a projection surface. For real-time environments, (Yang et al. 2005) describe 3D view deformations as postprocessing step to achieve multi-perspective views and nonlinear perspective projections. (Spindler et al. 2006) improves on this method by integrating the view deformation directly into the image formation process through a camera texture. (Glassner 2004, parts 1 and 2) describe an interesting non-interactive semiautomatic method to transfer the artistic Cubism style to computer generated images. Applications of multi-perspective views include, among others, story-telling, image processing with the goal of creating panoramas from multiple images or video footage

(Roman et al. 2004), recovery of 3D information (Li et al. 2004), and image-based rendering (Levoy and Hanrahan 1996).

Deformation is a well-established field in geometric modeling. (Barr 1984) is one of the first describing deformation operators. Such operators are used frequently in current modeling tools. Research topics include volume preservation, avoidance of self-intersection, or deformation control. Implementations can be classified as shape deformation or space deformation. Recent examples for the former approach are (Angelidis et al. 2004; von Funck et al. 2006), which result in interactive deformations for moderately sized models. The latter approach is useful for ray casting or ray tracing. (Kurzion and Yagel 1997) presents space deformations for hardware-assisted volume rendering.

A prerequisite for our implementation is rendering of spatial 3D environments, which includes terrain rendering (e.g., (Asirvatham and Hoppe 2005; Hwa et al. 2004; Lindstrom and Pascucci 2002)) and rendering of large scenes. Approaches for the latter include out-of-core algorithms (e.g., (Buchholz and Döllner 2005; Gobbetti and Marton 2005)), specialized visibility detection algorithms (e.g., (Chhugani 2005; Wonka et al. 2001)), and level-of-detail algorithms (e.g., (Sander and Mitchell 2006)). Additionally, interaction and navigation within a spatial 3D environment is necessary. (Buchholz et al. 2005) contains both, a survey of navigation techniques and improvements to common navigations.

## 3 Effective Presentation of Spatial 3D Environments

Multi-perspective views facilitate the implementation of effective presentation of spatial 3<D environments. They can add valuable cues by seamlessly integrating multiple perspectives in the resulting images and, therefore, make efficient use of the image space.

In the following, we present two related deformation techniques that implement multi-perspective views:

1. The bird's eye view deformation, which mimics Berann's panorama maps used to visualize mountain areas.
2. The pedestrian's view deformation, which swaps the role of foreground and background by presenting a low altitude perspective view in front of a top view of distant city parts.

In general, both deformations need to ensure the user's location awareness during navigation and interaction. Even experienced users get disoriented if the current perspective does not contain sufficient points of reference or if the image sequence does not provide spatio-temporal coherence. For these reasons, both techniques provide a seamless combination of different views in a single image and achieve interactive frame rates.

We describe both deformation techniques using a reference plane $T$, a usually horizontal plane. This plane can be elevated, e.g., to define the roof of the average building as the horizon or to reduce distortion artifacts. A point $P$ of the virtual 3D city model not lying in that reference plane is assigned a reference point $P_T$ in $T$. Deformation is then calculated using $P_T$ and applied to $P$. $P_T$ can be either a simple vertical projection of $P$ onto $T$ or – if an object's shape is to be kept free from distortion – a single reference point for the whole object.

## 3.1 Bird's Eye View Deformation

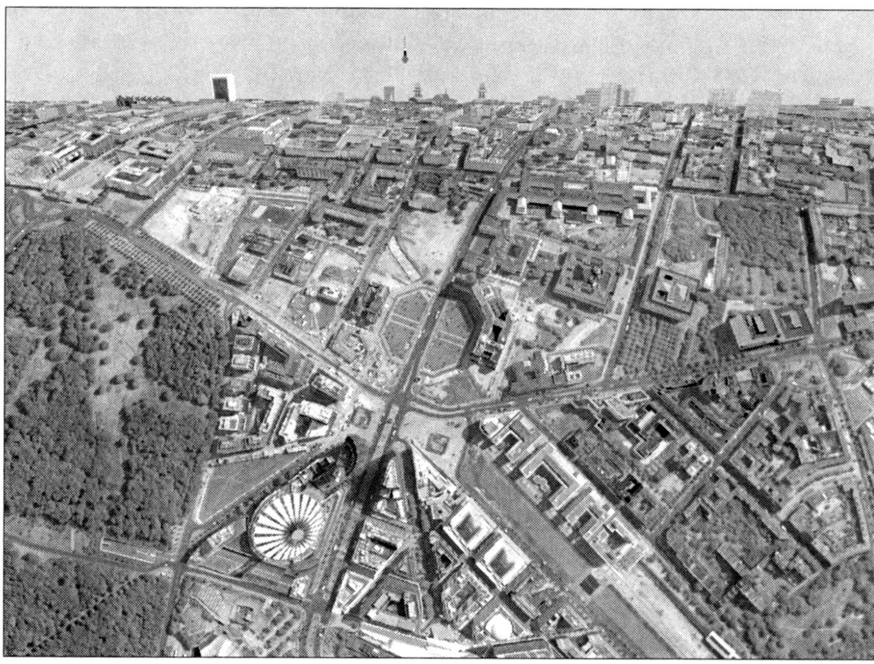

**Fig. 3.** The bird's eye view deformation shows a top view and the horizon simultaneously

Similar to Berann's panorama maps, this deformation is based on

- a depiction of the area of interest using a bird's eye view, which would not permit a visible sky,
- a view of the horizon and sky, and

- a smooth transition between both perspectives.

As a result, the landscape appears to be separated into two planar sections connected by a curved transition zone with the focus lying on the bird's eye view part in the foreground or lower image part (Fig. 3). Nevertheless, the area of interest is not strictly separated from the transition zone but often reaches into the curved section.

For a painted panorama map, the map designer decides on relevant parameters such as the two view points and the transition in between. In an interactive application the user can move the camera. To keep the three key properties of this multi-perspective view regardless of the camera's orientation or position, we define fixed image areas separated by horizontal lines for the bird's eye view, transition zone, and horizon (Fig. 4). This fixation results in a transition zone curvature that depends on the viewing angle, yet the fixed horizon provides strong temporal coherence and eases orientation tracking during navigation, whereas the ever-changing shape of the landscape does not lead to distraction. In addition, this implicit definition of the horizon's perspective permits the user to navigate relative to and interact with the focus area using standard metaphors for virtual environments while the visual context is adjusted automatically.

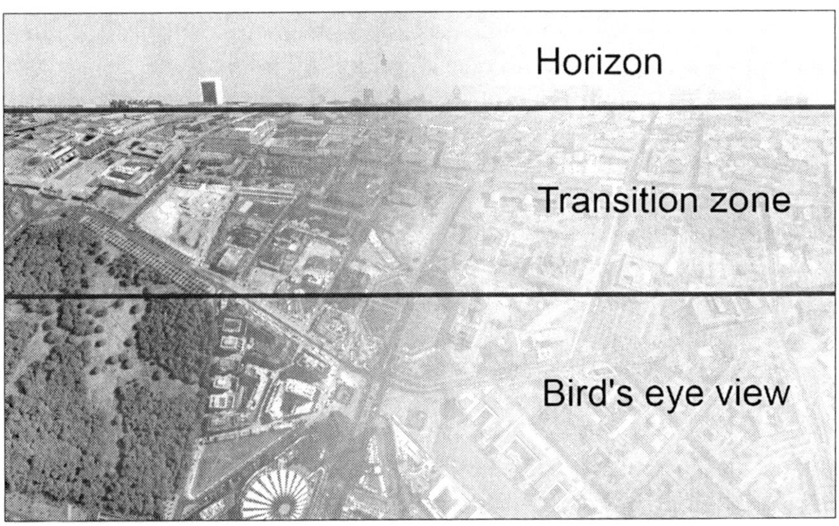


**Fig. 4:** Fixed image separation for the bird's eye view deformation

In general, painted panoramas exhibit a horizontal horizon. In contrast, an interactive application can permit rolling of the camera. In this case, the horizon should provide feedback about the roll angle. In the following descriptions, we assume no rolling.

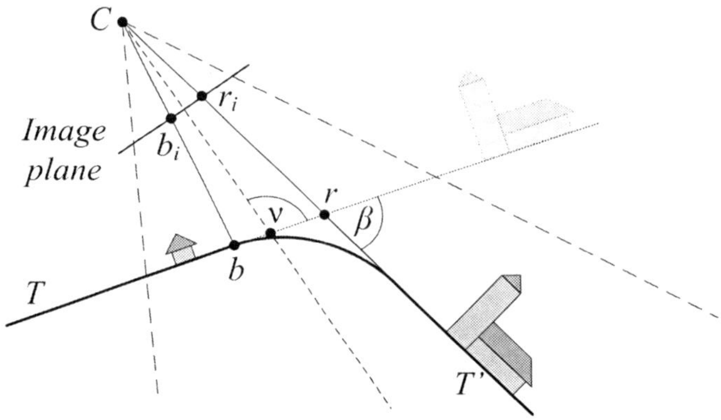


**Fig. 5.** Schematic side view of the bird's eye view deformation

The image subdivision results in the following set of viewing parameters:

- $C$ – camera position
- □ – viewing angle of the reference plane
- $b_i$ – line separating focus area and transition zone in the image
- $r_i$– line of the horizon in the image

Fig. 5 sketches a typical setting assuming a perspective projection. In our implementation the transition zone is guided by a quadratic Bézier spline due to its continuity properties at the borders. The exact computation is described in Section 4. The line $b$ – the projection of $b_i$ onto $T$ – marks the beginning of the transition zone. The half-plane following the transition zone, which leads to the horizon depiction, is referred to by $T'$. It is computed as a rotation of $T$ by an angle □ about the line $r$, the projection of $r_i$ onto $T$. Thus, an object's shape is maintained outside the transition zone. Within this zone an object's shape is preserved only if it uses a single reference point.

Typically, both $b_i$ and $r_i$ are rarely changed while $C$ and □ reflect the user's navigation. To make efficient use of the screen space, the amount of visible sky should be minimized while retaining a recognizable skyline. Placing $r_i$ in the upper quarter of the screen generally gives good results. The location of $b_i$ determines the curvature of the transition zone. Placing $b_i$ in the lower half of the screen gives a good compromise between smooth transition and visibility of the focus area.

## 3.2 Pedestrian's View Deformation

**Fig. 6.** The pedestrian's view deformation combines a realistic view of the user's vicinity with a top view of distant areas

The bird's eye view deformation supports answering questions such as "Which direction am I looking to?" without the need for a compass. For pedestrian's views, which occur in walk-through scenarios, the question changes to "Where am I going to?", e.g., if users want to look ahead the path along they are currently walking. Due to the low viewing angle, however, users can generally not obtain an effective overview without changing the perspective or navigation mode because large parts of the spatial 3D environment are occluded.

To counter this effect, the pedestrian's view deformation bends upwards distant parts of the reference plane (Fig. 6). Compared to the technique proposed in (Vallance and Calder 2001), which deforms the reference plane to fit the inside of a cylinder, the pedestrian's view deformation has the advantage of using a planar and, hence, clear and undistorted view of distant regions in the background. In terms of focus & context, the prominent sky in a pedestrian's view, which provides only little information, is replaced by a top view of the region ahead, resulting in a more efficient use of screen space.

Similar to the bird's eye view deformation, the landscape is separated into two planar sections connected by a curved transition zone, yet the image-based deformation definition is not appropriate as it does not lead to comprehensible context behavior. We observe a more effective visualization with the Pedestrian's view deformation when using a fixed orientation of *T'* relative to the reference plane *T* in world space. Particularly, this enables an intuitive "looking-up" operation to reveal more of the context information in the background. As a consequence, the curvature of the transition zone does not depend on the viewing angle but can be defined independently. Nevertheless, the deformation follows the camera such that the rotation axis *r* has a fixed distance and orientation relative to the camera.

With this definition, again the user is relieved from explicitly controlling the multi-perspective view. Standard navigation metaphors within the foreground remain applicable. Only interaction with the background, e.g., for the click-and-fly navigation (Mackinlay et al. 1990), needs to be aware of our deformation for correct object identification.

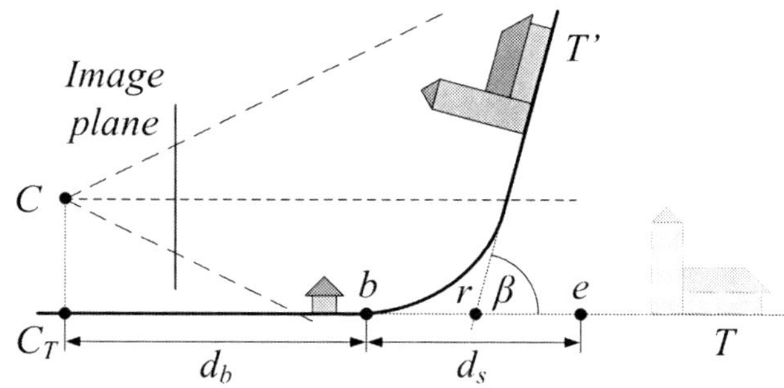


**Fig. 7.** Schematic side view of the pedestrian's view deformation

We use the following set of parameters to specify this multi-perspective view (cp. Fig. 7):

- *C* – camera position
- $\beta$ – angle between *T* and *T'*
- $d_b$ – distance between $C_T$ (*C* projected onto *T*) and *b*
- $d_s$ – width of the transition zone's source area

Analog to the bird's eye view deformation *T'* is a rotation of *T* about *r* and the transition zone follows a quadratic Bézier spline. The line *b* marking the beginning of the transition zone is always parallel to the image plane and keeps a distance $d_b$ from the camera's vertical projection $C_T$ onto *T*. We define the line *r* to be the center line of the transition zone's source area. Thus, it is parallel to *b* at a distance of $d_s$ / 2. This definition simplifies the implementation shown in Section 4.

The parameters except for *C* again change rarely. They control two main characteristics of the pedestrian's view deformation: the amount of available orientation reference in the focus area through $d_b$ and the amount of visible context information through □. Setting □ to values less than 90° trades magnification in the context area for visible space and allows for looking ahead a route farther. The parameter $d_s$ directly controls the transition zone's curvature where a small transition zone and thus rather high curvature shows good results.

### 3.3 Graphical Representation of Focus and Context

Both deformations presented in Section 3 smoothly and seamlessly combine focus and context. Due to the view dependent nature of both deformations, the user might loose distinction between geometrically correct information in the focus area and deformed information in the context during navigation. This might lead to misinterpretations, lost orientation, or erroneous navigation (Zanella et al. 2002). Specifically, the pedestrian's view deformation, permitting views without visible focus area and, hence, without navigation reference, is prone to such effects. Solutions require visual cues, e.g., iconic navigation aids or distinct rendering styles for focus and context such as context color desaturation.

Besides the more effective use of screen space, in focus & context visualization the two constituents can serve different purposes and thus are to display different information dimensions beyond change of rendering style (Keahey 1998; Stone et al. 1994). Whereas the focus gives core information, the context shows supporting information.

Panorama maps as inspiration for our bird's eye view deformation use this principle by adding thematic information such as trails to the focus area while the landscape depiction style is constant for the whole image. We demonstrate an extension showing a map with 3D landmarks in the focus area. The context remains a complete and photorealistic depiction since the skyline is required to be recognizable. Nevertheless, generalization techniques such as (Döllner et al. 2005) might prove useful. Additionally, context information can be enriched by labeling landmarks as seen in some of Berann's panorama maps.

The pedestrian's view deformation permits displaying more important information in the context. In fact, the focus area is limited to serve as navigation reference and location marker within the spatial 3D environment whereas the context generally receives the larger screen space and exhibits less occlusion. According to this observation, our sample visualization (cp. Fig. 1(c)) shows a photorealistic view in the focus and a map as context for visual distinction. On top, the current travel route is highlighted spanning both parts and thus allowing for a route preview.

Rendering such composite depictions does not require multi-pass techniques. Instead, the deformation implementation presented in Section 4 provides a vertex-based interpolation value $q \in [0;1]$ with $q = 0$ within the focus area, $q = 1$ within the context area, and a smooth transition in between. The rendering styles are then interpolated per pixel based on this value $q$.

## 4 Real-Time Deformation Implementation

The implementation shifts the deformation task to the GPU. Changing geometry on the GPU, however, has major consequences for standard application-based rendering optimizations such as occlusion culling or view frustum culling.
Our deformation scheme does not introduce new vertices. Rendering artifacts due to insufficient tessellation can only appear in the curved transition zone. This confined nature allows for a straightforward solution: a Level-of-Detail algorithm selects a more detailed object representation within the transition zone. Alternatively, on-demand tessellation using techniques such as generic mesh refinement (Boubekeur and Schlick 2005) or the newly introduced geometry shaders can be used.

The GPU is a highly parallel streaming processor, thus each vertex needs to be processed independently. This is achieved by formulating the deformation of an individual point $P$ as a function $f_D : P \mapsto P'$ which is computed by a vertex program. For efficient computation we want this function to perform an affine transformation $M_D$ on $P$, where the 4x4-transformation matrix $M_D$ depends only on the reference point $P_T$. Thus, we reformulate $f_D$ as $P' = M_D(P_T) \cdot P$. Both deformations described in Section 3 share the same underlying construction, allowing us to use a single function $M_D(P_T)$.

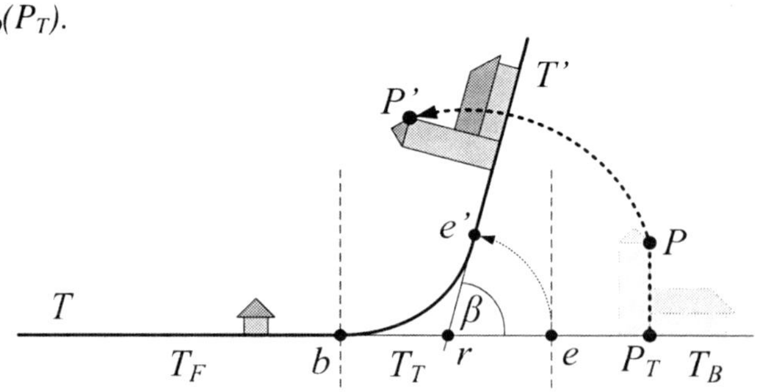


**Fig. 8.** Deformation parameters and definitions. Only objects located in $T_T$ become distorted

For our unified deformation, we divide the reference plane $T$ into three sections:

1. The undeformed part $T_F$ , which becomes the foreground or focus of the image,
2. The transition zone $T_T$ , which becomes curved, and
3. The remainder $T_B$ , which becomes the background or context of the image by rotating $T$ about $r$.

These three sections are separated by the line $b$ between $T_F$ and $T_T$ and the line $e$ between $T_T$ and $T_B$. The lines $b$, $r$, and $e$ are parallel and equidistant. Finally, □ denotes the angle between $T$ and $T'$. Fig. 8 sketches this setting. With these three sections, $M_D(P_T)$ can be formulated depending on the location of $P_T$. In addition, the rendering style interpolation value $q$ can be derived:

$P_T \in T_F$ : $M_D(P_T)$ is the identity matrix, since this section is not to be deformed; $q = 0$ .

$P_T \in T_T$ : $M_D(P_T)$ needs to specify a transformation that transforms $P_T$ and its frame of reference to the corresponding point $P_T'$ on a quadratic Bézier spline in the tangential frame of reference. A suitable transformation consisting of a scaling followed by a rotation based on the de Casteljau algorithm (Gallier 1999) is described in the following paragraph; $q$ equals the Bézier spline parameter $t$.

$P_T \in T_B$ : $M_D(P_T)$ is a rotation matrix about $r$ with an angle □; $q = 1$ .

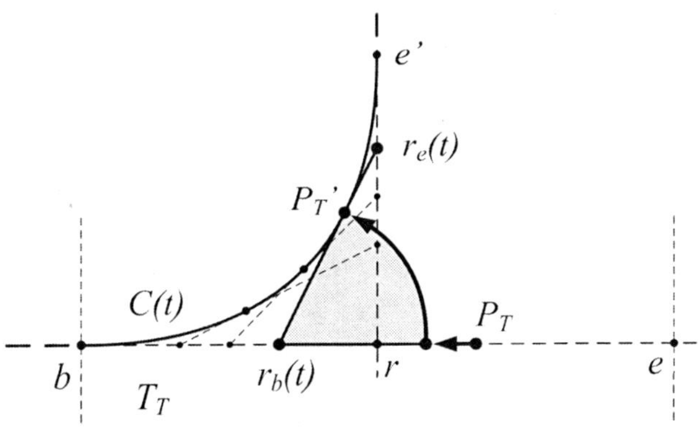


**Fig. 9.** The de Casteljau algorithm constructs a Bézier spline point through linear interpolations. It also provides the point's tangent

Fig. 9 shows the de Casteljau algorithm for the profile Bézier spline. The axes *b*, *r*, *e*, and *e'* appear as points in this side view, where *b*, *r*, and *e'* become control points of the spline. Since

$$\|b - r\| = \|e' - r\| \tag{1}$$

the resulting spline is symmetrical. For a quadratic Bézier spline $C(t)$ with $t \in [0; 1]$, the algorithm uses linear interpolations to construct two intermediate points

$$r_b(t) = (1 - t)b + tr \text{ and } r_e(t) = (1 - t)r + te \tag{2}$$

and the resulting point

$$C(t) = (1 - t)r_b(t) + tr_e(t) \tag{3}$$

For a given point $P_T \in T_T$, the corresponding point on the Bézier spline is found as:

$$P_T' = C(\|P_T - b\| / \|e - b\|) \tag{4}$$

This mapping does not define an arc length parameterization of *C* and thus introduces an unwanted flattening of objects within the transition zone. The suitable reparameterization of *C* can be achieved using a lookup table, which is left for future work.

For our purpose, the key property of the de Casteljau algorithm is the implicit tangent construction formed by the line through $r_b$ and $r_e$. To compensate for the variable length contraction along the curve, i.e., the missing arc length parameterization, a scaling along $e - b$ with a factor $\|P_T' - r_b\| / \|P_T - r_b\|$ centered at $r_b$ is necessary. Then, a rotation of the scaled $P_T$ about $r_b$ onto the tangent creates the correct tangential frame of reference for $P_T'$. This completes the definition of $M_D(P_T)$.

The parameters *T*, *b*, *e*, and  of this unified computation depend on the camera location. Thus, within an interactive application, they need to be derived from the original (camera-independent) deformation parameters described in Section 3 on a frame-by-frame basis. Also, especially for scene graph based systems, the current frame of reference needs to be taken into account. The most efficient and robust solution is to perform the deformation in the camera's frame of reference since it is constant during image generation.

# 5 Performance

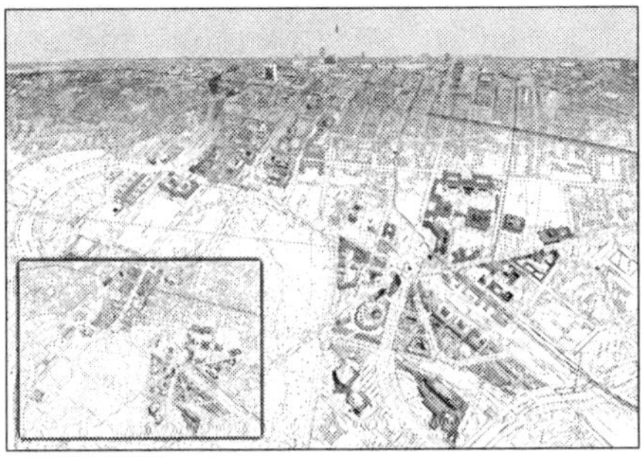

**Fig. 10a.** Bird's eye view deformation showing public transport lines

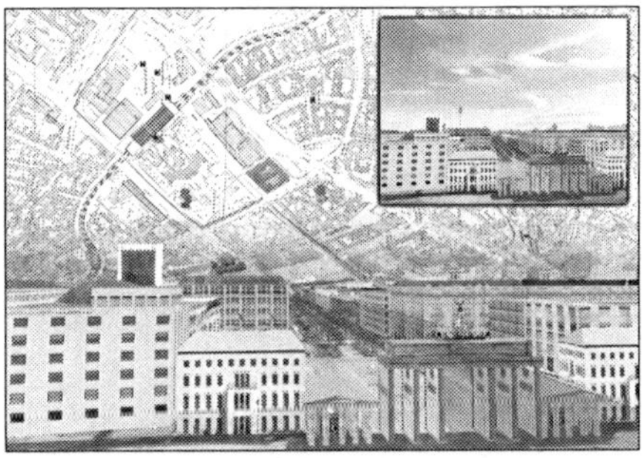

**Fig. 10b.** Pedestrian's view deformation with highlighted route

**Fig. 10:** Sample multi-perspective images. The insets show the corresponding standard perspectives

Fig. 10 shows sample images of the bird's eye view deformation and pedestrian's view deformation, respectively. For comparison, the insets show a standard perspective projection using identical camera settings to highlight the effects of our focus & context visualization.

We extended an existing 3D tourist information system for Berlin, Germany, with our technique. Despite the additional data handling overhead for two rendering styles, we were able to achieve interactive frame rates. Table 1 summarizes average frame rates for two sample camera paths per view deformation. The measurements were made on a PC with an AMD Athlon 64 X2 (2.3 GHz), 2 GB main memory, and a NVidia GeForce 7900GT with 256 MB video memory. The test application does not utilize the second CPU core. The sample dataset comprises the inner city of Berlin with about 16,000 generically textured buildings, about 100 landmarks, a 3 GB color aerial photo, and a 250 MB grayscale map image on top of a digital terrain model.

**Table 1.** Performance measurements for different screen resolutions and configurations

| Resolution | Configuration | Path | frames/ sec | frames/sec without bend. |
|---|---|---|---|---|
| 1600x 1200 | Pedestrian's view | 1 | 11.72 | 12.95 |
| | | 2 | 19.33 | 15.69 |
| | Bird's eye view | 3 | 8.35 | 29.86 |
| | | 4 | 6.73 | 17.64 |
| 1024x 768 | Pedestrian's view | 1 | 17.85 | 15.63 |
| | | 2 | 22.75 | 17.69 |
| | Bird's eye view | 3 | 8.87 | 27.24 |
| | | 4 | 5.42 | 16.07 |
| 800x 600 | Pedestrian's view | 1 | 20.54 | 16.49 |
| | | 2 | 23.94 | 18.42 |
| | Bird's eye view | 3 | 8.74 | 27.26 |
| | | 4 | 8.48 | 19.52 |

The frame rate without deformation is largely resolution independent suggesting texture access as main bottleneck in our test application. To deal with the texture amount, an out-of-core algorithm is used to load texture on demand in sufficient resolution from disk. Table 2 shows the average number of bytes read from hard disk per frame for our test setting at resolution 1600x1200.

**Table 2.** Average hard disk access per frame with / without deformation at resolution 1600x1200

| Configuration | Path | bytes/ frame | bytes/frame without bend. |
|---|---|---|---|
| Pedestrian's view | 1 | 260,207 | 407,822 |
| | 2 | 122,729 | 215,398 |
| Bird's eye view | 3 | 5,720,803 | 190,824 |
| | 4 | 2,555,602 | 243,067 |

The exceptionally high load rates for the bird's eye view deformation are caused by the visible horizon. Compared to the corresponding standard perspective projection, more terrain is visible and thus more texture requires loading – even though at low quality. At the same time, changing the view direction invalidates more texture. Hence, caching efficiency is reduced dramatically. With the pedestrian's view deformation, only a fraction of the terrain is visible compared to a standard view, but due to the deformation distant terrain requires a significantly higher texture resolution leading to only a slight reduction in texture load overhead.

## 6 Conclusions

We have demonstrated the concept and implementation of interactive multi-perspective views for spatial 3D environments. They are inspired by the well-known panorama maps and aim to increase the effectiveness of interactive applications by using the principle of focus & context visualization. Our implementation is based on a global space deformation processed by graphics hardware and permits the seamless combination of different graphical representations for focus and context areas. To verify its applicability we have successfully integrated our technique into an existing interactive 3D tourist information system.

The visual quality in the transition zone can be further improved by incorporation of on-demand geometry tessellation, e.g., through the use of geometry shaders, or by adaptation of a more advanced bending scheme. In contrast to the currently used simple static lighting, dynamic lighting and shadowing within a deformed 3D landscape model remains an interesting open question. While this contribution describes the underlying technology, user studies about the effectiveness and/or expressiveness of our visualization approach, different rendering style combinations, and navigation in a deformed 3D landscape model remain future work.

### Acknowledgements

We would like to thank 3D Geo GmbH (www.3dgeo.de) for providing the implementation platform LandXplorer, an authoring and presentation system for virtual 3D city models and landscape models. We also thank Matthias Troyer for his permission to use the panorama map shown in Fig. 1(a).

### References

The world of H.C. Berann (accessed 2007), url: http://www.berann.com

Angelidis, A., Cani, M.-P., Wyvill, G. & King S. (2004), Swirling sweepers: Constant-volume modeling, *in* Proceedings of the 12th Pacific Conference on Computer Graphics and Applications, IEEE Computer Society, Washington, DC, USA, pp. 10-15.

Asirvatham, A. & Hoppe, H. (2005), Terrain Rendering Using GPU-Based Geometry Clipmaps, *in* M. Pharr (ed.), GPU Gems 2, Addison-Wesley, pp. 27-45.

Barr, A. H. (1984), Global and Local Deformations of Solid Primitives, *in* SIGGRAPH '84: Proceedings of the 11th annual conference on Computer graphics and interactive techniques, ACM, New York, NY, USA, pp. 21-30.

Boubekeur, T. & Schlick, C. (2005), Generic Mesh Refinement on GPU, *in* Proceedings of ACM SIGGRAPH/Eurographics Graphics Hardware 2005, ACM, pp. 99-104.

Buchholz, H.; Bohnet, J. & Döllner, J. (2005), Smart and Physically-Based Navigation in 3D Geovirtual Environments, *in* IV '05: Proceedings of the Ninth International Conference on Information Visualisation, IEEE Computer Society, Washington, DC, USA, pp. 629-635.

Buchholz, H. & Döllner, J. (2005), View-Dependent Rendering of Multiresolution Texture-Atlases, *in* Proceedings Information Visualization 2005, pp. 215-222.

Carpendale, M. S. T. & Montagnese, C. (2001), A Framework For Unifying Presentation Space, *in* UIST '01: Proceedings of the 14th annual ACM symposium on User interface software and technology, ACM, New York, NY, USA, pp. 61-70.

Chhugani, J.; Purnomo, B.; Krishnan, S.; Cohen, J.; Venkatasubramanian, S. & Johnson, D. S. (2005), vLOD: High-Fidelity Walkthrough of Large Virtual Environments, *IEEE Transactions on Visualization and Computer Graphics* **11**(1), pp. 35-47.

Döllner, J.; Buchholz, H.; Nienhaus, M. & Kirsch, F. (2005), Illustrative Visualization of 3D City Models, *in* Robert F. Erbacher; Jonathan C. Roberts; Matti . T. Gröhn & Katy Börner, ed., Visualization and Data Analysis 2005, pp. 42-51.

von Funck, W., Theisel, H. & Seidel, H.-P. (2006), Vector field based shape deformations, *in* Proceedings ACM SIGGRAPH 2006, ACM, New York, NY, USA, pp. 1118-1125.

Gallier, J. (1999), *Curves and Surfaces in Geometric Modeling: Theory and Algorithms*, Morgan Kaufmann Publishers Inc., San Francisco, CA, USA.

Glassner, A. (2004), Digital Cubism, *IEEE Computer Graphics and Applications* **24**(3), pp. 82-90.

Glassner, A. (2004), Digital Cubism, Part 2, *IEEE Computer Graphics and Applications* 24(4), pp. 84-95.

Gobbetti, E. & Marton, F. (2005), Far Voxels: A Multiresolution Framework for Interactive Rendering of Huge Complex 3D Models on Commodity Graphics Platforms, *ACM Trans. Graph.* 24(3), pp. 878-885.

Hauser, H. (2003), Generalizing Focus+Context Visualization, *in* Scientific Visualization: The Visual Extraction of Knowledge from Data (Proc. of the Dagstuhl 2003 Seminar on Scientific Visualization), Springer, pp. 305-327.

Hwa, L. M.; Duchaineau, M. A. & Joy, K. I. (2004),Adaptive 4-8 Texture Hierarchies, *in* VIS '04: Proceedings of the Conference on Visualization '04, IEEE Computer Society, Washington, DC, USA, pp. 219-226.

Keahey, A. (1998),The Generalized Detail-In-Context Problem, *in* INFOVIS '98: Proceedings of the 1998 IEEE Symposium on Information Visualization, IEEE Computer Society, Washington, DC, USA, pp. 44-51.

Kurzion, Y. & Yagel, R. (1997), Interactive Space Deformation with Hardware-Assisted Rendering, *IEEE Computer Graphics and Applications* **17**(5), pp. 66-77.

Leung, Y. K. & Apperley, M. D. (1994), A Review and Taxonomy of Distortion-Oriented Presentation Techniques, *ACM Transaction on Computer-Human Interaction* **1**(2), pp. 126-160.

Levoy, M. & Hanrahan, P. (1996), Light Field Rendering, *in* Proceedings ACM SIGGRAPH 1996, ACM, New York, NY, USA, pp. 31-42.

Li, Y.; Shum, H.; Tang, C. & Szeliski, R. (2004), Stereo Reconstruction from Multiperspective Panoramas, *IEEE Trans. Pattern Anal. Mach. Intell.* **26**(1), pp. 45-62.

Lindstrom, P. & Pascucci, V. (2002), Terrain Simplification Simplified: A General Framework for View-Dependent Out-of-Core Visualization, *IEEE Transactions on Visualization and Computer Graphics* **8**(3), pp. 239-254.

Löffelmann, H. & Gröller, E. (1996), Ray Tracing with Extended Cameras, *Journal of Visualization and Computer Animation* **7**(4), pp. 211-227.

Mackinlay, J. D.; Card, S. K. & Robertson, G. G. (1990), Rapid Controlled Movement Through a Virtual 3D Workspace, *in* SIGGRAPH '90: Proceedings of the 17th Annual Conference on Computer graphics and Interactive Techniques, ACM, New York, USA, pp. 171-176.

Patterson, T. (2000), A View From on High: Heinrich Berann's Panoramas and Landscape Visualization Techniques For the US National Park Service, *Cartographic Perspectives* **36**, pp. 38-65.

Premoze, S. (2002), Computer Generated Panorama Maps, *in* Proceedings 3rd ICA Mountain Cartography Workshop. Mt. Hood, Oregon.

Roman, A.; Garg, G. & Levoy, M. (2004), Interactive Design of Multi-Perspective Images for Visualizing Urban Landscapes, *in* VIS '04: Proceedings of the conference on Visualization '04, IEEE Computer Society, Washington, DC, USA, pp. 537-544.

Sander, P. V. & Mitchell, J. L. (2006), Progressive Buffers: View-Dependent Geometry and Texture LOD Rendering, *in* SIGGRAPH '06: ACM SIGGRAPH 2006 Courses, ACM, New York, USA, pp. 1-18.

Spindler, M., Bubke, M., Germer, T. & Strothotte, T. (2006), Camera textures, *in* Proceedings of the 4th international conference on Computer graphics and interactive techniques in Australasia and Southeast Asia, ACM, New York, USA, pp. 295-302.

Stone, M. C., Fishkin, K. & Bier, E. A. (1994), The movable filter as a user interface tool, *in* Proceedings of the Conference on Human Factors in Computing Systems, ACM, New York, USA, pp. 306-312.

Vallance S. & Calder, P. (2001), Multi-perspective images for visualization, *in* ACM International Conference Proceeding Series, Vol. 147, ACM, New York, USA, pp. 69-76.

Whitfield, P. (2005), Cities of the World. A History in Maps, The British Library, London.

Wonka, Peter; Wimmer, Michael & Francois Sillion (2001), Instant Visibility, *in* A. Chalmers & T.-M. Rhyne, ed.,Proceedings of Eurographics 2001, The Eurographics Association and Blackwell Publishers, pp. 411-421.

Yang, Y.; Chen, J. X. & Beheshti, M. (2005), Nonlinear Perspective Projections and Magic Lenses: 3D View Deformation, *IEEE Computer Graphics and Applications* **25**(1), pp. 76-84.

Yu, J. & McMillan, L. (2004), A Framework for Multiperspective Rendering, *in* Alexander Keller & Henrik Wann Jensen, ed., Rendering Techniques 2004, Proceedings of Eurographics Symposium on Rendering 2004, EUROGRAPHICS Association, pp. 61-68.

Zanella, A., Carpendale, M. S. T. & Rounding, M. (2002), On the effects of viewing cues in comprehending distortions, *in* Proceedings of the second Nordic conference on Human-computer interaction, ACM, New York, USA, pp. 119-128.

# Scenario-Based Spatial Decision Support for Network Infrastructure Design

Gernot Paulus[1], Martin Krch[1], Johannes Scholz[1], Peter Bachhiesl[2]

[1]School of Geoinformation, Carinthia University of Applied Sciences, Europastrasse 4, A- 9524 Villach, Austria
[2]School of Telematics/Network Engineering, Carinthia University of Applied Sciences, Klagenfurt, Austria

**Abstract.** This paper describes an extended framework for scenario based spatial decision support for constructing new network infrastructure and its application in the domains of telecommunication, forestry and energy. There is an increasing need to provide new planning paradigms to support very expensive strategic investment decisions in new network infrastructure in these domains. The planning processes are still dominated by an expert approach based on empirical knowledge and manual implementation. With this conventional approach it is impossible to visualize and to consider different planning scenarios within a reasonable cost and time frame. We combined the powerful analytical and visualization capabilities of a Geographic Information System (GIS) with mathematical methods of graph theory and combinatorial optimization. This conceptual approach extends the basic spatial decision support model with a knowledge based module for scenario parameterization and graph generation, a module for geodata integration and processing, an operations research optimization module and a multi-level visualization module supporting the need of different communication channels within the decision making process.

**Keywords:** spatial decision support, knowledge base, network infrastructures, network analysis

## 1 Introduction

The major goal of this paper is the support of experts and decision makers responsible for planning and investing in the construction of new network infrastructures. Examples of such investments in new network infrastructures recently discussed in public are next-generation telecommunication networks, forest roads in Eastern Europe and Russia and new high capacity electricity transmission grids. German Telekom reported a planned investment volume of 3 billion Euro (Reuters 2005) for new fibre optic infrastructure. The Österreichische Bundesforste AG rent 176.000 ha of forested area in Russia to facilitate sustainable forestry. Therefore investments in forest accessibility of several million Euros are scheduled (ÖB*f* 2004).

Despite the significant investment volumes in these industries, intelligent computer based decision support in network planning and investment cost simulation is lacking. The planning process is still dominated by a manual expert approach based on empirical knowledge. With this approach it is impossible to consider, evaluate and visualize all potential route alternatives and its related investment costs within a reasonable cost und time frame. Furthermore, no simulations of scenarios for different route alternatives are possible in order to support strategic investment decisions e.g. based on best-case – worst-case scenarios.

In this paper we will discuss an extended SDSS framework based on the framework for spatial multicriteria decision analysis (MCDA). Malczewski (2006) gives a comprehensive survey of a well-established body of literature on GIS and MCDA (e.g. Diamond and Wright 1988; Carver 1991; Jankowski 1995; Malczewski 1999). Our approach discussed in this paper combines the powerful analytical and visualization capabilities of a Geographic Information System (GIS) with mathematical methods of graph theory and combinatorial optimization. This new approach offers promising perspectives for transparent strategic planning and comprehensible simulation of investment costs for constructing new network infrastructure.

## 2 Scenario-Based Spatial Decision Support (SDSS)

This chapter focuses on the generation of a scenario based spatial decision support structure based on the classical framework for spatial multicriteria decision analysis proposed by Malczewski (1999). From this starting point we elaborate on the combination of Geographic Information Science and Technology (GI S&T), Operations Research (OR) and Graph Theory (GT) to facilitate scenario based spatial decision support.

Malczewski's (1999) spatial multicriteria decision framework represents the process of decision making that starts with the problem recognition and definition. Once a problem is clearly identified sets of criteria are selected that support the design of alternatives to be generated in a following stage. These criteria are divided into two groups: evaluation criteria and constraints. Evaluation criteria and their combinations measure the performance of an alternative, whereas constraints limit the possible set of alternatives. According to Malczewski alternatives can be created using standard GI S&T methods. For evaluation purposes a decision variable or a set of decision variables is defined for each alternative, and calculated from evaluation criteria. With the help of decision rules, alternatives are ranked according to decision variables. The result of this process is a recommendation that can be altered by the user's preferences.

In the problem domain of network design, Paulus et al. (2003) state, that it is impossible to consider all possible alternatives with a manual "expert" design approach. In order to overcome the shortcomings of the manual planning approach a Geographical Information System (GIS) is coupled with methods of GT and OR to calculate the optimal solution based on given constraints out of a huge number of possible choices.

The proposed scenario based approach is divided into four different phases: scenario design phase, optimization phase, decision and feedback phase (Fig. 1). Scenario based spatial decision support starts with the generation of different planning scenarios that may have conflicting objectives: e.g. ecology vs. economy.

Each scenario is developed with the use of GI S&T and add-on applications that support the setting of evaluation criteria and constraints. Hence, evaluation criteria values are assigned to distinct spatial feature classes (e.g. cost factor for constructing fibre cable line) and constraints (e.g. for maximum slope for constructing forest roads) for each scenario.

The optimization phase calculates the optimal or near optimal result for each scenario by methods from OR and Graph theory. These include Graph optimization techniques, dynamic programming, simulated annealing, local search, etc, which have been investigated in literature (Jungnickel 2002; Kirkpatrick et al. 1983; Papadimitriou et al. 1998; Aerts and Heuvelink 2002).

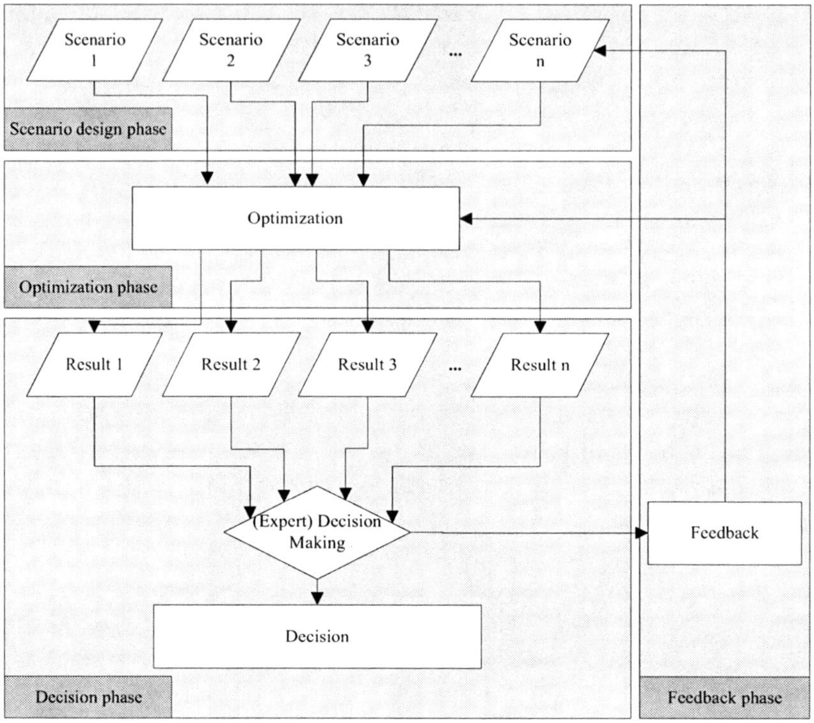


**Fig. 1.** Generic phases of a SSDS for planning network infrastructure

To calculate an optimal solution for the network design task the area of interest is modelled as Graph $G = (V, E, w)$, where $V$ denotes the set of vertices, $E$ the set of edges and $w: E \to R^+$ the weights or costs of each edge (Jungnickel 2002). This leads to graph optimization problems that have gained a lot of interest in the field of OR. The performance strongly depends on the complexity of the weighted graphs, i.e. the number of edges and nodes. For example, the complexity of specific optimization problems on the last mile in the telecommunication network domain comprises up to 300000 edges and 200000 nodes.

Within the decision phase the results and their corresponding decision variable or set of decision variables are calculated and presented in a GIS respectively. Subsequently the expert chooses the scenario that solves the problem best. The planning expert accomplishes that decision process with quantitative analysis of the decision variables and/or with a visual analysis of the generated maps, which serve as guidance in the decision process.

A comprehensive feedback control system is not implemented yet but has to be included in the feedback phase. It enhances the decision making process by altering the input parameters in the scenario design phase (evaluation criteria and constraints) and in the optimization phase (Vacik and Lexer 1999). The novel approach in the scenario based spatial decision support is the fact that scenarios and their results may be directly reused for the generation of new scenarios by changing the parameters. In addition, the system should calibrate itself by analyzing the results and decisions made, and adjust the parameters accordingly. Furthermore, an intelligent guidance system on which parameter values might be used should be available in order to avoid infeasible solutions on part of the user.

## 3 Extended SDSS Framework for Network Infrastructure Design

We propose an extended SDSS framework for network infrastructure design. Four complimentary extensions are added to the basic SDSS framework. We extend the model with a knowledge based approach for parameter definition and graph generation. Another improvement are modules for geodata integration and processing and visualization for result representation and parameter manipulation. The OR part is the kernel of the system where mathematical scenario optimization and result derivation takes place.

### 3.1 Knowledge

Expert domain knowledge is fundamental to acquire proper decision knowledge. This knowledge is acquired through various knowledge acquisition techniques like interviews, text analysis and technical domain report review. The major goal of the expert knowledge acquisition is to document and define a common understanding of the planning process and its involved stakeholders and processes. Structured one-on-one interviews based on specifically developed questionnaires and workshops with planning experts and decision makers in the various domains have been conducted and documented in text from. To integrate and organize it and for communication purposes the gained domain expertise is structured in a diagram based system which we call an informal ontology (Guarino and Giaretta 1995). This implies that the structure is a weak ontology similar to conceptual model (Obrst 2006) with focus on human and not machine interpretability.

In the next step, the informal ontology will be rewritten in a formal language like the Ontology Web Language (OWL) (W3C 2006) to suit as knowledge base (Gómez-Perez et al. 2004). This knowledge base includes all relevant network design parameters as well as rules for the graph generation algorithms. Beside that the ontology is the basis for geodata modelling and the framework for the evaluation criteria like the cost factors for cable laying or constraints like exclusion of specific landuse classes from the planning process (e.g. private property or environmental protection areas). The framework and parametrization of different scenarios in terms of evaluation criterias and constraints is directly derived from this knowledge base.

## 3.2 Geodata

The geodata model is derived from the informal ontology and has to satisfy the requirements of the network planning domain. During the process of geodata integration different data formats are integrated into one database format, while during data pre-processing semantic issues from the different data models are consolidated. The geodata processing delivers well structured data sets of planning relevant land cover and land use classes which are used as input for graph generation. However, in order to generate structured and well defined geodata sets, the following issues have to be solved:

- Semi-automatic data capture of important land use classes, which are not present in the available geodata provided by data vendors or the customer itself. One example is the missing road intersection polygons in cadastral data. These intersection polygons are important because these features represent edges in the graph offering the possibility of crossing roads. This geometric information is used in the optimization process in order to find the "best" connection of a given set of customers.
- Topological correct data. As the geometry of the selected geodata sets build up the graph structure, connectivity plays an important role in the geodata processing.
- Geodata quality: Selected geodata sets must be evaluated in terms of positional and attributive accuracy. This is especially important for integrating heterogeneous geodata sets from different sources and different scales.

In order to solve these issues and generate a valid input data for the graph generation, semiautomatic workflows have been developed using Feature Manipulation Engine FME (Prunner 2006). However, the process of evaluation and integration heterogeneous geodata sets is still a time consuming, domain dependent and therefore critical process in the presented workflow.

### 3.3 Operations Research (OR)

OR Methods and their combination are able to calculate the optimal solutions for a given problem due to a problem definition and a set of criteria and constraints. Possible strategies include: simulated annealing or dynamic programming. Kirkpatrick et al. (1983) proposed simulated annealing as a technique for optimizing combinatorial problems, without becoming trapped in local minima. Dynamic programming is a useful technique for optimization problems, where decisions are dependent on previous decisions. With graph theory it is possible to solve a number of relevant spatial problems like routing problems (Jungnickel 2002), shortest path, travelling salesman problem, or network flows (Ahuja et al. 1993).

In the case of infrastructure network design the input for optimization is a weighted graph that results from the transformation of the geodata. This transformation process is expert knowledge driven, e.g. the expert determines which land use or land cover class geometries and their corresponding weights should be used to build up the weighted graph. The optimization process itself relies on standard OR methods that are extended and adapted to solve complex problems related to network infrastructure design. Bachhiesl et al. (2003) demonstrated a successful approach of combining OR-methods with GIS for calculation of optimized fibre optic access networks.

### 3.4 Visualization

Visualization is important during the complete process of decision support, but there are some important process stages to emphasize.

Visualization is not limited to present simulation results but is also an integral part of the system development and system interaction process. As an example, the informal diagram oriented ontology is used for interdisciplinary communication not only between domain experts but between novice and expert users or technicians and managers (Mizoguchi et al. 1995). Effective workflow diagrams support the complexity of the model and an easy to handle user navigation as well (Alexander and Maden 2004).

For the validation of different steps in the SDSS process the integration of visualization of intermediate and final optimization results in conjunction with adequate reference data is indispensable (Fig. 4, 5, and 6). This visualization integration comprises (a) the geodata integration process and definition of spatial constraints (Fig. 2a); (b) the planning cost distribution, e.g. construction costs per meters (Fig. 2b) and (c) the evaluation of simulation results by the domain expert. Furthermore, such high quality and comprehensively visualized different steps of processing a scenario provide selective communication channels like (a) from system to expert; (b) from expert to expert; or (c) from expert to laymen, which is increasingly important in public participation processes.

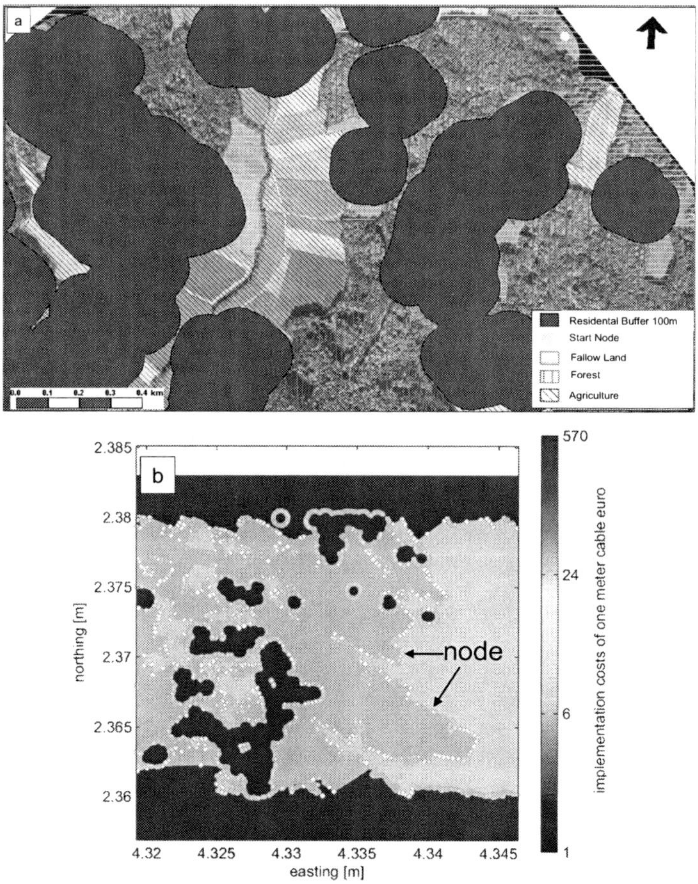


**Fig. 2.** Visualization of different steps in processing a planning scenario for a high voltage transmission line. (a) Spatial Constraints (100m Buffer around any residential area) and relevant land use classes; (b) Distribution of construction costs and nodes serving as input data for weighted graph generation.

### 3.5 Extended SDSS Process Model

The above defined four additions extend the SDSS process model as follows (Fig. 3):

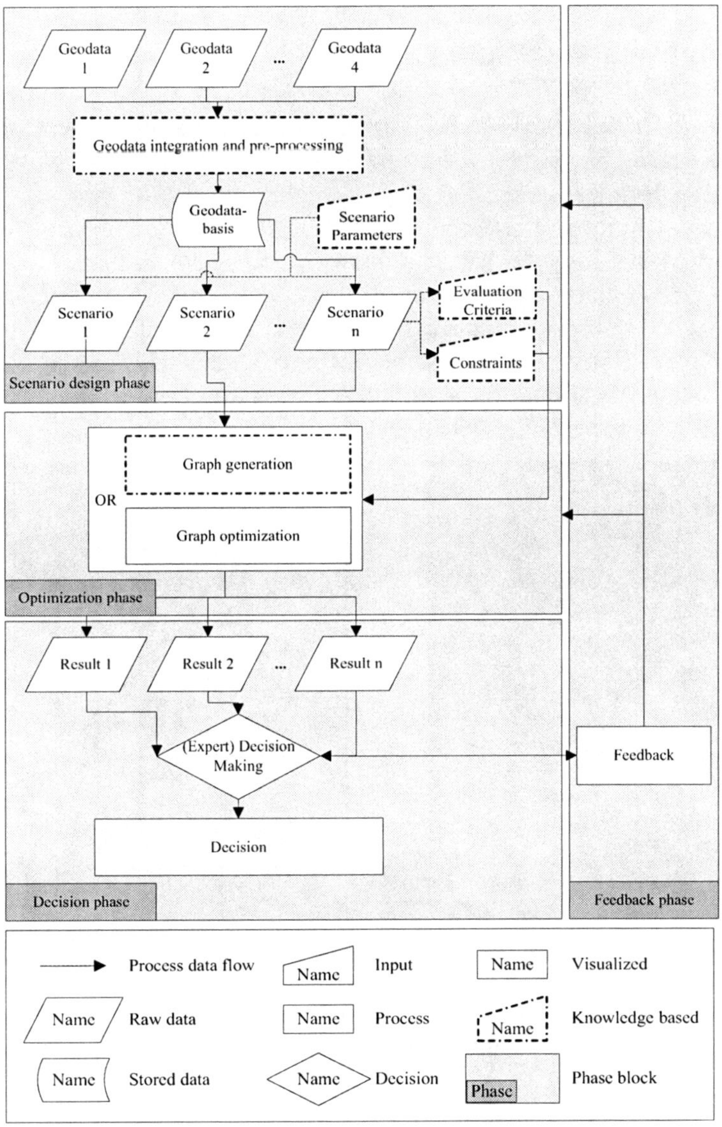


**Fig. 3.** Extended SSDS process model

Geodata are the initial input and serve as the basis of the decision support system. Visualization is applied in several phases of the process and guides the user through the decision making. All parameters are based on expert knowledge, which is also used during graph generation from planning-relevant land use classes.

## 4 Applications and Visualization

Scenario based spatial decision support can be used in forest road planning, which serves as basis for sustainable forestry. Together with the Österreichische Bundesforste AG (ÖB*f*) a pilot study has been conducted to evaluate the benefit of a GIS and spatial decision support in forest road planning (Gruber and Scholz 2005). This work shows the potential of sustainable forest road planning with methods of GI S&T and OR which build up scenario based decision support. Furthermore it is mentioned, that the developed project allows the forest engineer to create alternative forest road networks and evaluate them directly, which supports an interactive decision making process (Fig. 4, Tab. 1).

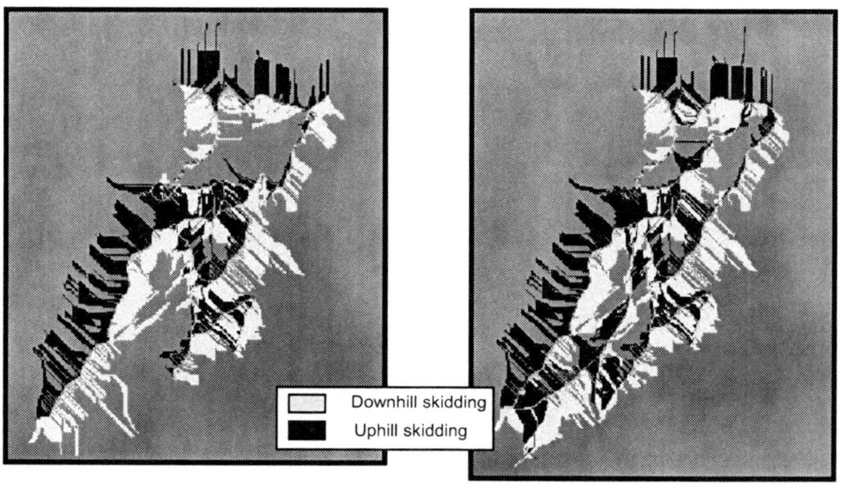


**Fig. 4.** Comparison of existing (left) and alternative (right) forest road network in test site T2. The accessible areas around the road networks are marked with grey and black color. Grey indicates the areas where downhill skidding and black indicates the areas where uphill skidding is assumed as most suitable harvesting methods.

**Table 1.** Results of existing and alternative, optimized forest road networks in test sites $T_1$ and $T_2$. The investigation period for the estimation of the benefit is 40 years

| | **Road Length [m]** | | **Construction Costs [EUR]** | | **Benefit [EUR]** | |
|---|---|---|---|---|---|---|
| | **Existing** | **Alternative** | **Existing** | **Alternative** | **Existing** | **Alternative** |
| $T_1$ | 13.234 | 13.876 | 422.000 | 466.000 | 899.000 | 1.294.000 |
| $T_2$ | 11.559 | 15.514 | 359.000 | 491.000 | 888.000 | 1.205.000 |

Currently the extended SDSS process model is applied in the FHplus project NETQUEST focusing on the planning of urban telecommunication networks with a special emphasis on fiber optic network infrastructure. Especially the facilitation of expert knowledge through ontologies and the voluminous data in urban planning scenarios are subject of ongoing research. First benchmark projects with industry partners have shown the relevance of the approach. Fig. 5 shows the results of a benchmark project for designing a network infrastructure for a given set of 37 customers based on planning relevant land use classes.

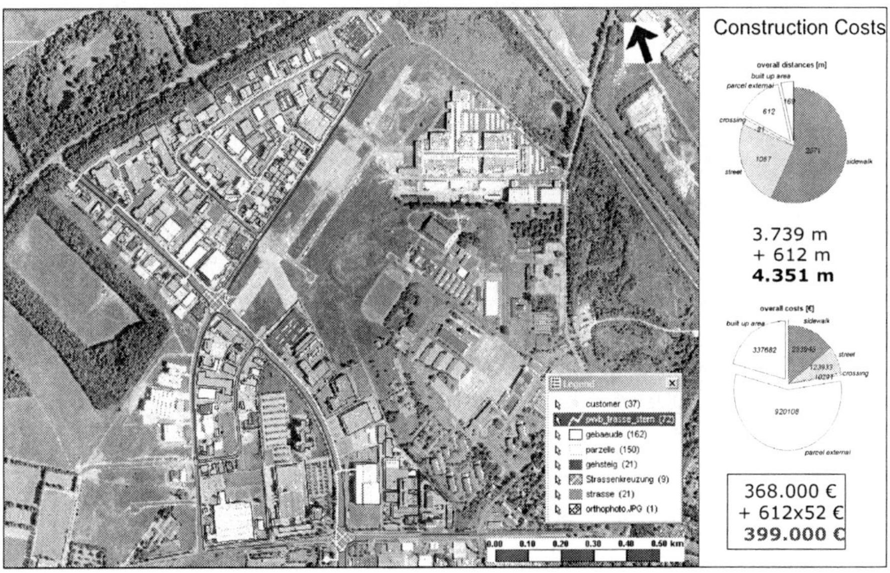


**Fig. 5.** Visualization of a cost optimized fibre optic network design together with expected construction costs for one specific scenario. (Benchmark project implemented for industry partner NetCologne in Köln-Ossendorf, Germany).

The land use classes for this particular benchmark project are street, sidewalk, street crossing and private property, each of them specific construction costs (e.g. street: 300 €/m; side walk: 150 €/m) are assigned.

The input data for the weighted graph generation are all customer locations to be connected, the geometry of the land use classes and their specific construction costs. The cost optimized network and the expected investment costs for this particular scenario are visualized using high resolution aerial views as reference data set.

We applied this framework successfully as an innovative transparent planning approach in a major 380 kV high voltage ring closure project in Austria. In this project it is impossible for a network planner to consider and evaluate all potential route alternatives with a manual approach based on empirical knowledge. The more customer nodes exist, and the larger the spatial extent of the project area is, the more alternatives for laying a transmission line can theoretically exist depending on the combination and consideration of individual factors and constraints like economic, ecologic or social issues in the planning process. The planner has to simplify his task to a feasible problem dimension, which results in most cases in non- optimized and non-transparent decisions. Furthermore, with this manual approach, no cost- and time effective simulations of investment and laying scenarios for different route alternatives are possible.

Various planning scenarios were simulated and visualized using aerial views and 3D animation. Planning relevant land use classes in this project are residential area, agriculture, fallow land, forest and willow trees, which are weighted using factors representing different paid compensation for crossing a land use class in €/m. Another constraint was that any residential area must be bypassed by the high voltage transmission line at the minimum distance of 50m. High quality geo data like high resolution aerial views and digital terrain models provide additional important realistic visualisation possibilities. This offers new opportunities in communicating the results to the public. Especially for any application dealing with visualizations at very detailed levels (e.g. single parcel) high resolution data are necessary. Figure 6 shows three scenarios visualized using a digital terrain model.

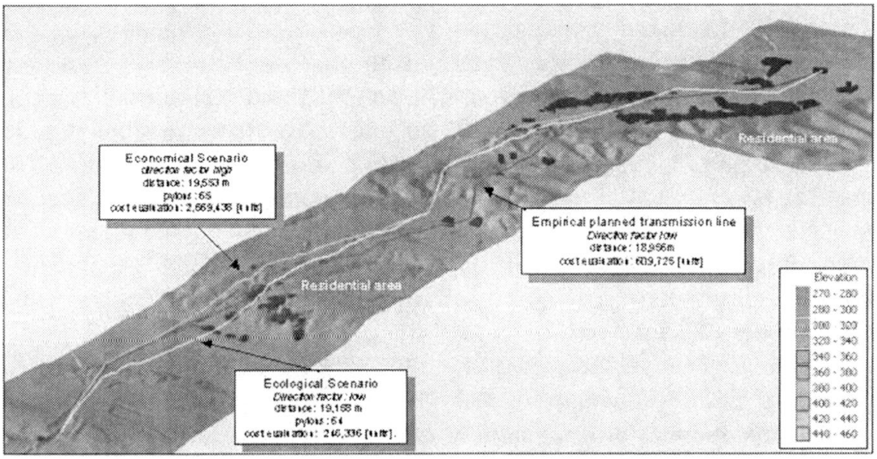


**Fig. 6.** Decision support by comparison three scenarios for the laying of the 380 KV high voltage transmission line. *Green:* Economic scenario (forest = "cheap"); *Yellow:* Ecologic scenario (forest = "expensive"); *Blue*: Empirical planned transmission line. (Paulus et al. 2003)

## 5 Conclusions and Future Work

The proposed extended SDSS planning framework provide the following opportunities for designing, simulating and presenting different scenarios and results of a planning process to its relevant stakeholders:

Different planning scenarios can be very efficiently calculated and compared. This will result in better understanding of the complex planning process by the public. Relevant scenarios to the public can be surveyed using e-voting, open discussions or other public participation processes. Different views on the project can easily be considered by changing the relevant land use classes and the cost factors to be assigned.

Increased principal availability of high quality geo data (e.g. cadastre data, orthophotos, high resolution digital terrain models) allows fully digital planning workflows. Limiting factor here are the still high costs for high resolution geo data, especially for DTM's or orthophotos. Furthermore, realistic visualisation of different results using 3D-animation is another important issue for public acceptance.

The level-of-detail of the calculated scenarios is totally flexible and can range from regional level down to a very detailed single-parcel level. Real investment costs as well fictive weighting units can be considered. Theoretically, it is possible to assign unique cost factors for every single feature in a project area under investigation. The finer the spatial planning resolution the more complex are the resulting weighted graphs.

Visualization interfaces at different levels of the decision making process are important in order to provide an interactive decision support process. We suggest three levels of visualization interfaces: (1) Scenario Design Phase Level, (2) at the network graph generation level and (3) at the Decision phase level, where the results of the different scenarios are compared. Future research will focus on how to visualize and analyze the characteristics of alternative planning scenarios especially when dealing with a large number of scenarios.

Another topic of future research is the linking of the extended SDSS framework with web mapping technology. This approach will integrate various stakeholders in the planning process providing a user-defined access to planning scenarios and personalized views on optimization results.

## Acknowledgments

This work is funded by the Austrian Research Promotion Agency (FFG) and is part of the FHplus project NETQUEST – Cost Simulation and Optimization of fibre optic networks on the 'last mile' (Reference Nr. 811378). The critical discussions with Josef Strobl and his remarks on this project are gratefully acknowledged.

## References

Alexander, I. F. and Maden, N. (Ed.) (2004) *Scenarios, Stories, Use Cases - Through the Systems Development Life-Cycle* (Chichester: John Wiley & Sons).

Ahuja, R. K., Magnati, T. L. and Orlin, J.B. (1993) *Network Flows: Theory, Algorithms, and Applications,* (Upper Saddle River, NJ: Prentice Hall).

Aerts, J. C. J. H. and Heuvelink, G. B. M. (2002) Using simulated annealing for resource allocation, *International Journal of Geographical Information Science*, **16** (6), pp. 571-587

Bachhiesl, P., Prossegger, M., Paulus, G., Werner, J. and Stögner, H. (2003) Simulation and Optimization of the Implementation Costs for the Last Mile of Fiber Optic Networks, *Networks and Spatial Economics*, **3** (4), pp. 467-482

Carver, S. J. (1991) Integrating multi-criteria evaluation with geographical information systems. *International Journal of Geographical Information Systems*, **5**, pp. 321–339

Diamond, J.T. and Wright, J.R. (1988) Design of an integrated spatial information system for multiobjective land-use planning. *Environment and Planning B*, **15**, pp. 205–214

Gómez-Perez, A., Fernández-López, M. and Corcho, O. (2004) *Ontological Engineering: with examples from the areas of Knowledge Management, e-commerce and the Semantic Web*, (London: Springer).

Gruber, G. and Scholz, J. (2005) GIS based planning of forest road networks. In *Angewandte Geoinformatik 2005. Beiträge zum 17. AGIT-Symposium Salzburg*, J. Strobl, T. Blaschke and G. Griesebner (Ed.), pp.218-223 (Heidelberg: Wichmann).

Guarino, N. and Giaretta, P. (1995) Ontologies and Knowledge Bases: Towards a Terminological Clarification, In *Towards Very Large Knowledge Bases: Knowledge Building and Knowledge Sharing,* N. Mars (Ed.), pp.25-32 (Amsterdam: IOS Press).

Jankowski, P. (1995) Integrating geographical information systems and multiple criteria decision making methods, *International Journal of Geographical Information Systems*, **9**, pp. 251–273

Jungnickel, D. (2002) *Graphs, Networks and Algorithms* (Berlin: Springer).

Kirkpatrick, S., Gelatt Jr., C.D. and Vecchi, M.P. (1983) Optimization by Simulated Annealing, *Science,* **220**, pp. 671-680.

Malczewski, J. (1999) *GIS and Multicriteria Decision Analysis* (New York: John Wiley & Sons).

Malczewski, J. (2006) GIS-based multicriteria decision analysis: a survey of the literature, *International Journal of Geographical Information Systems*, **20 (7)**, pp. 703–726

Mizoguchi, R., van Welkenhuysen, J. and Ikeda, M. (1995) Task Ontology for reuse of problem solving knowledge, In *Towards Very Large Knowledge Bases: Knowledge Building and Knowledge Sharing,* N. Mars (Ed.), pp.46-57 (Amsterdam: IOS Press).

Obrst, L. (2006) The Ontology Spectrum and Semantics Models. Available online at: http://ontolog.cim3.net/file/resource/presentation/LeoObrst_20060112/OntologySpectrumSemanticModels--LeoObrst_20060112.ppt (accessed August 2006).

Österreichische Bundesforste (2004) Bundesforste steigen in Russland ins Forstgeschäft ein. Available online at: http://www.dlv.de/sro.php?redid=46509 (accessed August 2006).

Papadimitriou, C. H. and Steiglitz, K. (1998) *Combinatorial Optimization: Algorithms and Complexity* (Mineola, NY: Dover Publications).

Paulus, G., Bachhiesl, P., and Pospischil, W. (2003) Strategic Planning and Simulation of Electricity Transmission Grid Infrastructure: New Opportunities for Public Participation – In *International Conference on Public Participation and Information Technologies*, 10-12 November 2003, Massachusetts Institute of Technology, Cambridge, USA.

Prunner, N. (2006) Evaluierung und Integration von heterogenen Geodaten für die Planung von urbanen Glasfaser-Leitungstrassen, Unpublished Master thesis, School of Geoinformation, Carinthia University of Applied Sciences, Villach, Austria.

Reuters (2005) Neues Glasfasernetz für Deutschland. Available online at: http://www.tkp.at/channel_telekom/news_21487.html (accessed November 2007).

Vacik, H. and Lexer, M. J. (1999) Spatial Decision Support Systems for Silvicultural Planning, In *IUFRO-Conference: The transformation of plantation forests*, 3-6 September 1999, Edinburgh, Scotland.

W3C (2004) OWL Web Ontology Language Overview W3C - Recommendation 10 February 2004. Available Online at: http://www.w3.org/TR/owl-features/ (accessed November 2007)

# Grouping of Optimized Pedestrian Routes for Multi-Modal Route Planning: A Comparison of Two Cities

Hartwig H. Hochmair

University of Florida, Geomatics Program
3205 College Avenue, Fort Lauderdale, FL 33314, USA
hhhochmair@ufl.edu

**Abstract.** The purpose of multi-modal route planners is to provide the user with the optimal route between trip start and destination, where the route may utilize several transportation modes including public transportation. The optimal route is defined over a set of evaluation criteria considered by the user during the route selection process. Especially in the case of multi-modal transportation, numerous evaluation criteria play a role in the traveler's route choice. Thus the number of requested search parameters in the route planner may be large, and the user interface is overcrowded easily. Based on a set of pedestrian routes that are optimized for various criteria in multi-modal, inner-urban transportation networks of two European cities, an exploratory study based on Principal Components Analysis (PCA) identifies underlying factors that capture the correlations among route selection criteria. The results show how the variability of routes can be parsimoniously described with a smaller set of components, and how these findings can be used to simplify the user interface design of multi-modal route planners.

**Keywords:** Multi-modal route planning, spatial decision support, user interface design, pedestrian navigation, network analysis

## 1 Introduction

Various existing route planners provide access to public transportation systems, which encourage their use and help reduce individual car traffic. Especially when it comes to multi-modal route planning, the number of optimization criteria involved in the decision making process is extensive, and user interfaces may be complicated to use. A simple user interface is crucial for the success of a route planning system. This research aims at finding underlying factors that explain the variability of routes between trip start and destination that are optimized for various criteria. The results can be used to simplify the route planner user interface by reducing the number of parameter settings input by the user.

Route selection problems commonly involve a set of route alternatives from which a choice of an alternative must be made under consideration of several evaluation criteria. This is also true for the use of a route planner if the user can specify route preferences. With this paper, we address route selection within the framework of multi-attribute decision making (MADM). The MADM framework involves a selection among a limited set of alternatives and has a single, implicitly defined objective (Malczewski 1999). The objective is functionally related to, or derived from the set of attributes, and alternatives are described by their attributes. Solving a MADM problem involves the sorting and ranking of alternatives according to an underlying decision rule. A decision rule is a procedure that integrates information on alternatives and the decision maker's preferences to produce an evaluation of the set of alternatives. Two classes of decision rules can be distinguished: compensatory and non-compensatory. The compensatory approach is based on the assumption that the high performance of an alternative achieved in one or more criteria can compensate for the weak performance of the same alternative in other criteria. Contrarily, under the non-compensatory approach a poor performance by an alternative in a criterion cannot be offset by another criterion's good outcome.

Evaluation criteria (which are also called attributes or decision variables in the MADM framework) include benefit criteria and cost criteria. For a benefit criterion, a higher attribute score is more attractive, whereas for cost criteria, a lower score is more desirable. Eliminatory constraints impose limitations on the set of decision alternatives. An alternative is feasible if it satisfies all eliminatory constraints.

## 1.1 Route Selection Criteria for Pedestrians in Multi-Modal Networks

Whereas pedestrian route choice has been studied for many years, research on route preferences in multi-modal trips is sparsely reported in literature. Findings from both areas are relevant, as a multi-modal route planner should provide the user with the necessary choice options for defining and selecting the optimal route. Generally speaking, the set of attributes must be complete to cover all relevant aspects of a decision problem (Keeny and Raiffa 1993).

An empirical study by Heye and Timpf (2003) investigated factors that influenced the complexity of transfer processes in public transportation networks. The four most frequently mentioned physical characteristics of transfer points in participants' responses were transfer distance, streets to cross, signage, and number of lines. Whereas minimizing time is one of the most widely used optimization criteria in multi-modal route planning, for budget tourists and the general public cost is often as important as time optimization (Chiu et al. 2005). Avoidance of uncertainties about travel conditions has also been found important. For example, studies have shown that public transportation for which travel time uncertainties are relatively low in cities where public transportation is favored, overpasses private transportation when travel times are rather similar (Peytchev and Claramunt 2001).

In pedestrian navigation shortest routes are often chosen, although pedestrians are seldom aware that they are minimizing distance as a primary strategy (Lausto and Murole 1974). The disutility of a route depends, besides distance or travel time, also on the proximity of obstacles or other physical obstructions, the number of sharp turns, the expected number of interactions with other pedestrians (Hoogendoorn and Bovy 2004), and on traffic, number of crossings, amount of crime, attractions, and weather protection (Senevirante and Morrall 1986). A study by Daamen et al. (2005) examined the influences of changes of vertical levels on passenger route choice and showed that stairs provided highest disutility, followed by ramps and escalators. Golledge (1995) identified relevant route selection criteria for pedestrian navigation in a known campus environment. In questionnaires, subjects rated shortest route, route taking the least time, and route proceeding in the direction of destination as the most important criteria. Fewest turns, first noticed, and usual route were the next important criteria. Muraleetharan and Hagiwara (2007) investigated the benefits of improving the overall level-of-service (LOS) at walkways and crosswalks. The results indicate that on longer travel paths, pedestrians divert from the shortest-path route and use high LOS sidewalks and crosswalks. On the contrary, when the destination is less than a few hundred meters away from the start, the probability that a pedestrian would utilize the shortest route becomes high, regardless of the route-LOS.

## 1.2 Designs of Multi-Modal Route Planners

A growing number of electronic multi-modal route planners that are freely available on the Internet provide evidence for the increasing popularity of these spatial decision support tools. To give an idea of frequent route planner functionalities this section reviews the pedestrian related route choice options provided in three selected multi-modal route planners for Europe (BayernInfo 2008; JPL 2008; SCOTTY 2008). All three route planners support non-compensatory decision making by allowing the user to set a single optimization criterion, and/or by setting eliminatory constraints. Thus the route planners do not provide importance weighting and the selection of compromise routes. JPL (2008), a journey planner for London, provides a selection option between fastest route, route with fewest changes, and routes with least walking between stops, whereas the other two route planners use a default optimization function that is hidden from the user. The latter two route planners allow, however, to set the maximum number of transfers as an eliminatory constraint instead. All three route planners provide an option to deselect undesired public transportation means, and to specify the maximum walking distance or time. With JPL (2008), the user can set mobility requirements, such as avoiding stairs. Further, it provides an option to search for walk-only routes if they are faster than with public transit. This function is, however, redundant if fastest route is chosen as optimization criterion. Other relevant route selection criteria, such as route with short transfer waiting times, are not included in any of the three planners. Chiu et al. (2005) suggest that users of multi-modal route planners should be offered with comparisons between fastest and cheapest route, as one optimization criterion may not provide the optimal route.

Related research found that route selection criteria used in bicycle trip planning can be grouped into four general criteria (Hochmair 2004; Hochmair 2007). If public transportation planning is combined with bicycle route planning, the criteria can be grouped into five general criteria, which are fast, simple, quiet, scenic, and safe (Hochmair 2008). Based on Principal Components Analysis (PCA) and the use of street and public transit datasets from two European cities, this research will expand previous research to pedestrian route planning in combination with transit transportation.

### 1.3 Structure of the Paper

The remainder of this paper is structured as follows. Section 0 describes the setup of the computer-based exploratory study, and section 0 formulates the challenges and computational approaches for seeking a Pareto optimal set of routes in a multi-modal transportation network. Section 0 analyses the retrieved route sets using PCA and discusses its results with respect to the user interface design of route planners. Conclusions and future work are presented in section 0.

## 2 Design of the Exploratory Study

### 2.1 Test Networks

The datasets represent part of the street and public transportation networks of Vienna, Austria (Fig.1a), and Bremen, Germany (Fig.1b).

(a) (b)

**Fig. 1.** Layout of street and public transportation networks for Vienna (a) and Bremen (b)

The Vienna test network has 631 links and 404 nodes, the Bremen test network 701 links and 470 nodes. A node is defined as intersection or end of a cul-de-sac, and a link is a roadway or transit segment between two nodes. In the Vienna test area, public transportation is provided through three metro lines (U2, U3, U6) (dashed lines), 15 tram lines (5, 6, 9, 18, 33, 37, 38, 40, 41, 42, 43, 44, 46, 49, J) (dark continuous lines), and two bus routes (13A, 48A) (light continuous lines). Metro-stops are marked with "U" symbols, and tram and bus stops with color-coded circles. The Bremen test area has 7 tram lines (1, 2, 3, 4, 6, 8, 10) and three bus routes (22, 24, 25). The transit density, which we define as total length of public transit edges over total length of street edges, amounts to 40.3% in the Vienna test network, and to 17.7% in the Bremen network. The timetables of public transportation used in the network modeling can be found at *http://efa.vor.at/wvb/index_en.htm* for Vienna, and at *http://www.bsag.de/eng* for Bremen. None of the two test areas contains pedestrian zones.

For the exploratory study, a set of 50 start-destination pairs was selected from each test network. For each of these pairs, a set of Pareto optimal paths was explored as described further below. The search algorithm was programmed in Delphi and run on a desktop PC.

## 2.2 Route Selection Criteria Included in the Optimization Process

As far as possible, the route selection criteria in pedestrian navigation and public transit use identified in previous work were implemented in the network structure and the algorithm for seeking Pareto optimal route sets (Table 1). A path is called Pareto optimal if no other path from the available route set has (a) a better value for at least one criterion and (b) at least equally good values for all other criteria. Such path is also called non-dominated. The set of non-dominated paths is the Pareto optimal route set.

**Table 1.** Route selection criteria used in the optimization process

| **Street level** | | **Public Transit level** | | **Combined level** | |
|---|---|---|---|---|---|
| *Link-related* | *Node-related* | *Link-related* | *Node-related* | *Link-related* | *Node-related* |
| Walking distance<br>Parks<br>Sights<br>Shopping streets | Traffic lights<br>Intersections<br>Turns<br><br>Street crossings during public transportation transfer | Transfers<br>Trip fare<br><br>Transfer waiting time<br><br>Public transportation portion | Choice options at transfer | Travel time | Turns and transfers combined |

The 15 route attributes are measured as ratio values. The Bremen test area is mostly residential and does not include sights, which makes 14 attributes for that network. As all transit vehicles in both public transit networks will be wheelchair accessible in the near future, mobility requirements were excluded from the set of optimization criteria in the study. The distinction into link-related and node-related criteria refers to whether the attribute values are stored with links or nodes in the network graph. Cost criteria are underlined, whereas benefit criteria are printed as plain text.

# 3 Modeling Approach

## 3.1 Problem Formulation

It has been found earlier by Vilfredo Pareto that even without making any multicriteria decisions the solution space of a multi-criteria problem is already partially ordered so that all dominated solutions can be eliminated from consideration before the multicriteria decisions are made (Pareto 1896). In consequence, we explore underlying factors among route characteristics based on a set of Pareto optimal routes. The methods used to seek the Pareto frontier will be described.

Single-criterion shortest path problems (SSP) find the shortest path using a single optimization criterion, e.g., travel time, whereas multi-criteria shortest path problems (MSPP) consider two or more independent criteria in evaluating the solution. Solving a SSP or MSPP can help to build the Pareto optimal route set. Historically many MSPP are reduced to a SPP by using a weighted linear combination of all criteria as the cost function. However, it may be difficult to compute an appropriate set of weightings for the criteria involved, and optimal solutions may be overlooked (Mooney and Winstanley 2006). Even the bicriteria path-problem in a graph is NP-hard, but pseudo-polynomial time algorithms are known that find all Pareto paths in a graph in time polynomial in the number of paths and nodes (Hansen 1980). The problem is that with a MSPP there may exist an exponential number of non-dominated solutions in the worst case. An SPP or MSPP path approach cannot solve problems that involve benefit criteria, as there exists no polynomial-time algorithm for the longest path problem if the network contains negative cycles (Hardgrave and Nemhauser 1962). The latter is generally true for street networks. This paper uses a genetic algorithm to optimize routes regarding benefit criteria, such as parks, and Dijkstra's algorithm to minimize cost criteria of a route, e.g., travel time.

Multi-modal route planning systems need to account for the transfer between different transportation modes. This involves modeling of the physical complexity of the transfers (Heye and Timpf 2003), or modeling of a dynamic waiting time that depends on the time the commuter arrives at the station and the departure time of the vehicle. For both public networks a fixed amount is charged per trip. The fare is independent of the number of transfers made or stops traveled.

### 3.2 Network Modeling and Graphs

The basic model of the road network is a weighted, directed node graph G=(V,E) which comprises a set of vertices V and edges E connecting these vertices. Edges carry values for cost and benefit criteria, such as distance or number of sights. In addition to this, vertex cost, i.e., cost associated with each pair of connected edges in the node graph, need to be included as well. For this purpose the node graph is mapped to a line graph $D(N_D, E_D)$, which allows cost functions for traversing a segment and a vertex in the node graph to be attached as attributes to graph elements in a line graph. For details the reader is referred to related work (Winter 2002).

The line graph is particularly useful for finding routes which minimize cost between connected edges, such as waiting time at traffic lights or intersections, turn cost, or cost associated with choice options at transfers. The SP algorithm is executed on the line graph of the original network graph. In search for routes that minimize a street bound cost criterion (e.g., turns), public transportation segments are excluded from the SP route search, and a route that is exclusively running on street segments is returned. A walking speed of 4 km/h is assumed for a pedestrian.

### 3.3 Modeling Travel on Public Transportation Routes

To model the transfer between transportation modes, a technique based on node explosion (Spiess and Florian 1989; Meng et al. 1999) is used. This approach takes the node graph of the street network and transforms it into the expanded node graph G'=(V', E') through the following steps (Fig.2):

For each directed public transportation route K, do the following:

- For each stop $v_i$ along the route add a new vertex $v_{i,K}$ to the expanded graph.
- Replace each directed edge $e_{i\text{-}j}$ along its route by three new directed edges, namely (a) an *access* edge $e_{i\text{-}i,K}$ that connects the access point $v_i$ to the transit route, (b) a *traveling* edge $e_{i,K\text{-}j,K}$ that represents the

commuter travel on the transit route from stop $v_i$ to $v_j$, and (c) an *alighting* edge for exiting the transit route K at stop $v_j$.

- Add *transfer* edges $e_{i,K-i,L}$ (dashed lines) to all other traveling edges.

After node explosion, the expanded node graph is mapped to a line graph to facilitate all necessary shortest path computations.

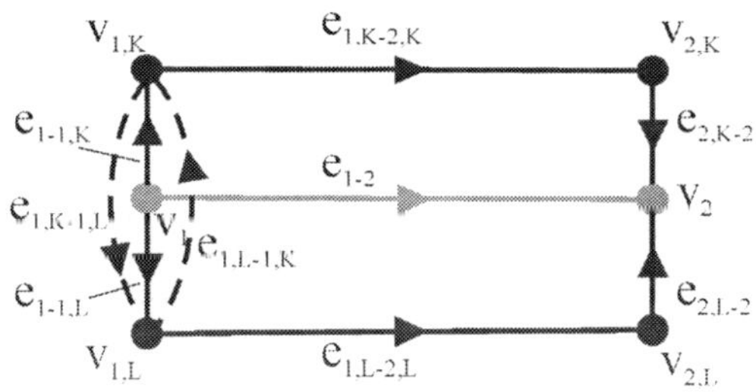


**Fig. 2.** Node explosion shown for two public transportation routes

### 3.4 Genetic Algorithm Framework

Over the last three decades genetic algorithms (GA) (Holland 1992), which are more recently also referred to as evolutionary algorithms (EAs), have gained high importance for exploring the Pareto front in multi-objective problems that are too complex to be solved by exact methods (Zitzler et al. 2000). Horn (1997) provides an overview of common methods that seek the Pareto front. Independent sampling (Fourman 1985) performs multiple single-criterion searches to optimize one criterion or a linear combination of criteria at a time where weights are varied from search to search. Simultaneous parallel search for multiple members of the Pareto optimal front includes among others criteria selection, aggregation selection, and Pareto selection. Pareto selection favors Pareto optimal solutions above others. Many of these efforts have incorporated some form of active diversity promotion, such as GA niching (Goldberg 1987), to find and maintain an even distribution sampling of points along the Pareto front.

Fig.3 depicts the structure of the genetic algorithm used in the exploratory network application.

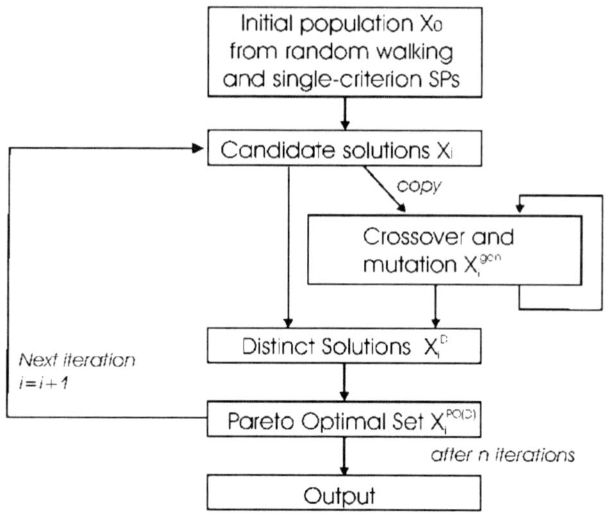


**Fig. 3.** Genetic algorithm for searching Pareto optimal routes

An initial population $X_0$ is created through random walking (Costelloe et al. 2001) on the line graph of the expanded node graph. In addition to this, initial routes are found through separate single-criterion SP computations that minimize walking distance, travel time, traffic lights, intersections, and turns. This gives the "corners" of the Pareto optimal surface for cost criteria. Next, a repeated standard one-point crossover and mutation are executed to find additional solutions on the Pareto front. A copy of $X_i$ ensures that good solutions are not destroyed during crossover and mutation. This is followed by the elimination of duplicate solutions from the intersection of $X_i$ and the modified set $X_i^{gen}$. Pareto elitist selection gives the Pareto optimal set of candidates $X_i^{PO(D)}$ which provides the population for the next iteration and grows with each of the 10 iterations used. No objectives are specified at this point, thus a fitness function and quality metric are not included in the framework.

Crossover describes the process where two chromosomes (the parents) line up and then swap the portions of their genetic code beyond the crossover point which creates two offspring. In the framework of this paper, candidate routes can be viewed as chromosomes, with the sequence of route segments being their genes (Fig. 4a).

Mutations make a random modification of the chromosomes. Whereas mutation is traditionally applied on one string (chromosome), the approach in this paper uses two parents to create a mutated offspring that replaces one parent. A random path is computed to connect the two randomly chosen points on both parent routes and to mutate the first parent (Fig. 4b).

Offspring with a walking distance more than twice the shortest path walking distance, or offspring with more than twice the fastest travel time were removed from the route pool, as these routes were presumed unacceptable by a user.

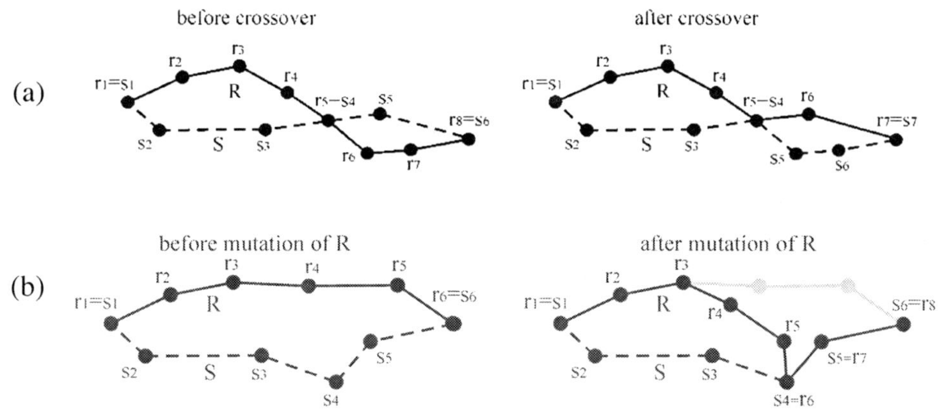


**Fig. 4.** Crossover (a) and mutation (b)

## 4 Results

913 (Vienna) and 1367 (Bremen) Pareto optimal paths were found in the search process for the 50 start-destination pairs. Some characteristics are summarized in Table 2.

**Table 2.** Descriptive values of the Pareto optimal route sets

| | **Vienna** | | **Bremen** | |
|---|---|---|---|---|
| Attribute | *Mean* | *SD (±)* | *Mean* | *SD (±)* |
| Total distance [m] | 2589 | 787 | 3498 | 1154 |
| Walking distance [m] | 1493 | 740 | 2483 | 994 |
| Trip time [min] | 29.6 | 7.4 | 44.5 | 13.2 |
| PT portion [%] | 35.5 | 34.8 | 23.3 | 29.3 |
| Waiting time [min] | 5.6 | 3.8 | 9.2 | 6.6 |
| Transfers | 2.4 | 1.2 | 2.0 | 1.1 |
| Choices at transfers | 10.8 | 7.0 | 7.8 | 4.5 |
| Turns | 5.4 | 4.0 | 7.8 | 4.4 |

Waiting time, transfers, and choices at transfers are derived from routes that use public transit. Entering a transit route from the street level is counted as one transfer. The table shows that a smaller transit density in the Bremen network (see section 2.1) leads to a significantly smaller portion of public transit use (thus a larger portion walked in trips), a longer waiting time, and a smaller number of transfers and choices at transfers (p=0.000 for all four criteria; Mann Whitney Test).

## 4.1 Principal Components Analysis

This section explores whether the variability of the Pareto optimal route set can be more parsimoniously described by a smaller number of components using Principal Components Analysis (PCA). PCA involves a mathematical procedure that transforms a number of (possibly) correlated variables into a smaller number of uncorrelated variables called principal components. The first principal component accounts for as much of the variability in the data as possible. That is, PCA chooses the first PCA axis as that line that goes through the centroid, but also minimizes the square of the distance of each point to that line. Each succeeding component accounts for as much of the remaining variability as possible. If as many components are retained as original variables specified, all variability will be accounted for. The trick is to retain the fewest number of components that explain the substantive amount of variance. Every axis has an eigenvalue associated with it, which is related to the amount of variation explained by the axis. The sum of eigenvalues amounts to the number of variables.

Except for the public transportation portion of a route (given in %) and trip fare, all route values, such as portion of parks along a route given in meters, were divided by the shortest path distance of the start-destination pair for cost criteria, or the actual path distance for benefit criteria, respectively. This scaling makes route characteristics for different start-destination pairs comparable, because attribute values, such as cultural landmarks passed by, generally increase with a longer trip distance. We model the complexity of a multi-modal route as a linear combination of turns and route transfers, because both the street and the public transit portion of a route contribute to route complexity. Based on the Pareto optimal route set, a combined complexity measure of $c=1*turns+3*transfers$ for Vienna, and $c=1*turns+1.3*transfers$ for Bremen was found to yield positive correlations between turns and transfers, and the combined complexity measure (Vienna: $r_{turn,c}=0.690$; $r_{transfer,c}=0.453$; p=0.000; Bremen: $r_{turn,c}=0.424$; $r_{transfer,c}=0.749$; p=0.000). These two linear combinations were used to calculate corresponding values for routes.

After the scaling, all attribute values were standardized using Z-scores. By standardizing, all variables have a standard deviation of 1, and the centroid of the whole data set is zero. A PCA on a standardized data set is an eigenanalysis of the correlation matrix, which must be applied if variables are measured in different units.

Table 3 lists the variance accounted for by successive components. The eigenvalues in the "Total" column describe the observed variance explained by each component. In the Vienna data set, for example, the first component with an eigenvalue of 7.382 accounts for about 49% of the variability of the 15 variables. Values are similar for Bremen.

Applying the Kaiser criterion, which suggests to retain all components with an eigenvalue of 1 or higher, yields four components for Vienna, and three for Bremen. In the Bremen data set, the variance explained by the fourth (6.7%) and fifth (5.8%) component are close to the third component (7.2%) so that consideration of four or five components to explain the variance seems justified as well. The first four components account for about 82% (Vienna) and 84% (Bremen) of the variance, as shown under "Cumulative %".

**Table 3.** Results of the PCA: Extracted components with initial eigenvalues and explained variances for Vienna and Bremen

| Compo- | **Initial Eigenvalues (Vienna)** | | | **Initial Eigenvalues (Bremen)** | | |
|---|---|---|---|---|---|---|
| nent | Total | % of Variance | Cumulative % | Total | % of Variance | Cumulative % |
| 1 | 7.382 | 49.211 | **49.211** | 6.634 | 47.387 | **47.387** |
| 2 | 2.501 | 16.677 | **65.888** | 3.156 | 22.541 | **69.927** |
| 3 | 1.376 | 9.171 | **75.059** | 1.011 | 7.223 | **77.150** |
| 4 | 1.093 | 7.284 | **82.343** | 0.941 | 6.723 | **83.874** |
| 5 | 0.639 | 4.263 | 86.606 | 0.818 | 5.845 | 89.719 |
| 6 | 0.611 | 4.074 | 90.680 | 0.501 | 3.579 | 93.297 |
| 7 | 0.525 | 3.500 | 94.180 | 0.380 | 2.711 | 96.008 |
| 8 | 0.270 | 1.799 | 95.979 | 0.213 | 1.521 | 97.529 |
| 9 | 0.246 | 1.637 | 97.616 | 0.161 | 1.147 | 98.676 |
| 10 | 0.176 | 1.172 | 98.788 | 0.091 | 0.653 | 99.328 |
| 11 | 0.074 | 0.491 | 99.278 | 0.061 | 0.439 | 99.767 |
| 12 | 0.061 | 0.404 | 99.683 | 0.030 | 0.212 | 99.979 |
| 13 | 0.041 | 0.276 | 99.959 | 0.003 | 0.021 | 100.000 |
| 14 | 0.006 | 0.041 | 100.000 | 0.000 | 0.000 | 100.000 |
| 15 | 0.000 | 0.000 | 100.000 | - | - | - |

Component loadings describe the correlation between components and variables, i.e., which of the original 15 (14) variables contribute to which component. To obtain a clear pattern of loadings, an orthogonal rotation that maximizes the variance on the new axes is obtained. The rotated component matrices (Table 4 for Vienna and Table 5 for Bremen) reveal what the different components represent. For example, the second column in Table 4 means that the value of a route along the second axis of PCA is

0.858 times the standardized walking distance plus 0.424 times the standardized number of traffic lights, etc.

The meaning attached to rotated components is subjective and sometimes ambiguous, however, some tendencies can be identified.

For the Vienna dataset, the first component is marked by high loadings on attributes associated with route complexity, such as transfers, and street crossings and choice options at transfers. Although this component receives high loadings on fare and waiting time, the correlation with public transit is smaller than 0.6. Based on this and on the fact that the first component—as the only component—receives a high loading on the combined route complexity measure (last row), this component can be mostly ascribed to route simplicity.

The second component receives high loadings on walking distance, turns, intersections, travel time, and a negative loading on public transit portion. Although turns contribute to route complexity, the combined measure shows only a small correlation with this component, thus finding a simple route is not strongly supported by this component. In summary, this component could either be ascribed to fast route, or to public transit use, or a combination of the two. As public transit will generally be associated with reaching one's destination faster, the public transit part is omitted in our description of the component.

The meaning of the third component can be ascribed to route scenery, whereas the fourth component has high loadings on shopping streets. It is interesting to see that the shopping component appears instead of a safety related component which has been identified as the fourth component for bicycle navigation in previous work (Hochmair 2007).

**Table 4.** Rotated component matrix for Vienna with four components retained. Component loadings > 0.6 or < -0.6 are printed in boldface

| | Component | | | |
|---|---|---|---|---|
| | 1 | 2 | 3 | 4 |
| Walking distance | -0.434 | **0.858** | 0.180 | 0.074 |
| Traffic lights | -0.441 | 0.428 | 0.412 | 0.411 |
| Turns | -0.115 | **0.882** | 0.001 | -0.197 |
| Parks | -0.118 | 0.172 | **0.805** | -0.153 |
| Sights | -0.098 | 0.084 | **0.871** | 0.090 |
| Shopping streets | -0.012 | 0.035 | -0.061 | **0.926** |
| Intersections | -0.417 | **0.854** | 0.193 | 0.041 |
| Travel time | 0.142 | **0.865** | 0.190 | 0.095 |
| Transfers | **0.911** | -0.335 | -0.106 | -0.019 |
| Fare | **0.649** | -0.598 | -0.061 | -0.038 |
| Waiting time | **0.817** | -0.288 | -0.029 | -0.040 |
| PT portion | 0.593 | **-0.753** | -0.166 | -0.048 |
| Street crossings | **0.746** | -0.023 | -0.115 | 0.054 |
| Choice options | **0.813** | -0.369 | -0.140 | -0.022 |
| Turns, Transfers | **0.788** | 0.474 | -0.103 | -0.198 |
| **Meaning attached to rotated components** | *simple* | *fast* | *scenic* | *shopping* |

With four components, the Bremen data set reveals a similar pattern of loadings as the Vienna data set (right part in Table 5). As the Bremen data set does not include sights the third component can be solely ascribed to parks along the route. Another noticeable difference is that the first component receives a comparably higher loading on the PT portion than the Vienna dataset.

When using three components (left part in Table 5), the loadings on the first two components show similar patterns as before. The last two components merge into a third component with a positive loading on shopping streets and a negative loading on parks. Reducing the weight for this component returns routes with fewer shopping streets and more parks. Thus this component can be referred to as quiet routes. When taking into account that shopping streets have not been identified as a prominent criterion in route selection in previous work, a three components solution would be as justified as a four components solution both for the Vienna and the Bremen data set. It would, however, be of interest to see whether the inclusion of pedestrian zones in the test networks would impact the component structures.

**Table 5.** Rotated component matrix for Bremen with three and four components retained. Component loadings > 0.6 or < -0.6 are printed in boldface

| | Component | | | Component | | | |
|---|---|---|---|---|---|---|---|
| | 1 | 2 | 3 | 1 | 2 | 3 | 4 |
| Walking distance | -0.432 | **0.870** | -0.090 | -0.412 | **0.869** | 0.162 | 0.035 |
| Traffic lights | -0.373 | 0.390 | 0.192 | -0.346 | 0.449 | -0.273 | -0.046 |
| Turns | -0.100 | **0.866** | 0.048 | -0.076 | **0.864** | 0.044 | 0.096 |
| Parks | -0.075 | 0.271 | **-0.753** | -0.111 | 0.149 | **0.948** | -0.013 |
| Shopping streets | 0.031 | 0.236 | **0.616** | 0.014 | 0.114 | -0.007 | **0.986** |
| Intersections | -0.399 | **0.852** | 0.064 | -0.372 | **0.868** | -0.004 | 0.072 |
| Travel time | 0.337 | **0.872** | -0.084 | 0.357 | **0.849** | 0.183 | 0.045 |
| Transfers | **0.965** | -0.162 | 0.025 | **0.961** | -0.183 | -0.021 | -0.011 |
| Fare | **0.800** | -0.284 | 0.081 | **0.791** | -0.313 | -0.032 | 0.077 |
| Waiting time | **0.901** | -0.139 | 0.015 | **0.898** | -0.159 | -0.012 | -0.015 |
| PT portion | **0.781** | -0.542 | 0.082 | **0.765** | -0.568 | -0.063 | 0.052 |
| Street crossings | **0.794** | -0.074 | 0.032 | **0.796** | -0.083 | -0.051 | -0.037 |
| Choice options | **0.957** | -0.139 | 0.015 | **0.954** | -0.161 | -0.010 | -0.015 |
| Turns, Transfers | **0.850** | 0.428 | 0.045 | **0.863** | 0.406 | 0.019 | 0.048 |
| **Meaning attached to rotated components** | *simple* | *fast* | *quiet* | *simple* | *fast* | *parks* | *shopping* |

## 4.2 Implications for the User Interface Design

When using the 15 (14) factor loadings in each of the four (three) components as coefficients in a weighted linear combination of the z-scored criterion values, a large percentage (between 75% and 84%) of all Pareto optimal routes can be accessed. Without going into detail on how the optimal route, based on user preferences and settings, can be computed (Hochmair 2008), Fig. 5 shows some user interface designs that reflect the findings of PCA.

The design in Fig. 5a allows the user to weight all 15 original multimodal route selection criteria, which can be simplified to three or four slider bars based on the PCA results (Fig. 5c,e). Searching for a route that is optimized for exactly one of the three (four) components can be implemented through radio buttons (Fig. 5d,f), which simplifies the selection process compared to a comprehensive design (fig. 5b). Additional check boxes to set eliminatory constraints (e.g., avoid unpaved streets), and edit fields to specify thresholds (e.g., set maximum cycling distance accepted) will be helpful in specifying the optimal route search parameters (Hochmair and Rinner 2005).

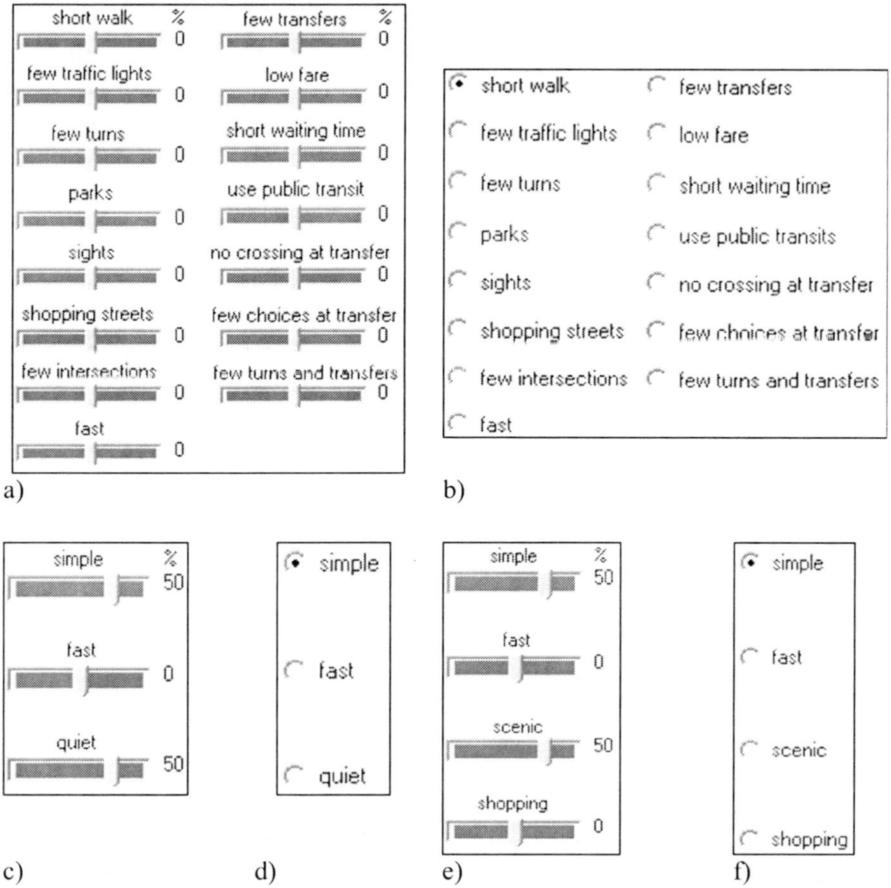


**Fig. 5.** User interface designs: Slider bars and radio buttons for all 15 original criteria (a,b), and slider bars and radio buttons for three (c,d) and four (e,f) components

## 5 Conclusions and Future Work

The results of this explorative study show that a high number of route selection criteria in multi-modal pedestrian trips can be more parsimoniously described through three or four components. Although both analyzed networks reveal a significant difference in transit density, PCA gives similar results for both networks. This suggests that for other cities, as long as they provide public transportation, a similar grouping of route selection criteria and, in consequence, a similar simplification of user interfaces for route planners seems plausible. A possible variation in results may, however, be found if analyzed travel routes become longer (both in terms of distance and travel time), or when the type of environment changes, such as on an inter-urban trip with rural and urban regions involved.. This, however, is largely speculative at this point and must be subject to future research.

Another aspect of future work is to compare the presented results to real world travel behavior, i.e., how travel routes observed in an urban environment can be grouped based on PCA. Real world data collection is feasible but challenging. Harvey et al. (2008), for example, analyze bicycle commuter behavior that relies on the use of Global Positioning System (GPS)-based data collection. For multi-modal transportation modes, GPS may not suffice due to signal blockage inside buildings. Independent of the method used, future work should also take into consideration upcoming transportation means for pedestrians, such as the growing number of public bicycle rental programs in European cities, which offer free (or nearly free) access to bicycles for inner-city transportation.

## References

BayernInfo (2008). Trip Information for all Modes of Transport. Retrieved 01/25/08 from http://www.bayerninfo.de

Chiu, D. K. W., Lee, O. K. F., Leung, H.-F., Au, E. W. K., and Wong, M. C. W. (2005). A Multi-Modal Agent Based Mobile Route Advisory System for Public Transport Network. 38th Hawaii International Conference on System Sciences.

Costelloe, D., Mooney, P., and Winstanley, A. C. (2001). From Random Walks to Pareto Optimal Paths. In 12th Irish Conference on Artificial Intelligence and Cognitive Science (pp. 309-318). Maynooth.

Daamen, W., Bovy, P. H. L., Hoogendoorn, S. P., and Reijt, A. V. d. (2005). Passenger Route Choice concerning Level Changes in Railway Stations. Transportation Research Board - 84th Annual Meeting, Washington, D.C. Transportation Research Board of the National Academies.

Fourman, M. P. (1985). Compaction of symbolic layout using genetic algorithms. In J. Grefenstette (Ed.), Proceedings of the First International Conference on Genetic Algorithms (pp. 141-152). Hillsdale, NJ: Lawrence Erlbaum Associates.

Goldberg, D. E. (1987). Genetic Algorithms in Search Optimization and Machine Learning. Reading, MA: Addison-Wesley.

Golledge, R. G. (1995). Path Selection and Route Preference in Human Navigation: A Progress Report. In A. U. Frank and W. Kuhn (Eds.), Conference on Spatial Information Theory (COSIT'95) (LNCS 988, pp. 207-222). Berlin: Springer.

Hansen, P. (1980). Bicriterion Path Problem. In G. Fandel and T. Gal (Eds.), Multiple Criteria Decision Making: Theory and Application (Lectures Notes in Economics and Mathematical Systems 177, pp. 109-127). Heidelberg: Springer.

Hardgrave, W. W. and Nemhauser, G. L. (1962). On the Relation between the Traveling-Salesman and the Longest-Path Problems. Operations Research, 10 (5), 647-657.

Harvey, F., Krizek, K. J., and Collins, R. (2008). Using GPS Data to Assess Bicycle Commuter Route Choice. Transportation Research Board - 87th Annual Meeting, Washington, D.C. Transportation Research Board of the National Academies.

Heye, C. and Timpf, S. (2003). Factors influencing the physical complexity of routes in public transportation networks. Electronic Proceedings of the 10th International Conference on Travel Behaviour Research, Lucerne, Switzerland.

Hochmair, H. H. (2004). Towards a classification of route selection criteria for route planning tools. In P. Fisher (Ed.), Developments in Spatial Data Handling (pp. 481-492). Berlin: Springer.

Hochmair, H. H. (2007). Dynamic Route Selection in Route Planners. Kartographische Nachrichten, 57 (2), 70-78.

Hochmair, H. H. (2008). Effective User Interface Design in Route Planners for Cyclists and Public Transportation Users: An Empirical Analysis of Route Selection Criteria. Transportation Research Board - 87th Annual Meeting, Washington, D.C. Transportation Research Board of the National Academies.

Hochmair, H. H. and Rinner, C. (2005). Investigating the Need for Eliminatory Constraints in the User Interface of Bicycle Route Planners. In A. G. Cohn and D. M. Mark (Eds.), Conference on Spatial Information Theory (COSIT'05) (LNCS 3693, pp. 49-66). Berlin: Springer.

Holland, J. H. (1992). Adaptation in Natural and Artificial Systems (2nd ed.). Cambridge: MIT Press.

Hoogendoorn, S. and Bovy, P. (2004). Pedestrian Route-Choice and Activity Scheduling Theory and Models. Transportation Research Part B, 38, 169-190.

Horn, J. (1997). Multicriterion decision making. In T. Bäck, D. Fogel and Z. Michalewicz (Eds.), Handbook of Evolutionary Computation (pp. F1.9:1-F1.9:15). Oxford, England: Oxford University Press.

JPL (2008). Journey Planner for London. Retrieved 01/25/08 from http://journeyplanner.tfl.gov.uk/

Keeny, R. L. and Raiffa, H. (1993). Decision Making with Multiple Objectives: Preferences and Value Tradeoffs. Cambridge, UK: Cambridge University Press.

Lausto, K. and Murole, P. (1974). Study of Pedestrian Traffic in Helsinki: Methods and Results. Traffic Engineering and Control, 15 (9), 446-449.

Malczewski, J. (1999). GIS and Multicriteria Decision Analysis. New York: John Wiley.

Meng, F. H., Lao, Y., and Leong, H. W. (1999). A Multi-Criteria, Multi-Modal Passenger Route Advisory System. Proceedings 1999 IES-CTR International Symposium.

Mooney, P. and Winstanley, A. (2006). An evolutionary algorithm for multicriteria path optimization problems. IJGIS, 20 (4), 401-423.

Muraleetharan, T. and Hagiwara, T. (2007). Overall Level-of-Service of the Urban Walking Environment and Its Influence on Pedestrian Route Choice Behavior: Analysis of Pedestrian Travel in Sapporo, Japan. Transportation Research Board - 86th Annual Meeting, Washington, D.C. Transportation Research Board of the National Academies.

Pareto, V. (1896). Cours d'Economie Politique (1). Lausanne, Switzerland: F. Rouge.

Peytchev, E. and Claramunt, C. (2001). Experiences in building decision support systems for traffic and transportation GIS. In W. Aref (Ed.), 9th International ACM GIS Conference (pp. 154-159). New York, NY: ACM Press.

SCOTTY (2008). Multi-modal door-to-door routeplanner for Austria. Retrieved 01/25/08 from http://fahrplan.oebb.at/bin/query.exe/dn?L=vs_addr

Senevirante, P. N. and Morrall, J. F. (1986). Analysis of factors affecting the choice of route of pedestrians. Transportaton Planning and Technology, 10, 147-159.

Spiess, H. and Florian, M. (1989). Optimal Strategies: A New Assignment Model for Transit Networks. Transportation Research Part B, 23B, 83-102.

Winter, S. (2002). Modeling Costs of Turns in Route Planning. GeoInformatica, 6 (4), 345-360.

Zitzler, E., Thiele, E., and Deb, K. (2000). Comparison of Multiobjective Evolutionary Algorithms: Empirical Results. Evolutionary Computation, 8 (2), 173-195.

# Spatial Decision Support in the Pedagogical Area: Processing Travel Stories to Discover Itineraries Hidden Beneath the Surface

Pierre Loustau, Thierry Nodenot, Mauro Gaio

DESI Research Group, LIUPPA Laboratory, 64012 Pau University Cedex, France - Contact: Pierre.Loustau@univ-pau.fr

**Abstract.** Local cultural heritage documents are characterized by contents strongly attached to a territory (i.e. Geographical references). Numerous corpora of such local documents become available and a challenging task is to process them automatically in order to retrieve and to make explicit the geographical information that they contain. The research reported in this paper aims at developing a toolset that teachers could use, first to retrieve travel stories from these corpora, and then to study the itineraries reported in these travel stories. To provide an adequate support to teachers, we propose two computational models from which we have built a Geographical Information Retrieval toolset in tune with travel stories characteristics. The paper demonstrates that these quite simple computational models are well fitted to process automatically (at discourse level) travel stories and to make explicit the geographical itineraries reported in such texts.

**Keywords:** Spatial Decision Support, itinerary, retrieval, travel stories

## 1 Introduction

The work presented in this article deals with the processing of large digital documentary cultural heritage holdings, which are made available by a local media center which is composed of 200.000 pages of text-document.. Most documents that make up our corpus have contents that very frequently refer to a local territory: they strongly refer to specific geographical locations and their surrounding area (Casenave *et al.* 2004).

Our research consists in designing methods and tools allowing the contents of such documents to be revitalized and exploited by non-expert users (e.g. tourists, teachers and learners). One user must be provided with adequate tools to consider such documents according to a focus, which takes into account his/her geographical interests, and which allows him/her to access the relevant document's contents.

In this paper, we present some computational models and a toolset that we designed to address some needs of teachers trying to make use of particular localized documents called "travel stories". We focus on the toolset required during the preliminary design steps of a Computer-Aided Learning (CAL) application which aims at teaching about a high level geographical information (travel itineraries) embedded in a travel story / a set of travel stories. We already have developed and tested in the real world (with learners) a prototype of such CAL applications.

From the automatic processing standpoint, the aforementioned uses require a phase of Information Extraction (IE) and one of Information Retrieval (IR). Because of the field of our application, we should better talk about Geographic IE (GIE) and Geographic IR (GIR). GIR is concerned with providing effective access to huge quantities of information that relate in some way to particular geographical locations. Some of its techniques derive from discipline of Geographic Information Systems (GIS), where the focus is on the accessing to geometrically structured data based on digital maps.

In case of a use requiring an unguided interaction (like in tourism), the GIR phase can only be based on the processing of local contents. The next phases process at the scale of a noun syntagm gravitating around a toponym; we call that a Spatial Feature (SF) (Lesbegueries *et al.* 2006). On the other hand, if the expected use requires a more guided interaction (cf. teaching activities), then a more elaborate interpretation and an extra effort towards formalization are necessary. Here, GIR processing must allow one to discover a pattern of a granularity that is greater than a syntagm. The extraction of these patterns raises a discourse problem.

Extracting itineraries is our approach to raise this discourse level in route narratives. Much research has been done on itineraries, some of the first of which dealt mainly with concept, to the exclusion of all other considerations (Kuipers 1977), (Fraczak and Lapalme 1999). Other researchers (Wunderlich and Reinelt 1982), (Mathet 2000), (Przytula-Machrouh, et al. 2004) studied how the itinerary was expressed in an oral or written description.

In this paper, we first describe the context of our research and the global architecture for our toolset. In the next sections (3 and 4), we address the main topic of our contribution: the itinerary characterization and the automatic way we extract itineraries from travel stories. Finally, we show in section 5 some examples that demonstrate that our multi-structure and multi-granularity approach is suited to do pedagogical "bricolage" on travel stories.

## 2 Context and Proposed Infrastructure

### 2.1 Context of Our Work

Travel stories have intrinsic characteristics that make them good teaching-resource candidates (CRDP 1997). A travel story is a sort of text whose author tells what he discovered while travelling through a territory or a country. On the one hand, the author tries to very precisely present the places he visited; on the other hand, he tries to tell the events that occurred during his trip, he reports on his activities and explorations. Indeed, the travel story aims at fitting words and travel reality (Granier and Picot 2004): the travel story is told day after day, the duration of the trip is explicitly written in the text in conjunction with the travelled locations.

Geographic Information Systems (GIS) provide generic functionalities to browse, to query and to create map data. Indeed, a GIS can be used to process and represent the itineraries if the input is a set of structured data. In our pedagogical context, such GIS use comes up against three problems.

First, a GIS requires structured data while our corpus consists of unstructured documents. We got them from MIDR as text files and we need first to process text at the different semantic level to identify Geographic Information. We already worked on identifying low level Geographic Information by extracting and interpreting absolute locations (e.g. "*Lourdes*" or "*Argelès*") or relative locations (e.g. "*South of Lourdes*"). These previous works were part of the PIV project. The research presented in this paper aims at completing the PIV project available results (Lesbegueries *et al.* 2006; Gaio *et al.* 2007). Our current works reach a higher level of structuration involving relation between locations (e.g. enumeration, comparison, description, itinerary, etc.). Displacements between such locations (e.g. "*I walked from Argelès to the South of Lourdes*") are the subclass more precisely described in this paper.

Secondly, the MIDR has provided us with a lot of local documents. Morevover, we can imagine to complete this corpus with documents put at internauts disposal by Digital Libraries which either are already available or will be available soon (CENL 2007; Europeana 2007; Gallica 2007;, Google 2007), etc. Yet, wherever they come from, these documents are not necessarily travel stories and it is impossible to manually check whether they are travel stories or not. So to help teachers retrieve useful travel stories, we need to retrieve such documents at conceptual level (the itinerary), passing by different simpler levels (Geographic Named Entity, Spatial Features, Displacements). Then we need to automatically process the query to retrieve candidate documents from the extended/available corpus.

Thirdly, behaviour and technical capabilities of teachers are very different from those of computer scientists. From the technical viewpoint, one can provide them with a smart user-interface (Marquesuzaà and Etcheverry 2007) that mainly hinders the technical complexity of spatial and temporal queries. From the behavioral viewpoint, different works have shown that the way teachers tackle design tasks has certain similarities to "Bricolage" (Perrenoud 1983; Caronet al. 2005) : here, "Bricolage" means to somehow manage to design a piece of educational software even if success requires that one diverts available resources (e.g. texts, photos, etc.) and tools' functionality from their primary purpose. Thus, an educational designer doing some "bricolage" with a travel story has the power to appropriate such text property. That is why we have to design a system giving him the possibility to retrieve relevant documents from the corpus but also to rewrite some passages to better fit his pedagogical goals. The teacher must also have the possibility to resubmit the modified document to the system to see if it is well understood.

## 2.2 Spatial Decision Support Infrastructure

To tackle these three problems, we have decided to investigate the development of an infrastructure that could, on the one hand, benefit from the power of GIS functionality and could, on the other hand, help teachers doing some educational "bricolage" from a corpus of documents. Such an infrastructure provides educational designers with three complementary sets of services: (i) a set of services to automatically retrieve from the corpus travel stories that could be good candidates for the next design stages; (ii) a set of services to help a teacher both studying the itineraries told in the selected travel stories and evaluating their adequacy as regards to his pedagogical aims; (iii) a set of services to exploit the geographical information embedded in the selected travel story/stories while designing a piece of educational software.

Figure 1 (cf. next page) details the set of services that we have already developed (cf. section 4.1) to support each stage of the design process. Our infrastructure is internet-based; the front-office provides tools to support teachers' bricolage while the back-office consists of tools dealing with the contents of the documents and the extracted semantics. For example, to check the adequacy of candidate documents as regards to pedagogical aims, a teacher is provided with 4 services (green shapes labelled 2) : a service to query the itineraries reported in the documents, a service to visualize such itineraries, a service to select adequate documents and a service to rewrite inadequate paragraphs of such selected documents.

Since the upper part of the figure 1 (cf. yellow shapes with the label 1) was tackled by our previous works, the main concern of this article is on the second part (cf. green shapes with the label 2). Here we first recall these previous works and results. They focused on low level Geographical Information and they are based on a linguistic hypothesis (Borillo, 1998) that considers that a SF is recursively defined from one or several other SFs and spatial relations are part of the SFs' definition. The target/landmark principle (Vandeloise 1986) can be defined in a recursive manner. For instance, the SF "*north of the Biarritz-Pau line*" is first defined by "*Biarritz*" and "*Pau*" that are well known named places, the term "line" creates a new well-known geometrical object linking the two landmarks and cutting the space into two sub-spaces, finally, an orientation relation creates a reference on the target to focus on.
Thus, we defined two types of SF: the Absolute one (ASF) and the Relative one (RSF). We also defined 5 relations: an adjacency ("*nearby Laruns*"), an inclusion ("*centre of Laruns*"), a distance ("*at about 10 kms of Laruns*"), a geometric form ("the *Laruns-Arudy-Mauleon triangle*") or an orientation ("*in the west of Laruns*").

To sum-up, we can model a SF with both:

1. an XML tree representing its eventual relations with other SF,
2. a projection of this XML tree on a geometric structure which is compliant with GIS structures.

This projection is done by distorting the original shapes of the ASF according to the extracted relations. A prototype supporting this approach has been developed. It starts from text documents and is able to extract SF's in order to create a spatial index in a GIS. This index will be consulted in a GIR based search engine.

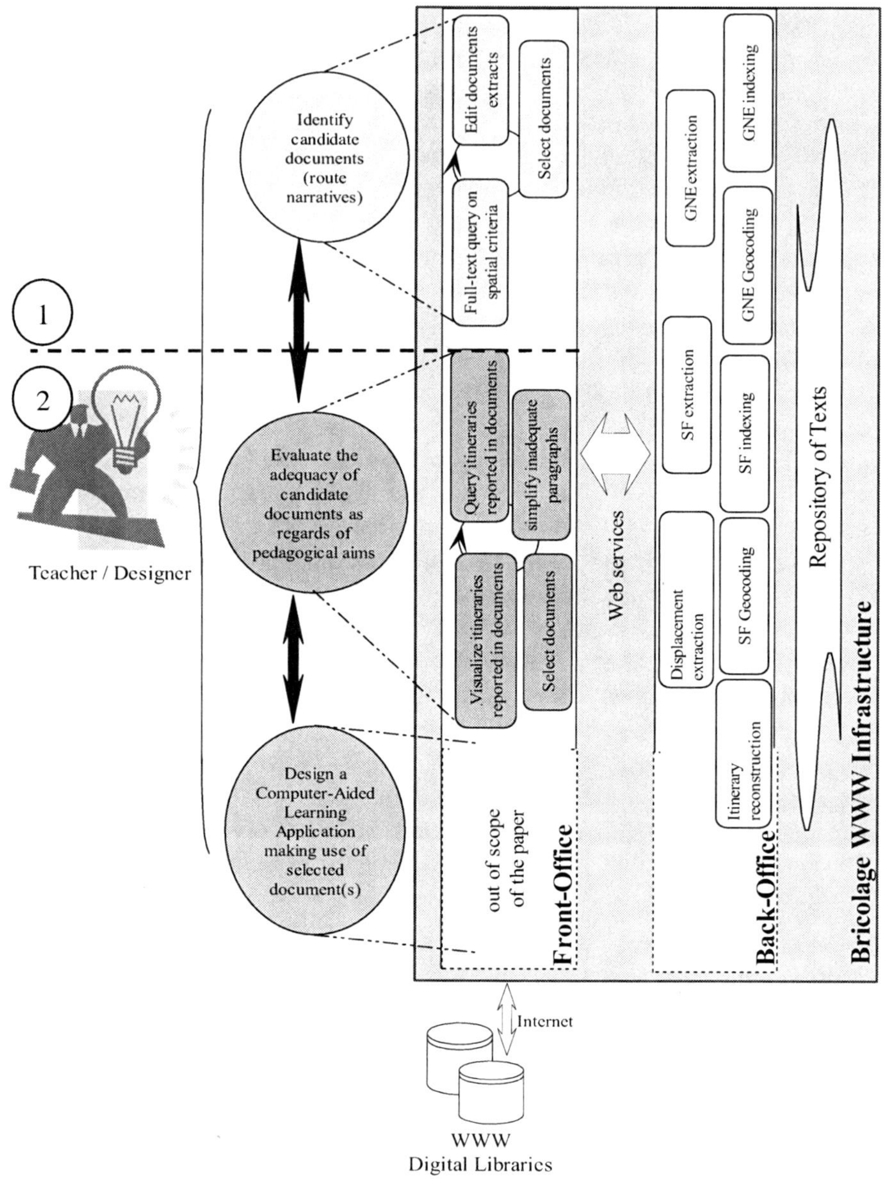


**Fig. 1.** The Infrastructure (services and data) used to process Travel Narratives.

# 3 Itinerary Characterization

Lot of work has been done in the field of itineraries. However, we have both to fine down the generic concept of itinerary and also to specify it. In this section, we present related works done in the field of itineraries. Considering the results of this study, we propose the principles of a model which is simpler than those presented, but can be filled automatically to address the problem of routes narratives retrieval.

## 3.1 Related Work on the Generic Concept of an Itinerary

Most researchers in the field agree that, at a high level of abstraction, an itinerary is composed of a starting point, intermediary stages, and an ending point. However, the way these stages are defined varies according to the works that we have studied. For Wunderlich and Reinelt (1982), the intermediary stages are landmarks, which are defined as having certain particular characteristics that make it easily recognizable. However, Przytula-Machrouh *et al.* (2004) and Denos (1997) explain that describing an itinerary consists essentially in describing its actions and reference points; so an object can become a reference point if it has certain outstanding properties (this converges with the viewpoint of Wunderlich and Reinelt (1982)). Fraczak and Lapalme (1999) write about relays and segments. The segment is a fragment of an itinerary in which one or more characteristics remain constant, whereas a relay changes in characteristics. These characteristics can be an orientation, a direction, or a type of path, etc.

## 3.2 Formalizing Textual Descriptions of an Itinerary for a Pedagogical Use

There are two reasons for the simplification of the model. The first one is due to our starting point. Let us recall that we want to automatically instantiate our model starting from text by analysing documents. That is why the information we need to fill the model has to be automatically extracted or deducted from the analysed text-documents. Some existing conceptual models of itineraries don't take into account this facet of the itinerary and that's why they are too complicated to be automatically instantiated, that is to say that we are not able to automatically extract from documents some elements to fill such models. The second reason is due to the application itself: the extracted itineraries have to be simple enough (Nodenot *et al.* 2006) to match the pedagogical goal of being understood by learners.

So in our case, the intermediary stages of an itinerary appear when the traveller passes through a place and describes it in its narration. However, the traveller does not necessarily describe every place he visits. Thus, only the most important places make up the intermediary stages of the itinerary. This brings us back to the previously mentioned notion of outstanding properties (Wunderlich and Reinelt 1982; Denos 1997; Przytula-Machrouh *et al.* 2004). What seems important is the approach of (Fraczak and Lapalme 1999) that aims at discovering a change in the characteristics when going through the intermediary stages. Indeed, we frequently noticed in our texts that the intermediary stages were marked when the traveller made a change (to stop for visit, to eat, to get means of transportation, etc.). Therefore, we propose to take up the notions of relays and segments.

***The relay:*** It is a part of geographical space occupied by the itinerary at one point in time. However, only taking account of the spatial aspect would be inadequate since the notions of space, time and thematic (activities, objectives, etc.) intervene. The latter three notions combine to make up the concept of Geographic Entity (GE). Indeed, in several works taken up by (Usery 2003), an GE has a Spatial Marker (SM), a Time Marker (TM) and a thematic or Phenomenon Marker (PM). These markers can be explicit (the chapels of 17th century Laruns), or they may be implicit and vague (the city of Pau). Each marker has its own representation (like a spatial area for the SM, a time interval for the TM or a knot in the ontology for the PM) and they are linked to the relays. The concept of GE is born from their unification. Indeed, two relays that refer to the same GE can appear in two different itineraries and have a different SM, TM, and/or PM (especially from a scale standpoint) : *I left Pau at 8 am [...]I came back to Pau at 10 pm.*

In this article, the centre of attention is the spatial aspect of GEs, but the thematic and temporal aspects make up some of our future perspectives.

***The segment (or fragment of an itinerary):*** this is the path that links together two consecutive relays. It is noteworthy that the path is of a virtual nature, meaning that it is only one possible representation. In the cognitive map (Kuipers 1977) that the reader mentally visualizes, the path that links two relays is not necessarily the one that the itinerary's traveller really took. Yet, this virtual path approaches the true path more or less accurately. This depends on some hints that help the reader to follow along the real path taken by the traveller. These hints can be classified into two categories. On the one hand, the information given by the traveller in the story (transportation, speed, places, etc.) and on the other hand, the knowledge the reader has about the region being travelled (topology of the land, difficulty of the trail, the existence of a means of communication, etc.). We will show further how these hints can be exploited in an automatic interpretation of the itineraries.

## 3.3 The Case of Route Narratives

The model we proposed in the previous paragraph is the ideal conceptual model of an itinerary. Even if it takes into account the information available in the documents and the uses of the extracted itinerary, it can't be instantiated directly. In the case of route narratives, authors rarely textually describe the segments and the relays of their itinerary saying *segment PAU-LARUNS / two relays PAU and LARUNS*. The relays and the segments of the described itinerary are hidden in the narration (*"I left Pau at 8 AM with my brother [...] I arrived in Laruns at 9 AM"*). Hereafter, we are showing with two experiments how the itinerary appears in a specific subclass of textual itinerary description. We are concluding by the necessity of intermediary different level models.

### *3.3.1 Experiments on the Corpus*

Of all the documents made available to us by the media centre, three were chosen to undertake two experiments that would more accurately validate our initial observations. These three texts tell the story of an explorer who leaves the big cities of Paris and Bordeaux to discover the Pyrenees Mountains for a trip of several days.

Experiment 1: The first experiment consisted in finding the clauses that referred to the author's location during his travels (LS) and then calculating how many of these contained Spatial Features (SF) to carry out the (CCSF).

The results of the experiment can be found in Figure 2. The use of SF to refer to displacements appears pretty clearly. An 80% order rate is reached, which varies according to the author and to the kind of travels that are related. Indeed, the absence of the SF's that refer to displacements appears most commonly when the author evokes shorter displacements (*I went to the park, I took a walk on the port, I left the hotel, etc.*). This mainly occurs when the author is at an intermediary stage of his trip. Take for example Ann Lister's text, in which the author essentially refers to the activities during the longer stages of the trip.

Experiment 2: This second experiment aims at evaluating the impact of verbs. It consists in counting the number of clauses (LS) containing movement verbs (CCV). Indeed, with an average of 87.7%, this study shows their predominance. Results per document are given in Figure 2.

| | | Experiment 1 | | | Experiment 2 | | |
|---|---|---|---|---|---|---|---|
| Text | Words | LS | CCSF | % | LS | CCV | % |
| J.-D. Forbes | 3662 | 75 | 63 | 80% | 75 | 63 | 80% |
| A.Lister | 3225 | 40 | 20 | 50% | 40 | 35 | 87,5% |
| J.-R. Bals | 19015 | 213 | 170 | 80% | 213 | 190 | 89% |

**Fig. 2.** Experiments 1 and 2 : The predominance of clauses containing SF (CCSF) and movement verbs (CCV) in Sentences that refer to the Location of the author (LS).

These two studies require further analysis. However, by increasing the amount of documents, one may already get a good idea of the high proportion of SF and movement verbs in our corpus. These observations also collaborate research done in reference to displacements in language such as that of (Laur, 1991), (Sarda, 1992), (Muller & Sarda, 1999). That is why we have to propose a model for the characterization of the displacements in the language.

### *3.3.2 Characterization of the Displacements in the Language*

We essentially uphold the criteria of a verb's aspectual polarity that was introduced by Boons, and further developed by Laur (Boons 1987; Laur, 1991).

Thus, we model displacements in the language by taking account of movement verbs that are necessarily associated with SF's, and that are optionally associated with spatial clauses. We have named this triplet formed by the verb, the Clause (optional so we use "?") and the Entity: the triplet (V,P?, E).

In the model we propose here, the triplet is discriminate to bring out the spatial meaning of certain polysemic verbs (*to leave someone* is of little interest to us, whereas *to leave Bordeaux* attracts our attention). The same is true for *to get out of a tough spot* vs *to get out of Bordeaux*. So, in this model we uphold Boon and Laur's notions of aspectual polarity.This means that the extracted displacements will be initial (*to leave*), median (*to cross*) or final (*to arrive*).

The original notions of destination and median position also come into play. The construction of movement verbs result in the surfacing of the following: one (*to leave Pau*), two (*to leave Pau for Bordeaux*) or three (*to leave Pau for Bordeaux on the RN 134*).

These notions meet the "Lieu de Référence Verbal" (LRV[19]) thesis of Sablayrolles (Asher and Sablayrolles 1995) who also studied the displacement verbs. The LRV is the argument of a displacement verb which is compatible with his polarity. If the LRV is explicitly mentioned, it fixes the narrator's position on it. This can be seen in example like (*to go out of Laruns, to arrive in Bedous from Pau, etc.*). The LRV are underlined.

## 4 Automatically Filling Our Conceptual Model of an Itinerary

Our general approach is first to extract the SF's from text documents, then to extract the displacement of the narrator and finally to reconstruct the itinerary.

### 4.1 Spatial Features Extraction

The SF's extraction is done by the previously mentioned PIV prototype (Lesbegueries *et al.* 2006). We can briefly describe it telling it is based on the discovering of GNE in text documents. By exploring the syntagms around this GNE, it can make a semantic interpretation of the more generic concept of Spatial Feature.

The SF extraction is done by a processing stream in the Linguastream platform. The following analyses are carried out : a tokenization and a splitter (to split the raw text flow in words and to cut it in paragraphs), a morphosyntactic analysis made by the Tree-Tagger part-of-speech tagger, several pattern matching recognitions with regular expressions to detect possible GNE (beginning with a capital letter for example), then a DCG grammar is processed by Prolog to discover the SF and their syntactic links with spatial modifiers like "*au sud de*" ("*at the south of*"), "*près de*" ("*near*"), etc.

### 4.2 Extraction and Structuration of Displacements

Concerning the extraction of displacement, we design movement verbs thanks to transducers, or a finite state machine based device that transforms one language into another by correlating two regular languages. In-

[19] LRV : Lieu de Référence Verbal (Verbal Reference Location)

deed, we believe that as a first approach the construction of movement verbs can be considered a regular language. So, it may be designed by finite state machines like those in Fig. 3.

Displacement extraction is therefore based on the same principles as those of the SF's. The transducers are translated into grammar rules in which one can find the triplet (V,P? ,E) . This rule based analysis is founded on the initial results of analysis. The first is a morphosyntactic analysis capable of processing the form and the lemma of each lexical unit (@lemma=sortir in Figure 3). This enables to work on the canonical forms of the words, especially those of conjugated verbs. The second analysis is the extraction of the SF's (@sem=sf in Figure 3).

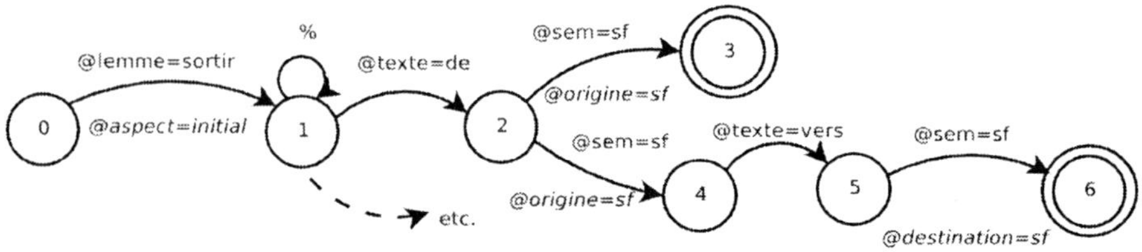


**Fig. 3.** Simplified extract of a transducer. The % symbol represents the "joker". The transition in dots shows that other states bring other meanings to surface.

The extraction of displacements is a linguistic processing chain in and of itself. It is made up of the main linguistic treatment phases and it is implemented through a linguistic processing platform called Linguastream (Widlöcher and Bilhaut 2005). The latter contains Tree-Tagger, which is a time-tested morpho-syntactic analyser (Schmid 1994). The verb transducers of the displacements are translated into DCG[20] grammar rules and the grammar based analysis is in turn handled by the Prolog language, in order to take advantage of its mechanisms of deduction and unification.

## 4.3 Itinerary Reconstruction

Going from the displacements to the itinerary reaches a discourse problem. Starting from the displacements that are local in the document (syntagms), we want to reach a higher level of semantics that appears on the entire paragraph if not on the whole document. We are using Spatial Reasoning and Spatial Resources to bridge this gap, as well as a reader can do it using his spatial knowledge of the world and his spatial intelligence.

[20] DCG : Definite Clause Grammar

As mentioned above, the resources needed to interpret an itinerary that appears in a text document can be split in two categories.

On the one hand we have knowledge of the world, especially of the geographic world (topology of the area, distance between two places, existence of a road, etc.). On the other hand, we have reasoning and especially, spatial reasoning. This spatial reasoning gives the reader the ability to construct mental representation of the path described by the narrator. This second-category resource obviously needs the first one, that is to say the knowledge of the geographic world. Typically, the first category resources (knowledge of the geographic world) can be given to an automated system thanks to GIS layers, Gazetteers or ontologies. The second one is part of our contribution.

### 4.3.1 Relay Discovering

First, we are stating the hypothesis that the text is recounting the itinerary chronologically. This hypothesis can be made because route narratives which have this particularity. Raising this hypothesis is part of our perspectives.

Secondly, we have to identify the intermediary stages (the relays) of the trip. This is done by analysing the extracted displacements which are put into a sorted list according to our hypothesis. We are stating that a relay appears when the LRV of a displacement is explicitly mentioned. Let us exemplify this point.

*(1)J'ai quitté Pau à 8h00 pour Laruns*[21]
*(2)J'ai traversé Toulouse à 10h00*[22]
*(3)Je suis arrivé de Pau à 13h00*[23]

The LRV of the displacement mentioned by the quitter (to leave) verb is the origin of the displacement. In (1) the origin is explicitly mentioned: Pau. This reveals a relay. The second part of the sentence pour Laruns (to Laruns) is the destination of the displacement but we cannot say that the narrator will reach his destination or if pour Laruns (to Laruns) is only the direction of the initial displacement. We are only able to say that à 8h00 (at 8 o'clock) he was in Pau. In (2), the LRV of the traverser (to go across) verb is mentioned, a relay appears as well as in (1). In (3) the LRV of the arriver (to arrive) verb is not mentioned. We are not able to say anything.

[21] I left Pau at 8 AM to Laruns
[22] I went across Toulouse at 10
[23] I Arrived from Pau at 1 PM

The (3) sentence only give us an origin of a displacement but reading (3) not help us to know the narrator's localisation.

When the intermediary stages are discovered, we need geographic resources to assign geocodes to them. This will be done using the approach of the PIV prototype (Lesbegueries *et al.* 2006) which is the results of previous works of our research team.

### 4.3.2 Segments Building

The second phase of the itinerary reconstruction is to build the path from a relay to another. The paths we are going to build are not necessarily the paths taken by the narrators but we want to be as close as possible. An important clue to determine this path is the modality of the transport. Going from A to B by train is not the same as doing it by car or walking. This kind of deduction is part of our contribution. Our system is capable to build paths according to an initial point, an end point and a modality of transportation.

For an unknown modality of transportation, the path is the right line. This is what the reader have in his mind when he read *J'ai quitté Pau à 8h00, [...] Je suis arrivé à Laruns à 10h00 (I left Pau at 8:00, [...] I arrived at Laruns at 10:00).* This is the easiest case: we just need the geocoding of the start point, the geocoding of the end point and we have to build the geometrical segment with these two extremities.

For a car displacement, the reader imagines a displacement on the road. There are a lot of other modalities that can be imagined on a road (bus, truck, motorcycle, horse-drawn carriage, etc.). To be the more precise possible, we should examine all of these modalities separately as they convey other information (for example, a horse-drawn can't take a motorway). In a first approximation we have chosen to group all the road modalities in a single category. We need more geographic resources to be able to construct the path with such a modality. The road network and algorithms that can find shortest paths in this network are required.

This approach can be generalized to other displacements. We just need to know which path can be taken by which modality. For example, with a bike or a by foot displacement, we need the dirt road network, with a train displacement, we need the rail-road network, etc. We call this approach the network-modality approach.

Two other modalities can be distinguished as they don't use networks. This is the case of a by-plane or a by-boat displacement. The by-plane displacement could be interpreted as a right line, like the unknown modality. For the by-boat one, it is a little bit more complicated. We have two solutions. The first one is to consider the maritime routes and to reason with a

network. The second one is to build a path that connects two ports with one constraint: staying onto the water.
The whole approach presented here has been implemented in a prototype. By submitting a raw-text route narrative, it can extract, interpret and store the itinerary embedded in the narration.

## 5 Multi-Level and Multi-Structure Querying System: Application to the Choice of the Right Documents

Let us recall that a specific task of a teacher consists in choosing documents according to teaching activities in a huge document repository. Here we are showing how our multi-structure and multi granularity approach is particularly adapted to this pedagogical "bricolage".

Fig. 4 presents the increasing complexity of the GI. The first column shows a raw-text example where the focused GI is underlined. The second one shows a semantic structure corresponding to the GI considered and the third one shows its projection on a GIS structure.

The first level of granularity can be used in several GIR based applications. The higher levels are more specific. In our study case we are interested in route narratives so we have to look at the displacements with displacement verbs. If we were interested in narrative descriptions, we should have had to deal with local descriptions based on perception verbs. But the lowest levels of granularity have to be taken into account in both approaches.

We are going to show this by simulating a query a teacher could ask to our system. For each type of query we are showing which granularity is addressed and which structure is the more appropriate to answer it. Moreover, the treatments of the corpus, the treatments of the query and how the matching is done are considered for each query.

**Query 1:** *Find documents which talk about Barèges and Argelès*

In this query, only low-level GI is addressed, a simple GNE comparison between the document repository and the query is needed.

Corpus treatments: GNE Extraction, GNE geocoding, GNE indexing (in a GIS for example).

Query treatments: if the query is in natural language, the same treatments have to be applied (GNE Extraction and GNE geocoding). However, the query could be formulated on a rich interface (on a map for example) and then we directly obtain its geocoding.

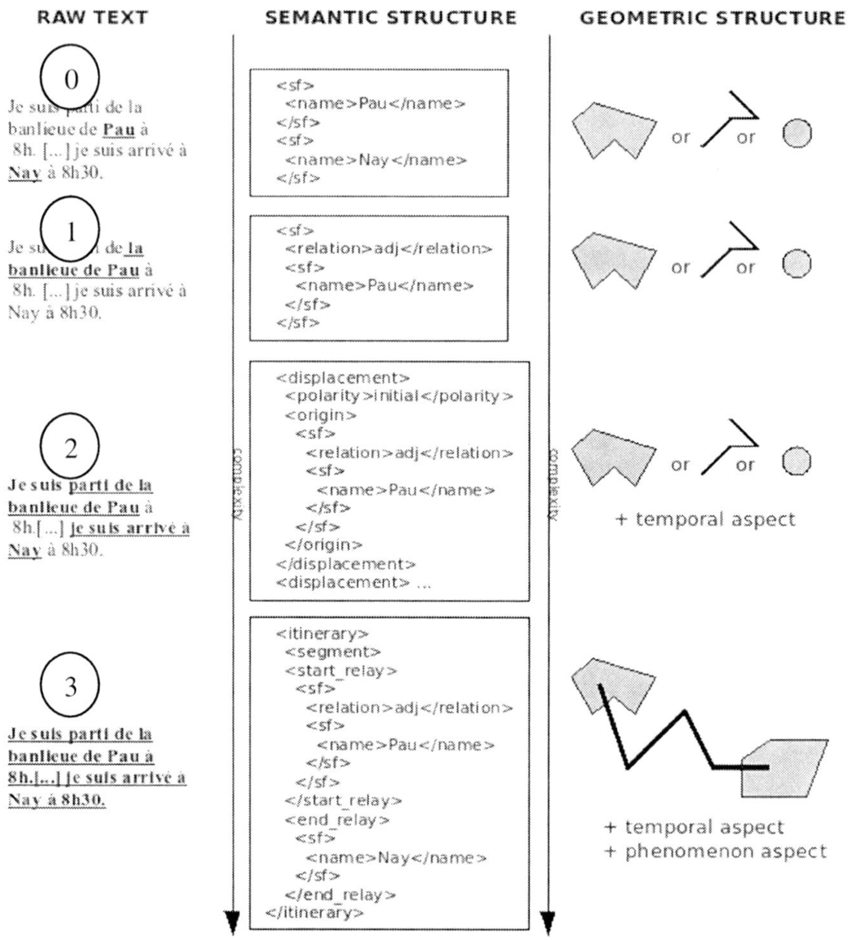


**Fig. 4.** The increasing complexity of the Geographical Information. The example english-translation is *I left the suburbs of Pau at 8. I arrived in Nay at 8:30.*

Matching query vs corpus: it's a strict geometric comparison between the geocoding of the query and the geocoding of the GNE extracted from the documents. For this kind of query, the geometric approach is not really improving a classic IR approach (fulltext for example).

**Query 2:** *Find documents which talk about the region of Barèges and Argelès*

In this query, the GI level is increased, a simple GNE comparison is not sufficient. The system needs to understand *Argelès* and *Barèges* but he also need to interpret *the region of*, that is to say it has to interpret the SF's.

Corpus treatments: SF Extraction, SF geocoding, SF indexing (in a GIS for example).

Query treatments: if the query is in natural language, the same treatments have to be applied. However, the query could be formulated on a rich interface and then we directly obtain its geocoding.

Matching query vs corpus: it's a strict geometric comparison between the geocoding of the query and the geocoding of the SF extracted from the documents.

**Query 3:** *Find documents in which the narrator arrives in Argelès*

In this query, the GI level is again increased. The system needs not only to understand the SF but also the relation between the narrator and the SF. In other words, it has to know if *Argelès* appears in a final polarity displacement of the narrator. Here the approach begins to be more specific: the relation we are interested in is the relation of the displacement but we could imagine other relation like description, enumeration, etc.

Corpus treatments: SF Extraction, SF geocoding, SF indexing (in a GIS for example) + Displacement Extraction and Interpretation.

Query treatments: if the query is in natural language, the same treatments have to be applied. However, the query could be formulated on a rich interface and then we directly obtain its geocoding.

Matching query vs corpus: it's not only a geometric comparison between the geocoding of the query and the geocoding of the SF extracted from the documents. The system also query the semantic structure (coded as an XML tree) to find the semantic of the displacement (located in *Argeles* in our example) that is to say if it is initial, final, etc.

**Query 4:** *Find documents in which the narrator starts from Argelès, crosses Luz and arrives in Barèges.*

In this query, the GI level is again increased. The system needs to understand the whole itinerary Agelès – Luz – Barèges.

Corpus treatments: SF Extraction, SF geocoding, SF indexing (in a GIS for example) + Displacement Extraction and Interpretation + Itinerary reconstruction.

Query treatments: if the query is in natural language, the same treatments have to be applied. However, the query could be formulated on a rich interface and then we directly obtain its geocoding.

Matching query vs corpus: it is a geometric comparison between the geocoding of the query and the geocoding of the extracted itineraries from the documents.

## 6 Conclusion

In this paper, we have presented a computational approach and a toolset that implement an innovative geographic semantics approach on a full process of Geographic Information (GI) Extraction and Retrieval. Such a toolset focuses on the semantics of travel stories which are particular documents commonly found in local cultural and heritage documents corpora. Such a corpus is too big to allow ones to manually analyse its contents.

This article suggests a complete processing method to extract high level GI from text document. The approach we proposed starts from low level GI Extraction like Geographic Named Entity (that appears in the textual contents as toponyms) and tries to infer more and more complex GI by looking around this anchor.

We have also shown that the low level treatments (GNE or SF recognition) can be rather generic whereas higher level analyses are more specific. This latter point has been emphasized in the pedagogical field where more and more teachers are doing pedagogical "bricolage" to find relevant route narratives documents for their teaching activities.

The specific treatments that we have proposed consist in extracting itineraries from low-level GI (ie GNE and SF extraction) retrieved a document. Then, thanks to two experiments, we have shown that we need to first extract the displacements of the narrator before to understand the whole document. This kind of discourse analysis is also addressed in this paper by proposing a method and a prototype that uses GIS knowledge and GIS technology to simulate Spatial Reasoning.

Future work will try to go further by mixing the Spatial facet of the GI with the Temporal facet. While the described PIV prototype (section 4.1) has been developed for the low level Spatial Features extraction, another prototype developed in our team can handle the Temporal Features (TF) in the same way. We plan to exploit these extracted TF to assign timestamps to our displacements so that we can better understand the chronology of the displacements.

Concerning the evaluation of our works, we have evaluated the strength of the automatic extraction of the displacements. This evaluation has shown that our prototype is able to extract and interpret around 70% of the displacements of the narrator. To go further, we need to measure the correctness of the interpreted itineraries. This evaluation is in progress: we have selected a set of documents embedding itineraries. And we are in the process of comparing the itineraries drawn by different people from such documents with the itineraries that our toolset can interpret. Such an evaluation will lead us to determine the current efficiency of our toolset and further improvements to be made.

## References


Asher, N., & Sablayrolles, P. (1995). A Typology and Discourse Semantics for Motion Verbs and Spatial PPs in French. Journal of Semantics, 12(2), 163-209.

Boons, J.-P. (1987). La notion sémantique de déplacement dans une classification syntaxique des verbes locatifs. Langue Française, 76, 5-40.

Borillo, A. (1998). L'espace et son expression en français. L'essentiel, Ophrys.

Caron, P.-A., Le Pallec, X., & Derycke, A. (2005). Bricolage and Model Driven Approach to design distant course. Paper presented at the E-learn 2005, Vancouver (Canada).

Casenave, J., Marquesuzaà, C., Dagorret, P., & Gaio, M. (2004, 13-15 octobre 2004). La revitalisation numérique du patrimoine littéraire territorialisé. Paper presented at the Colloque EBSI-ENSSIB : Le numérique : impact sur le cycle de vie du document pour une analyse interdisciplinaire, Montréal (Canada).

CENL. (2007). The European Digital Library. Retrieved october 2007, from http://www.theeuropeanlibrary.org

CRDP. (1997). Enseigner le document. Revue pédagogique d'histoire-géographie, n°42.

Denos, N. (1997). Modélisation de la pertinence en recherche d'information : modèle conceptuel, formalisation et application. Université Grenoble 1.

Europeana. (2007). La contribution française à la bibliothèque numérique européenne. Retrieved october 2007, from http://www.europeana.eu

Fraczak, L., & Lapalme, G. (1999). Utilisation de stratégies cognitives dans la génération automatique de descriptions d'itinéraires. TALN'99, 10 pages.

Gaio, M., Sallaberry, C., Etcheverry, P., Marquesuzaà, C., & Lesbegueries, J. (2007). A Global Process to Access Documents' Contents from a Geographical Point of View Journal Of Visual Languages And Computing. Special Issue on Spatial and Image-Based Information Systems. Elsevier 18(6).

Gallica. (2007). Projet Gallica2 de la bibliothèque numérique de France. Retrieved october 2007, from http://gallica2.bnf.fr/

Google. (2007). Google Books. Retrieved october 2007, from http://books.google.fr

Granier, G., & Picot, F. (2004). La place des documents dans l'enseignement de l'histoire et de la géographie. Paper presented at the Actes du Colloque "Apprendre l'histoire et la géographie à l'école", Versailles.

Kuipers, B. (1977). Modeling Spatial Knowledge. IJCAI, 292-298.

Laur, D. (1991). Sémantique du déplacement et de la localisation en français : une étude des verbes, des prépositions et de leur relation dans la phrase simple. Université de Toulouse II.

Lesbegueries, J., Gaio, M., Loustau, P., & Sallaberry, C. (2006). Geographical Information access for non structured data. Paper presented at the ACM SAC 2006, track on Advances in Spatial and Image-based Information Systems (ASIIS), Dijon.

Marquesuzaà, C., & Etcheverry., P. (2007). Implementing a Visualization System suited to Localized Documents. Paper presented at the Fifth International Conference on Research, Innovation and Vision for the Future (RIFV 2007), Hanoi (Vietnam).

Mathet, Y. (2000). Etude de l'expression en langue de l'espace et du déplacement : analyse linguistique, modélisation cognitive, et leur expérimentation informatique. Doctorat en Informatique de l'Université de Caen (France).

Muller, P., & Sarda, I. (1999). Représentation de la sémantique des verbes de déplacement transitifs du français. revue TAL, 39(2), pp. 127-147.

Nodenot, T., Loustau, P., Gaio, M., Sallaberry, C., & Lopistéguy, P. (2006). From Electronic Documents to Problem-based Learning Environments: an ongoing Challenge for Educational Modeling Languages. Paper presented at the 7th International Conference on Information Technology Based Higher Education and Training (ITHET 2006), Sydney (Australia).

Perrenoud, P. (1983). La pratique pédagogique entre l'improvisation réglée et le bricolage. Essai sur les effets indirects de la recherche en Education. La formation des enseignants entre théorie et pratique, Education et Recherche, 1983(2), pp. 198-212.

Przytula-Machrouh, E., Ligozat, G., & Denis, M. (2004). Vers des ontologies transmodales pour la description d'itinéraires. Revue Internationale de Géomatique : n° Spécial sur les ontologies spatiales, 14(2), 285-302.

Sarda, L. (1992). Syntaxe et sémantique - Sémantique du lexique verbal. Presses Universitaires de Caen.

Schmid, H. (1994). Probabilistic Part-of-Speech Tagging Using Decision Trees. International Conference on New Methods in Language Processing.

Usery, E. L. (2003). Multidimensional Representation of Geographic Features: U.S Geological Survey (USGS).

Vandeloise, C. (1986). L'espace en français: Seuil.

Widlöcher, A., & Bilhaut, F. (2005). La plate-forme LinguaStream: un outil d'exploration linguistique sur corpus. Actes de TALN 2005, 517-522.

Wunderlich, D., & Reinelt, R. (1982). How to Get There from Here. Speech, Place, and Action, pp. 183-201.

# Ownership Definition and Instances Integration in Highly Coupled Spatial Data Infrastructures

Alberto Belussi[1], Federica Liguori[2], Jody Marca[2], Mauro Negri[2], Giuseppe Pelagatti[2]

[1]Dipartimento di Informatica, Università di Verona, Strada Le Grazie, 15, Verona, Italy, alberto.belussi@univr.it
[2]Dipartimento di Elettronica e Informazione, Politecnico di Milano, Via Ponzio 34/5, Milan, Italy, giuseppe.pelagatti@polimi.it

**Abstract.** Building and maintaining Spatial Data Infrastructures (SDIs) in large networks of public bodies is a great challenge, in particular for European countries since the INSPIRE project has become a EU directive in the last months. However, SDIs can have different goals and thus different requirements depending on which type of processes they aim to serve.

In this paper we consider a highly coupled SDI where different actors join the SDI at different times, and each actor is responsible for providing geographical data during the joining phase and for maintaining them updated afterwards. The main result is the definition and the experimental validation of an SDI architecture where the well-known problems of conflation and semantic harmonization can be approached in a consistent way; moreover, two less-evident problems, called instance integration and ownership definition are precisely identified and integrated in the same architecture.

**Keywords:** Spatial Data Infrastructure, interoperability, geodata fusion, instance ownership, instance integration

## 1 Introduction

The notion of Spatial Data Infrastructure (SDI) is widely known and adopted by different countries around the World. In Europe the idea of SDI is tightly related to the EU project INSPIRE, that has became a EU directive in the last months.

The notion of SDI is subject to different interpretations in particular with regard to the level of cooperation of the different providers of spatial information, which can vary from "loosely coupled" to "highly coupled".

The context of this paper is a highly coupled SDI (HC-SDI), characterized by the goal to provide a shared representation of the territory that is adequate to:

- satisfy the needs of a given set of actors that govern it,
- support their cooperating processes, including the continuous update,
- guarantee adequate levels of quality (in particular consistency of the data).

It should be mentioned that a highly coupled SDI can be part of a higher level loosely coupled SDI in an SDI hierarchy, however this aspect is not further explored in this paper.

This paper is based on an experimental project sponsored by the regional administration of Lombardy in order to explore the technical feasibility of a highly coupled SDI at the regional level. Eight public administrations at Regional, Municipal and Provincial level (in Italy a Province is at intermediate level between Region and Municipality) have participated to the project in 2 ways: by discussing the spatial data requirements of the cooperative processes involving these administrations and by providing their spatial databases for experimentation.

Some of the problems which have emerged in the project are well-known in literature, although still not generally solved, in particular the problems of geometric harmonization of spatial data and the semantic harmonization due to the different data models adopted by different data providers.

Examples of geometric harmonization are:

- shared objects and topological relations on the border between two adjacent municipalities;
- the snapping of the local roads to the graph of the regional roads.

Examples of situations requiring semantic harmonization are:

- inside the same administration body, among different offices,;
- between administration bodies at different levels that govern the same area;
- among administration bodies of different types that work on overlapping territories;
- between two adjacent administration bodies.

This project has applied existing solutions to the above problems; its main contribution therefore has not been in the development of new strategies for performing geometric and semantic harmonization, but in the following aspects:

- to develop an integrated architecture in which all the harmonization problems can be solved in an ordered way;
- to characterize and solve two related problems which have been insufficiently considered in literature, called here ownership definition and instance integration, explained below;
- to deal with the geometric and semantic harmonization not only during the initial phase of SDI creation, but also through subsequent update operations, thus aiming to save in time the investment done at SDI creation.

The problems of **ownership definition and instance integration** arise in highly coupled SDIs because it must be clearly stated who is the actor who should provide a given kind of spatial information and perform updates on it; moreover, since different parts of some feature instances may be provided by different actors, these instances must be integrated. Indeed, we claim that ownership definition is a fundamental issue in highly coupled SDIs, since it establishes the responsibility concerning the data management and update.

**Reference example.** The following very simplified example, extracted from the real project, will be used throughout this paper. There are two actors providing data: a local municipality (actor PE) and the regional administration (actor RE).

The application context is the road administration, dealing with 3 feature classes: Road Elements (ES) and Main Roads (TR), both represented by curve geometries, and Road Junctions (RJ), represented by point geometries. There are also spatial integrity constraints, stating that (1) every RJ must *BelongTo* the boundary of some ES, (2) the boundary of every ES must be *ComposedOf* RJs, and (3) each TR geometry must be *ComposedOf* a set of underlying ES. The keywords BelongTo and ComposedOf are taken from the GeoUML language, their meaning is obvious in this context and has been formally defined in (Belussi et al. 2006).

Road Elements and Road Junctions are provided by both actors, while Main Roads are provided only by actor RE.

In Figure 1, the Road Elements provided by the two actors are shown (before any harmonization). Supposing that we have already converted data in a shared structure, defined by a common Application Schema, it is now necessary to establish: (i) the common geometry (geometric harmonization) of those Road Elements that are provided by both providers and (ii) the ownership of these geometries and the subsequent update activity.

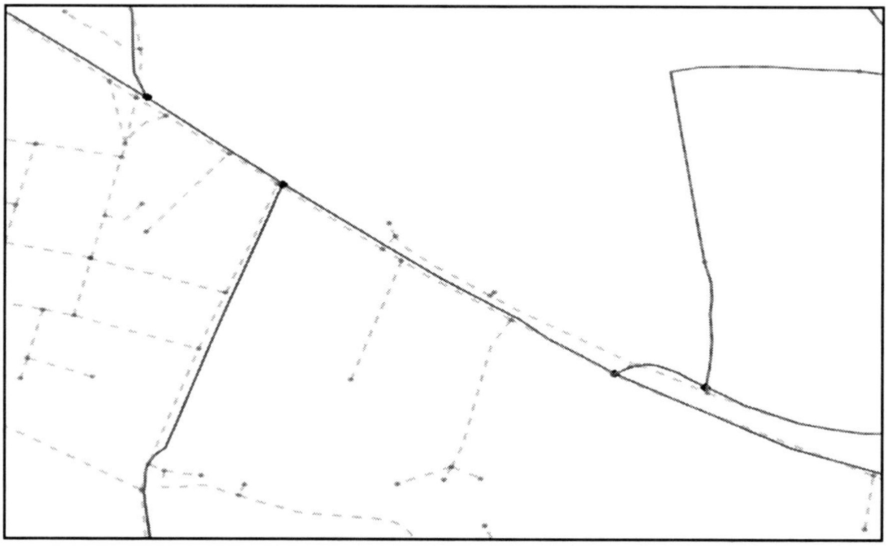

**Fig. 1.** Road Elements provided by two different actors (dashed lines represent data provided by PE, continuous lines represent data provided by RE).

Figure 2 shows the feature class Main Roads and, since there is the spatial composition integrity constraint between Road Elements and Main Roads, the ownership of instances of the different classes must be treated consistently and conflicts must be solved.

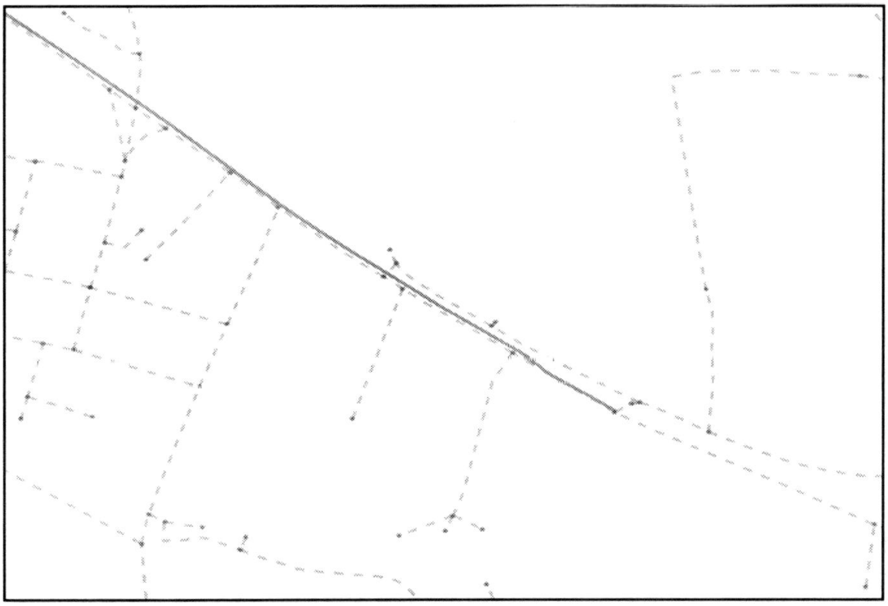

**Fig. 2.** with respect to Figure 1, here only the Main Roads provided by RE are shown as continuous lines (dashed lines represent data provided by PE).

**Related works**. In the last few years, in the research area of spatial databases and geographic information systems some new approaches have been proposed for the integrated representation (Duckham and Worboys 2007; Hariharan et al. 2005; Belussi et al. 2003) and integrated querying (Boucelma et al. 2007; Essid et al. 2004; Zaslavsky et al. 2000) of different spatial data sources. Considering in particular the area of SDI building and maintenance many works have recently been carried out on spatial data conflation, among them we recall (Chen et al. 2003; Volz and Bofinger, 2002; Rahimi et al. 2002; Cobb et al. 1998) and on spatial data semantic integration (Duckham and Worboys 2007; Jang and Kim 2007; Torres and Levachkine 2007; Gnaegi et al. 2006); however, to the best of our knowledge the problems of instances integration (or instances fusion) and ownership definition for spatial data in the context of SDI have not been analyzed yet.

The paper is organized as follows: in Section 2 the basic requirements regarding the building of an HC-SDI are presented. In Section 3 the proposed architecture is illustrated in details. Finally Section 4 presents a formal definition of the HC-SDI Participation Commitment and illustrates the proposed technique for handling the problems of ownership definition and instances integration. Section 5 outlines conclusions and future work.

## 2 Basic Requirements and Reference Architecture of an HC-SDI

A HC-SDI is an infrastructure that produces and maintains, through the cooperation of a variable but well-defined set of *actors*, a *virtual database* that integrates in a consistent way geographical information of *adequate quality* and is suitable for obtaining the goals of the administration bodies by supporting a set of cooperative processes.

This definition contains some key concepts that have to be further described.

A *well-defined set of actors*: the set of actors is well-defined if there exists a procedure for the explicit join of an actor to the SDI. Through this procedure the actor assumes explicitly the responsibility related to the join. This is done by defining the *participation commitment* of each actor, in which the actor declares the set of data types and the quality level that it will provide.

*Virtual database*: the SDI content is handled like a federated database, which means that a user can access this database as she will do with a traditional one, but it is the SDI manager that, given a user query, generates the correct set of queries on local systems. However, query optimisation is performed only inside each local system. This is indeed the characteristic that distinguishes a federated database from a distributed database. Following ISO TC 211 terminology, we call *Application Database* (*ADB*) the content of the virtual database and *Application Schema* (*AS*) its schema (in fact, the global virtual database consists of several thematic application databases, but we avoid to deal with this additional complexity in this paper).

*Adequate quality*: the concept of quality is important in the HC-SDI and it forces the actors to produce data that are compliant to the declared quality levels. This regards in particular the following aspects: (i) the accuracy, consistency and completeness of data; (ii) the temporal update level of data; (iii) the harmonization of the provided data with those provided by other actors of the HC-SDI.

An important issue that the HC-SDI architecture should face is its incremental development, since it is not feasible to build such a complex system in one shot.

A HC-SDI should have mechanisms to support the incremental development with respect to different aspects:

- Functional progressiveness: the SDI content should start by including data covering a part of the whole schema it will cover in the final state.

- Temporal progressiveness: actors will join the SDI in different instants of time.
- Space progressiveness and subsidiarity: since at a given time some geographical areas may not be covered by the actor at the most appropriate level (as required by recital 6 of the INSPIRE directive), other bodies at higher level could provide data in a subsidiary way.

In order to implement the functional progressiveness the global schema of the virtual database is subdivided in different parts, each one describing an independent subset of data, called Theme.

In order to implement the temporal progressiveness it must be possible to handle the case in which different actors join the HC-SDI in different moments and with different themes. Thus the participation commitment of an actor can evolve including new themes that the actor provides.

Finally, some cooperative processes require data covering the whole geographical area of the SDI, even if not all providers have joined the SDI. Thus, it should be possible for an administration body at higher level to provide less accurate data while waiting the joining of an actor that can provide more detailed information. This last situation is very common and should be managed by introducing a method for handling adequately the change of ownership.

Another important requirement refers to the possibility for each actor to use the hardware and software that the actor prefers, as far as they conform to the general standards of interoperability. The technological interoperability is based on the main ISO, OGC and W3C standards:

- XML of W3C.
- Service Oriented Architecture (WSDL and SOAP).
- Web Map Service (WMS) e Web Feature Service (WFS) of OGC.
- Geography Markup Language (GML) of OGC and ISO 19136,
- The "Spatial Schema" (ISO 19107), the "Rules for application schemas" (ISO 19109) for geographical data conceptual modelling.

The architecture of an HC-SDI should take into account the requirements described above and solve the critical issues that they pose.

Let us consider a data request from a final user, which is one of the basic operations that a HC-SDI has to perform. This operation is composed of the following steps:

1. The user requires data referring to the Application Schema AS and thus she expects to receive data coming from the Application Database ADB.

2. The SDI manager computes the set of Local Internal Databases (LIDB) that contain the requested data.
3. Data are extracted from the LIDBs, transformed and integrated, so that they appear to the user like they were extracted from a virtual global database (the ADB), and they are sent to the user.

The main issue in the above described process concerns the words "*transformed and integrated*" in the third step. Indeed, four are the fundamental transformations that must be applied to data in order to include them in the ADB:

1. Local systems are based on different technologies, thus a **Physical Mapping** is required, that should map the local data in a common data model (the Physical Mapping allows one to obtain the Local Conceptual Database LCDB from the Local Internal Database LIDB).
2. There exist many different approaches for describing and viewing data, thus a successive step for translating the data in the structures of the global shared schema is necessary. This is the **Semantic Mapping** that translates the LCDB into the Local Application Database LADB.
3. **Geometric Harmonization** is the necessary step for producing consistent geometries between different LADBs
4. In order to produce the content of the ADB also a phase of **Instances Integration** is necessary; it contributes, together with the geometric harmonization, to obtain the ADB from the LADBs.

Each of these transformations could be executed in two different moments: (i) when data are required by the user; (ii) when data are created.

An analysis of the four transformations described above has allowed us to verify the possibility of applying them in sequence during the data extraction, and this is a fundamental result in order to build an effective architecture for an HC-SDI.

The reference architecture that supports the above operations is described in Figure 3. This architecture is composed of 4 layers, where the three lower layers are associated to different representations of the local databases, while the fourth one represents the global and integrated database of the SDI.

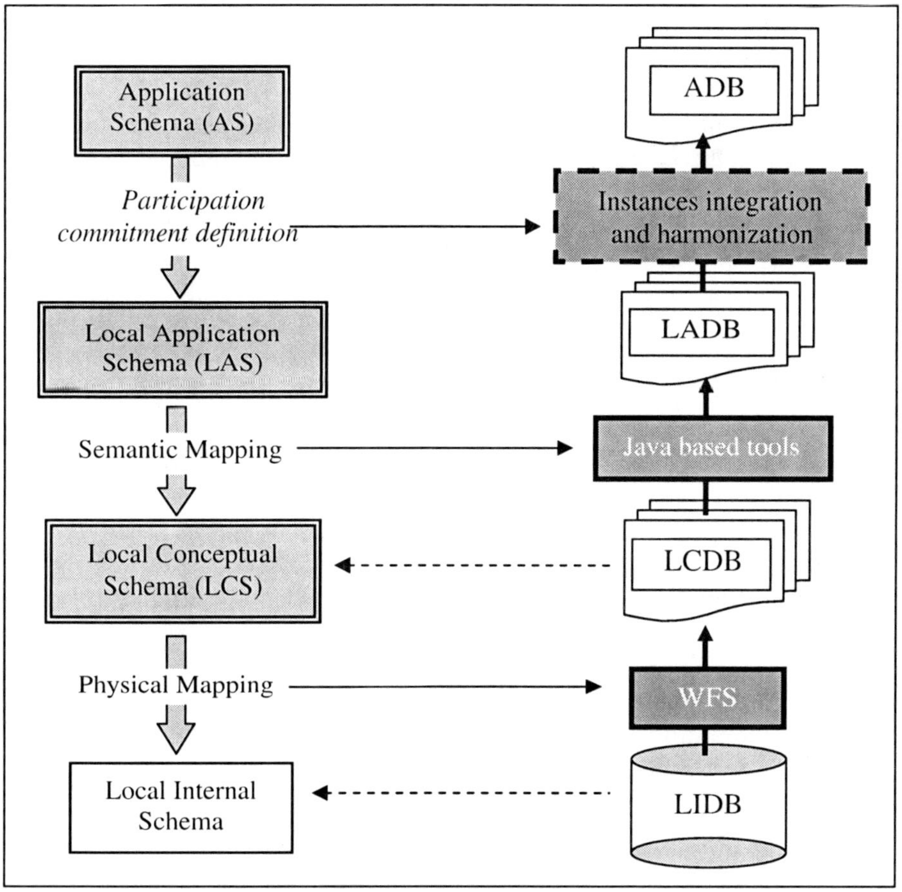


**Fig. 3.** The schema of the reference architecture: on the left the process of configuration of the joining phase is shown, starting from the participation commitment definition, while on the right we show the transformations applied on data during the extraction for feeding the SDI.

If we consider the intermediate results that each layer produces during the export of data from the Local Databases to the Application Database, we can identify the following three representations, each one obtained from the previous one:

- The *Local Internal Database* (LIDB): representing data in its native format, i.e. with a physical structure depending on local systems.
- The *Local Conceptual Database* (LCDB): representing data resulting from the Physical Mapping with a schema described in a shared conceptual model.

- The *Local Application Database* (LADB): representing data resulting from the Semantic Mapping, which are consistent with the global schema AS.

Each database has an associated schema: thus we have the *Local Internal Schema* (LIS), the *Local Conceptual Schema* (LCS) and the *Local Application Schema* (LAS).

By applying the *model driven approach* the first two transformations can be done by specifying the mapping among the corresponding schemas and then by applying automatic procedures guided by the defined mappings.

In the "Regione Lombardia" project, the following "road map" for building the SDI has been successfully tested and can be extended to other cases.

- Initially the Application Schema of the SDI is defined, then the actors perform their individual joining phases; since the Application Database is initially empty and the joining of the different actors is strictly sequential, every time that a new actor joins the SDI, it has to perform the harmonization of its data with the existing Application Database, which contains the data provided by those actors who have already joined the SDI.
- During the joining phase of each actor:
    - Some automatic tools are configured for executing the Physical and Semantic Mapping (model-driven approach).
    - Data of the joining actor are harmonized and instances integrated with the existing Application Database (with semi-automatic tools); at the same time ownership of instances is established.
    - Finally, the harmonized geometry and the metadata describing instances integration and ownership definition are stored back in the LIDB of the joining actor.
- During the subsequent update activity performed by the actor on its LIDB the update is harmonized with respect to the content of the SDI and the ownership of instances. For new instances also the instances integration and ownership definition are repeated.

In the following section the joining phase of an individual actor is analysed in detail with respect to the ownership determination and instance integration problems.

## 3 Joining the SDI

After the definition of the Application Schema (AS), which is the result of the initial interaction among the interested actors for building the HC-SDI, the joining of the first actor starts with the definition of its *Participation commitment*. Indeed, this is the first activity of the joining process.

The *Participation commitment* is a formal specification that defines the contribution of a actor and in particular it defines:

- The UML classes, associations and, for each class, the properties that the actor will provide; this is specified by assigning a portion of the whole AS, which could include also integrity constraints.
- The set of metadata describing how the instances provided by the actors have to be handled in the phase of instance integration.
- The set of rules describing how the ownership of the instances have to be managed in the phase of instance ownership definition.
- The portion of the AS which defines data that have some integrity constraints involving data from the provider and data coming from other providers of the HC-SDI.

### 3.1 Participation Commitment

Each actor $s_i$ participating to the HC-SDI with schema AS defines its *Interoperability Interface*, that is composed of a pair of specifications: the *Provider Interoperability Interface* and the *Consistency Requestor Interoperability Schema.*

The ***Provider Interoperability Interface*** is composed of:

- A ***Provider Interoperability Schema*** ($PIS(s_i)$ or $PIS_i$): it is the portion of AS that the actor $s_i$ promises to provide joining the HC-SDI. This schema is written in UML:

$$PIS_i = (C, A, V) \tag{3.1}$$

  where $C$ is a set of classes, $A$ is a set of associations among classes and $V$ is a set of integrity constraints. Moreover, each class (association) has a set of attributes. Finally $PIS_i$ must satisfy the following inclusion constraint with respect to $AS$:

$$\begin{gathered}\forall c \in PIS_i.C{:}\ \exists c' \in AS.C{:}\ (c.name = c'.name \wedge \\ \forall a \in c.attributes{:}\ \exists a' \in c'.attributes{:}\ a.name = a'.name \wedge \\ a.domain = a'.domain)\end{gathered} \tag{3.2}$$

where *c.name* is the class name, *c.attributes* is the set of its attributes, *a.name* is the name of the attribute *a*, and *a.domain* represents its domain. The actor $s_i$ is responsible for providing instances of the set of classes and associations declared in $PIS_i$ satisfying the integrity constraints declared in $PIS_i$.

- A ***Provider Interoperability Metadata*** *($PIM(s_i)$ or $PIM_i$)*: it is a set of metadata describing the quality of the data provided by $s_i$, in particular they state: how the instances of the LADB are obtained from the LIDB (they can be preserved or generated applying some transformations: for example, a merge or a split operation may be applied) and which is the accuracy of the provided geometries. Many other metadata descriptors can be added. We limit our attention to these ones, because they are used in the joining phase. Given the schema $PIS_i$ the following metadata are defined in terms of functions:

$$PIM_i = (type, gen, acc, \ldots) \tag{3.3}$$

where:

- $type: PIS_i.C \rightarrow \{ det, geo \}$ is a total function where:
  *"det"* indicates that the instances of the class *c* have an existence which is independent from their geometry; we say also that c is intrinsically determined.
  *"geo"* indicates that the instances of the class *c* can be determined by applying a geometric algorithm (possibly non deterministic) that given a scene produces the instances.
- $gen: PIS_i.C \rightarrow \{ generated, native \}$ is a total function where:
  *"generated"* indicates that the instances of the class *c* do not exist in the LIDB of the provider, but they have to be generated starting from other instances available in the LIDB;
  *"native"* indicates that the instances of the class *c* exist in the LIDB of the provider.
- $acc: PISi.C \cup \{a \mid \exists c \in PISi.C: a \in c.attributes\} \rightarrow \Re \cup \{derived\}$ is a partial function that indicates the accuracy by specifying the average error of the absolute position of the provided geometries. The special value *"derived"* is used when the geometry is built by using geometric values of other attributes. This function is extended also to attributes representing geometric properties. Given a class *c* if *acc(c)* is defined and *acc(a)* is defined (where *a* is a geometric attribute of *c*) then the value of *acc(a)* wins. If *acc(c)* is defined and *acc(a)* is not defined, then the value of *acc(c)* is propagated to *a*.

- An ***Ownership Interoperability Rule*** *($OIR(s_i)$ or $OIR_i$)*: it is a set of rules stating how the ownership of an instance is established in case of conflicts during the joining phase of $s_i$. Given the schema $PIS_i$ the ownership rule is defined by means of the following function:

$$
\begin{gathered}
own: PIS_i.C \cup \{a \mid \exists c \in PIS_i.C: a \in c.attributes\} \rightarrow 2^{Priority} \\
where \\
priority = (0..1) \cup \{ bottom, top \} \cup (0..1)(R) \cup \\
\{ bottom, top \}(R)
\end{gathered}
\tag{3.4}
$$

  *own* is a partial function that can be applied to the classes of $PIS_i$ and to all the attributes of these classes. Its co-domain is the set of real numbers from zero to one union the set of the following special values:
  *"bottom"* indicates that the instances of this class (or values of the attributes) are provided by the actor $s_i$ with the lowest priority, i.e. in any possible conflict $s_i$ looses.
  *"top"* indicates that the instances of this class (or values of the attributes) are provided by the actor $s_i$ with the highest priority, i.e. in any possible conflict $s_i$ wins.
  *x(R)* indicates a priority x that is applied only when the instance is spatially completely contained in the region R.
  The function own is partial, thus there could be an attribute *a*, such that *own(a)* is not defined. In this case the attribute *a* inherits the priority level from the class *c*.
  Notice that the function *own* gives only a general method for resolving ownership conflicts. The user can always modify the result of the automatic processing by assigning the ownership with ad hoc criteria.

The ***Consistency Requestor Interoperability Schema*** *($CRIS(s_i)$ or $CRIS_i$)* is the subset of the AS that the actor $s_i$ is going to request in order to guarantee the consistency of the data provided by $s_i$ itself.

The union of the $PIS_i$ e $CRIS_i$ schemas produces the schema of *the Interoperability Interface* of $s_i$ and is indicated as *Local Application Schema* *($LAS(s_i)$* oppure *$LAS_i$*).

**Examples**. The following 2 Participation Commitments refer to the actors and application context described in the Reference Example of the Introduction.

***Participation commitment of the local municipality PE:***

$PIS_{PE}.C = \{ Road\ Elements\ (ES), Road\ Junction\ (RJ) \}$

*$PIS_{PE}$.V = { RJ BelongsTo ES.boundary (VJBel),*
*ES.boundary ComposedOf RJ (VEComp)*
*$PIM_{PE}$ : type(ES) = geo; gen(ES) = native; acc(ES) = 0.6 m;*
*type(RJ) = det; gen(RJ) = native; acc(RJ) = 0.6 m;*
*$OIR_{PE}$: own(ES) = 0.9(MunicipalityTerritory)*
*own (RJ) = 0.9(MunicipalityTerritory)*
*$CRIS_{PE}$.C = { Main Road (TR) }*
*$CRIS_{PE}$.A = { roadElementMainRoad (ES-TR) }*
*$CRIS_{PE}$.V = { TR ComposedOf ES (VComp) }*

In the *CIS* of PE the class Main Road, the association *ES-TR* and the constraint VComp are included; indeed, the integrity constraint VComp involves the class Road Element and thus also the providers of this class. However, since the class Main Road and the corresponding association are not provided by PE, they are not contained in the *PIS* of PE, but in the *CIS* of PE.

***Participation commitment of the Regional Administration RE:***

*$PIS_{RE}$.C = { Road Elements (ES), Road Junction (RJ),*
*Main Road (TR) }*
*$PIS_{RE}$.A ={ roadElementMainRoad (ES-TR) }*
*$PIS_{RE}$.V ={ RJ BelongsTo ES.boundary (VJBel),*
*ES.boundary ComposedOf RJ (VEComp)*
*TR ComposedOf ES (VComp) }*
*$PIM_{RE}$: type(ES) = geo; gen(ES) = derived; acc(ES) = 2 m;*
*type(RJ) = det; gen(RJ) = native; acc(RJ) = 2 m;*
*type(TR) = det; gen(TR) = native; acc(TR) = derived;*
*$OIR_{RE}$: own(ES) = bottom; own(RJ) = 0.9; own(TR) = top*
*$CRIS_{RE} = \varnothing$;*

These two commitments state that:

- The municipality PE is responsible for providing the Road Elements and Junctions on its territory.
- The Regional Administration is providing the Road Elements only for the portion of the territory where no local municipality has joined the SDI. Moreover, notice that the road elements are not native instances in the local system of the Regional Administration.
- The Regional Administration is providing data about main roads and there is a spatial integrity constraint that requires that each main road is spatially composed of road elements.

- There are additional spatial constraints *VJBel* and *VEComp* for specifying the bindings among road elements boundaries (start and end points) and road junctions.
- Notice that in the participation commitment of the local municipality there is a not empty *CIRS*, which contains the class Main Road, the association between Road Elements and Main Roads and the constraint *VComp*.

## 3.2 Instance Integration and Ownership Definition

In Figure 4 we show in more details the joining process of a new actor to the HC-SDI, in particular we describe the instances integration and ownership definition sub-phase.
Considering Figure 4, notice that:

- The semantic mapping can either transform the instances of the *LCDB* (producing the *LADB*) or generate new instances starting from those contained in the *LCDB* (producing the *LADBgen*); thus new *IDs* should be generated for new instances, that are not stored in the *LIDB*.
- The input of the geometric harmonization phase includes not only the *LADB* and the *LADBgen* but also the current state of the *ADB* (denoted as *ADBpre*), representing the SDI state before the join of the new actor.
- Moreover, after the instances integration and ownership definition, we can identify two sets of instances with harmonized geometry: the *LADBpost*, that contains the set of instances that have preserved the ownership of the new provider joining the HC-SDI, and the *LADBgen-post* that contains the new instances generated in the semantic mapping or in the instance integration sub-phases.
- Finally, the *LADBpost*, the *LADBgenpost* and the *ADBpre* with harmonized geometry are merged to produce the new state of the HC-SDI, called *ADBpost*. Since the joining of a new actor can have impacts on other providers in terms of harmonized geometry and instances ownership, some data will be propagated to the other providers of the HC-SDI. Moreover, since part of the work done during the joining phase cannot be easily replicated on the fly (many human interaction occurred for producing a correct harmonization), at least the harmonized geometry has to be stored in the *LIDB* of the provider (*LIDBpost*).

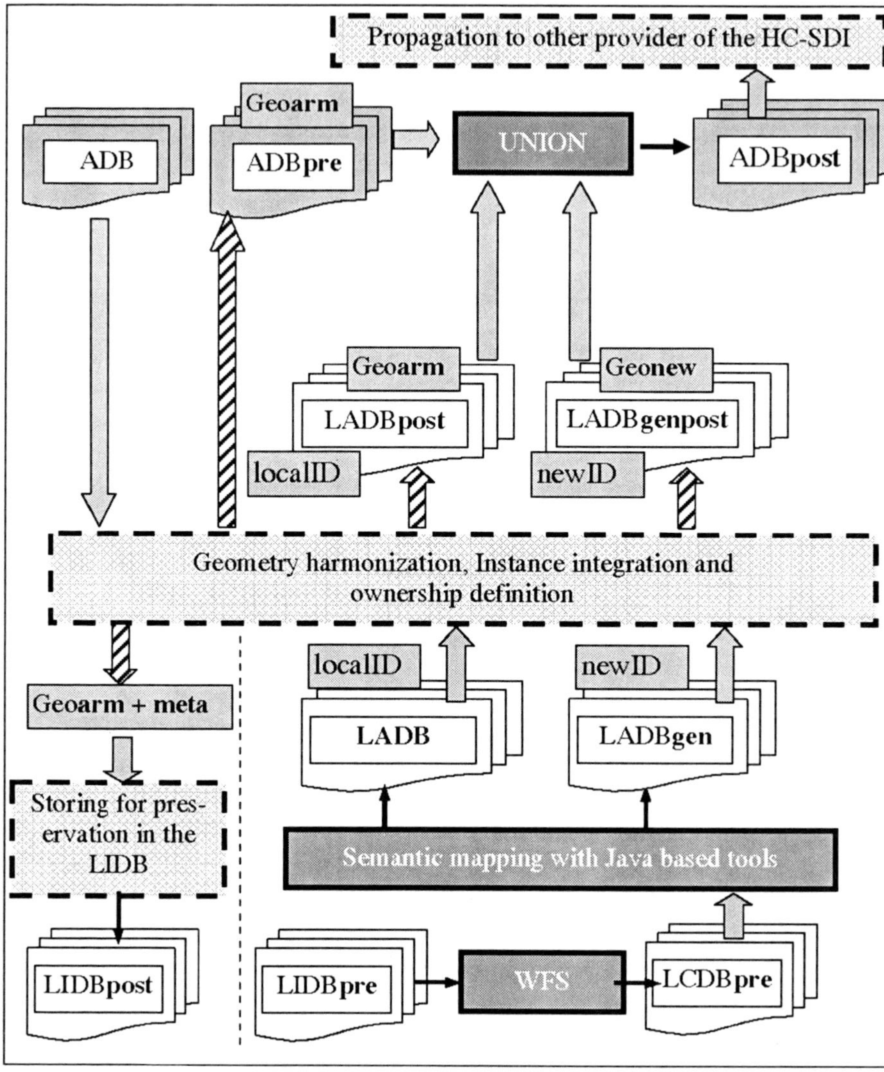


**Fig. 4.** The schema of the instance integration and ownership definition sub-phase.

In order to present the proposed method for instance integration we start considering a generic instance $x$ of the class $C$ contained in the *LADB* that has to be transferred into the *LADBpost* during the joining phase of $s_i$. Different cases can occur:

- *type(C) = det*: if $x$ is not contained in the *ADBpre*, then $x$ is copied in the *LADBpost* and the ownership of $s_i$ is preserved; if $x$ is present in

*ADBpre* (this test is based on global IDs) then a conflict of ownership is raised and the function *own* can be used to define the "winner". If the winner is $s_i$ the ownership is preserved and $x$ must be delete from the database of the "looser", otherwise $s_i$ looses the ownership of $x$. However, other possible cases can occur: a subset of attributes becomes ownership of $s_i$ and other ones remain ownership of other providers of the HC-SDI (***shared-ownership***); or, in particular for a geometric attribute, we can also have a ***multi-ownership*** commitment. The multi-ownership can be used in different situations: (i) when the geometry of the instance is very simple (like a point) and it is totally shared; (ii) when the geometry of the instance is very complex and widely extended on the territory of the SDI, in this case usually each provider has a piece of the whole geometry. This second case is possible only due to the fact that $x$ is intrinsically determined (*type(C) = det*), thus the instance is one but its geometry is divided in different pieces, that are owned by different providers (for example, this is the case of A4 highway, which is one of the main roads of Lombardy, or the Po river).

- *type(C) = geo*: if $x$ is not contained in the *ADBpre*, then $x$ is copied in the *LADBpost* and the ownership of $s_i$ is preserved; if x is present in *ADBpre* (this test is based on geometry) then a conflict of ownership is raised and the function *own* can be used to define the "winner". If the winner is $s_i$ the ownership is preserved and $x$ must be delete from the database of the "looser", otherwise $s_i$ looses the ownership of $x$. Also the cases described in the previous point can occur (shared-ownership and multi-ownership). Moreover, another situation is possible for geometric instances, indeed, the conflict can be relative to the whole geometric value, but also to a portion of it. In this last case the ownership of the overlapping geometry should be decided and as a consequence the geometry of $x$ should be cut or extended according to the decided ownership assignment.

In order to show an application of this method we consider the example proposed in Figure 1 and 2, supposing that first the Lombardy administration completes its joining phase and then the local municipality PE performs its joining to the HC-SDI.

In Figure 5 the situation after the joining of the Lombardy administration is shown. Figure 6 shows the state of the SDI and the *LADB* of the local municipality PE that is joining the SDI before the geometry harmonization and instances integration. Finally, in Figure 7 the situation after the joining of the local municipality PE is shown.

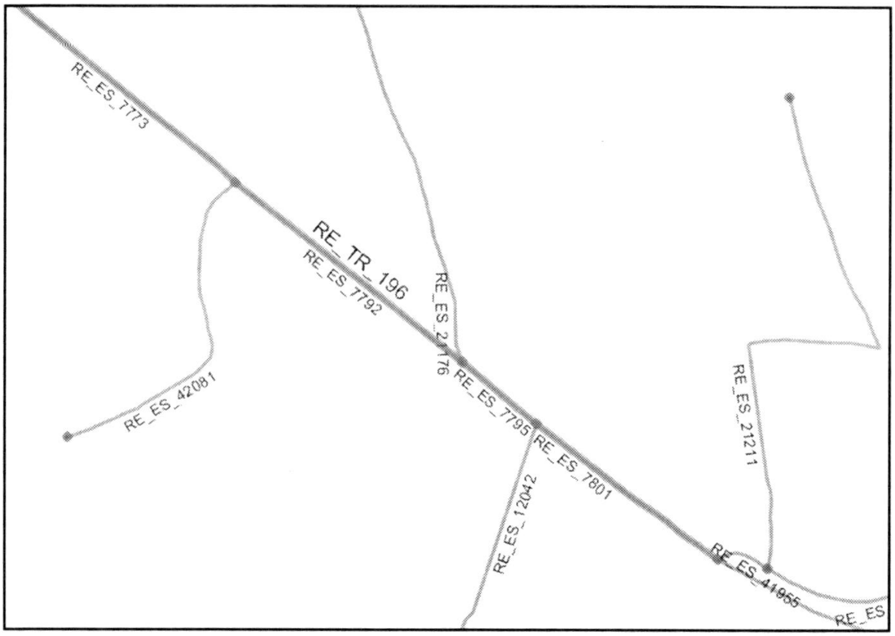


**Fig. 5.** The HC-SDI content after the joining of Regional Administration: the ID shows the provider of the object (RE for regional administration and PE for the local municipality), the class the instance belongs to (ES for Road Elements and TR for Main Road) and finally the local ID.

Notice that:

- The simple union of the two providers cannot become the new state of the SDI.
- The harmonization process without an ownership definition could lead to inconsistent updates, since the providers could modify the geometry of their local copy of the shared instance.
- The SDI content after the joining phase of PE gives the ownership of the Road Elements to the local municipality PE, but the ownership of the Main Road remains of the Lombardy Administration.
- The integrity constraint *VComp* becomes an inter-provider constraint, this means that any geometry modification of a Road Element (or of a Main Street) must be handled with a mechanism based on distributed transactions, otherwise some transitory inconsistent states have to be accepted.

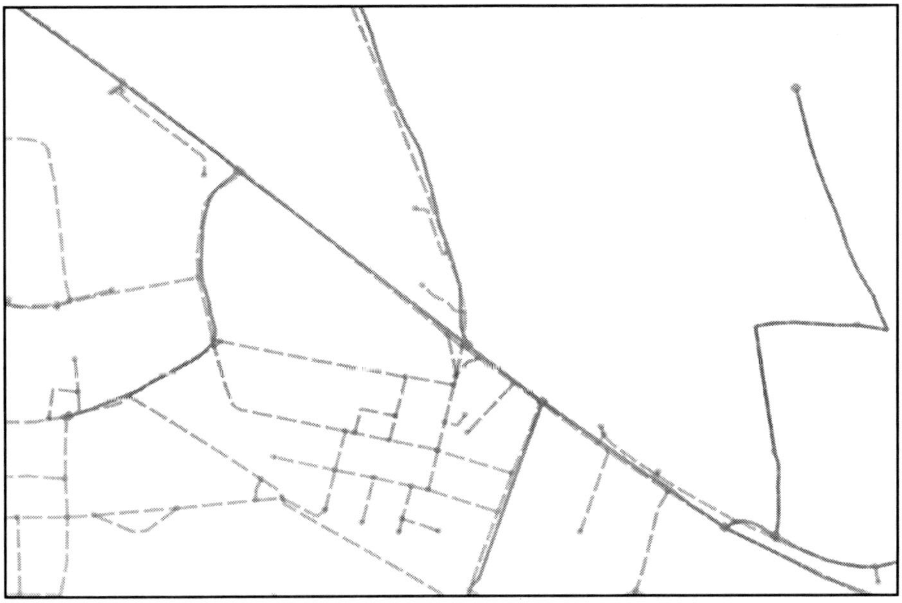

**Fig. 6:** The HC-SDI content before the geometric harmonization and instances integration of the municipality PE (dash lines represent the Road Elements of PE).

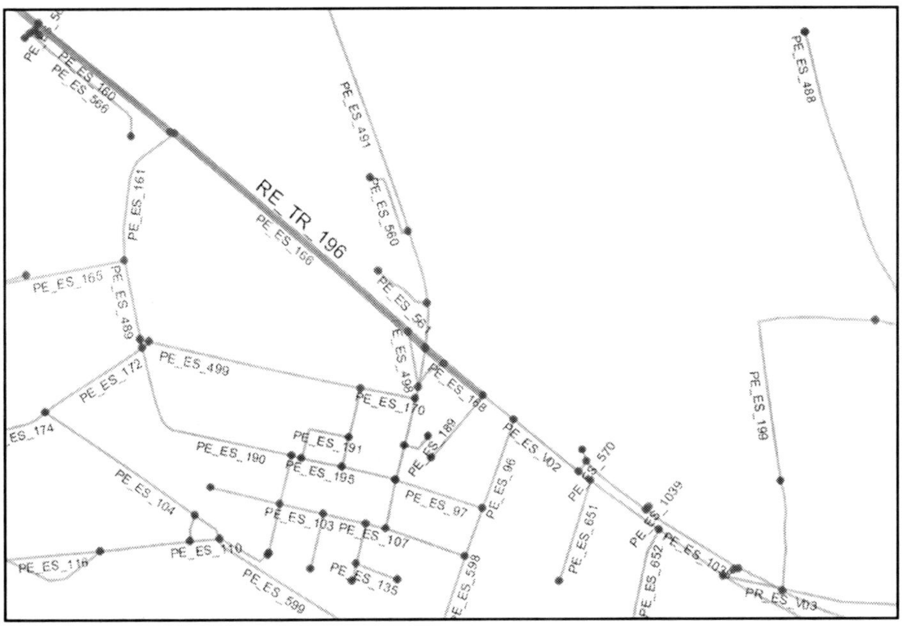


**Fig. 7:** The HC-SDI content after the joining phase of the municipality PE.

## 4 Conclusions and Future Work

The project described in this paper has analyzed the requirement of a set of real actors for the development of a Highly-Coupled Spatial Data Infrastructure, has defined a reference architecture for satisfying them and has implemented a prototype of this architecture. The experience gained with this prototype has shown that it is possible to effectively combine in a unique architecture the algorithms which have been or are being developed by the research community in order to solve some difficult problems like geometric harmonization and semantic mapping.
Moreover, the requirements have put in evidence the necessity of dealing also with two rather neglected problems: the instances integration and the ownership conflict resolutions. An approach for performing them has been proposed.

## References

Abdelmoty A.I., Smart P.D., Jones C.B., Fu G., Finch D. (2005) A critical evaluation of ontology languages for geographic information retrieval on the Internet, Journal of Visual Languages & Computing, 16, 331-358.

Belussi A., Catania B., Bertino. E (2003) A reference framework for integrating multiple representations of geographical maps. In Proceedings of the 11th ACM GIS'03 (New Orleans, Louisiana, USA, Nov. 07-08, 2003), pp 33-40.

Belussi A., Negri N., Pelagatti G., (2006) GeoUML: an ISO TC 211 compatible data model for the conceptual design of geographical databases. In Proceedings of the ER 2006 (CoMoGIS), Tucson, AZ, USA, 6-9 Nov. 2006.

Boucelma O., Essid M., Lassoued Y. (2007) A Quality-enabled Spatial Integration System. In Spatial Data on the Web: modeling and management, Springer Verlag, pp 133-157.

Chen C.C., Thakkar S., Knoblock C., Shahabi C. (2003) Automatically Annotated and Integrating Spatial Datasets, In Proceedings of te 8th International Symposium, SSTD 2003 (Santorini, Greece, July 24-27, 2003), pp. 469-488.

Cobb M.A., Chung M.J., Foley III H., Petry F.E., Shaw K. B. (1998) A Rule-based Approach for the Conflation of Attributed Vector Data, GeoInformatica, Vol. 2, N. 1, pp 7-35.

Duckham M., Worboys M. (2007) Automated Geographical Information Fusion and Ontology Alignment. In Spatial Data on the Web: modeling and management, Springer Verlag, pp 109-132.

Essid M., Boucelma O., Colonna F., Lassoued Y. (2004) Query Processing in a Geographic Mediation System, In Proceedings of the 12th ACM-GIS'04 (Washington DC, USA, Nov. 12-13, 2004), pp 101-108.

Gnaegi H.R., Morf A., Staub P. (2006) Semantic Interoperability through the Definition of Conceptual Model Transformations, In Proceedings of the 9th AGILE'06 (Visegràd, Hungary, Apr. 20-22).

Hariharan R., Shmueli-Scheuer M., Li C., Mehrotra S. (2005) Quality-driven approximate methods for integrating GIS data. In Proceedings of the 13th ACM GIS'05 (Bremen, Germany, Nov. 04 - 05, 2005), pp 97-104.

Zaslavsky I., Marciano R., Gupta A., Baru C. (2000) XML-based Spatial Data Mediation Infrastructure for Global Interoperability, 4th Global Spatial Data Infrastructure Conference, Cape Town, South Africa, 13-15 March 2000.

# Spatial Data Integrability and Interoperability in the Context of SDI

Hossein Mohammadi, Abbas Rajabifard, Ian Williamson

Centre for Spatial Data Infrastructures and Land Administration, Department of Geomatics, University of Melbourne, Parkville, Victoria, 3010, AUSTRALIA

**Abstract.** The number of multi-sourced heterogeneous spatial datasets continues to grow and the fragmentation of organizational arrangements has caused much technical and non-technical heterogeneity. Spatial Data Infrastructures aim to facilitate spatial data use and sharing, and can be an effective platform to aid in data integration. This paper discusses the technical and non-technical heterogeneities of multi-sourced spatial data within the holistic framework of Spatial Data Infrastructure. The paper capitalizes on research and case studies undertaken within Australia. The paper also introduces Geo-WebServices as a means of facilitating spatial data integration and interoperability. Geo-WebService can provide a platform to assess the level of Integrability and readiness of multi-sourced datasets. The results of this research aim to assist practitioners in developing the necessary technical tools including geo web-services and guidelines for effective data integration.

**Keywords:** spatial data, Spatial Data Infrastructure, integratability, interoperability, Geo Web-Service

## 1 Multi-Source Data Integration and Interoperability

Multi-source data integration and interoperability has become a significant issue as it ensures effective access and reuse of spatial data by many spatial users and applications. This has created many opportunities and possibilities for using and applying spatial datasets in a range of services.

Many spatial applications and services try to model and analyze some aspects of the environment utilizing different criteria. These applications rely highly on multi-sourced spatial data to meet the requirements of diverse criteria. For example, the Emergency Information Coordination Unit (EICU) of New South Wales (NSW)-Australia utilizes different spatial data ranging from fundamental datasets including cadastre, topography, roads and imagery to locational data including police, fire and points of interests to socio-economic and infrastructure data including demography, valuation, public transport and utilities (Colless 2005). Many of these datasets are managed by different custodians in NSW. For example cadastre and topography is managed by local councils and Department of Lands, roads by local councils, Roads and Traffic Authority (RTA) and National Parks (Baker and Young 2005), and fire data by Department of Land, National Parks and Wildlife Service and Royal Botanic Gardens.

Table 1 summarizes some spatial datasets which are necessary for emergency management purposes and also their potential sources within the state of NSW-Australia.

**Table 1.** Example of spatial data with potential sources used in NSW-Australia

| Spatial data | Source |
|---|---|
| cadastre and topography | Local councils and Department of Lands |
| roads | Local councils, Roads and Traffic Authority (RTA) and National Parks etc. |
| imagery | Department of Lands, RTA, and Department of Agriculture |
| vegetation | Department of Land and Water Conservation, National Parks and Wildlife Services, Forests NSW, Department of Defense |
| fire | Department of Land, National Parks and Wildlife Service and Royal Botanic Gardens |
| threatened species | Department of Land, National Parks and Wildlife Service and Royal Botanic Gardens |
| waste | Environmental Protection Agency, Waste Service and Local Councils |

As shown in the above table, different organizations are responsible for different datasets. Organizations develop their own strategies and policies in regards to capturing, managing and sharing data. The diversity of approaches utilized by these organizations leads to many technical and nontechnical inconsistencies and heterogeneity among datasets.

## 2 Spatial Data Integration Challenges

Despite the importance of spatial data integration for many applications and services, the fragmentation of the institutions that are responsible for the production and management of different datasets has caused heterogeneities and inconsistencies from different technical and non-technical aspects, as illustrated in Fig 1. These inconsistencies can be classified into institutional, policy, legal and social categories as suggested by Mohammadi et al. (2006a).

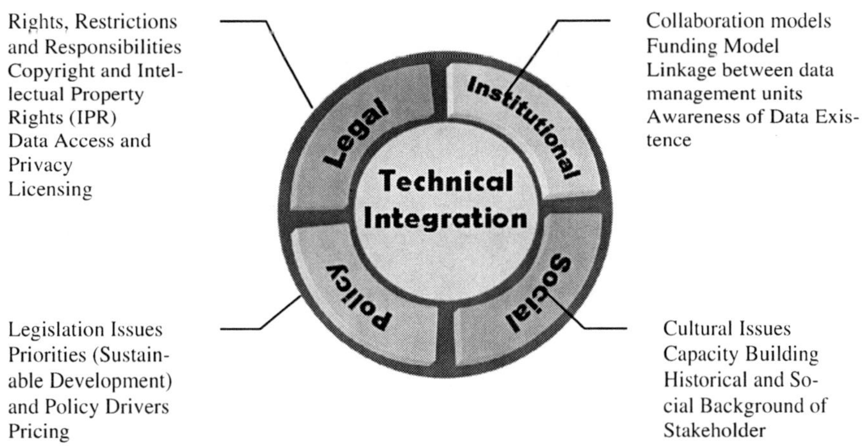


**Fig 1.** Technical data integration and associated non-technical issues

Technical issues including inconsistent standards, semantic heterogeneity, poor metadata/no metadata and inconsistency in data models hinder effective data integration. However the problem is not only technical in nature, with non-technical issues also hindering effective integration.

The non-technical problems including inconsistency of institutional arrangements and policies, different understanding and knowledge, capacity building, lack of regulation and lack of efficient metadata are also a concern, as has highlighted by Syafi'I (2006).

The collaboration between stakeholders, business models together with data management approaches are also key institutional issues which act as barriers against data integration. Policy drivers and priorities of nations, pricing and legislation have been found to be major issues from a policy view point. Cultural differences, capacity building and the social background of spatial data stakeholders are also paramount in the social category. From a legal perspective, the following issues are prominent:

- Rights, Restrictions and Responsibilities (RRR)
- Copyright and intellectual property rights (IPR),
- Data access and privacy, and
- Licensing

Table 2 has summarized technical and non-technical issues associated with spatial data integration and integratability.

**Table 2.** Technical and non-technical issues associated with spatial data integration and interoperability (adopted from Mohammadi et al 2006b)

| **Technical issues** | **Non-technical issues** | | | |
|---|---|---|---|---|
| | Institutional issues | Policy issues | Legal issues | Social issues |
| inconsistent standards<br><br>semantic heterogeneity (attribution etc)<br><br>poor metadata/no metadata<br><br>inconsistency in data models<br><br>bounding box<br><br>projection system | utilizing inconsistent collaboration models<br><br>funding model differences<br><br>lack of awareness of data integration | lack of supporting legislations<br><br>inconsistency in policy drivers and priorities (sustainable development)<br><br>pricing | definition of rights, restrictions and responsibilities<br><br>inconsistency in copyright and intellectual property rights (IPR) approaches<br><br>different data access and privacy policies | weakness of capacity building activities<br><br>different background of stakeholder |

## 3 Spatial Data Coordination in Australia

In terms of Australia, there are several national level organizations responsible for different aspects of spatial data coordination, including Public Sector Mapping Agency (PSMA), Geoscience Australia (GA), Office of Spatial Data Management (OSDM) and ANZLIC – the Spatial Information Council. PSMA produces national coverage maps from best available data sources. GA produces and maintains national level spatial data which is mostly small scale maps. OSDM provides spatial policy and guidelines for use of data within federal governmental organizations, while ANZLIC is the peak body for development and coordination of Australian SDI (ASDI) guidelines. Collaboration between organizations at federal and state levels is mostly done through PSMA and ANZLIC. Access to data and pricing strategies are developed at different jurisdictional levels.

Access to spatial data at a national level is done on a case-by-case basis. OSDM develops federal governmental spatial data pricing policy, while PSMA has its own pricing policy for spatial data in place. Data integration is also done in-house as GA and PSMA hold almost all national level spatial datasets.

At state level, states communicate to each other through certain channels including PSMA and ANZLIC and also on a project basis. Pricing policies differ from state to state, ranging from the state of Victoria, which sells spatial data based on cost recovery policy to the state of Western Australia (WA), which provides spatial data to users at the cost of distribution.

Within states, due to large numbers of data custodians without well-established collaboration, integration and interoperability of spatial data is problematic and is a time consuming and costly process. Standards are developed by every state individually, while where applicable and if in line with state priorities, they adopt ANZLIC's guidelines. An example of a nationally consistent initiative is a metadata standard developed by ANZLIC and adopted by the states. Access channels are not singular and there are different access channels for spatial data. States produce their own spatial data with their own policies and guidelines.

Local councils produce large scale maps but follow state standards and policies on spatial data. Access to local council's data is done on a case-by-case basis. Pricing on spatial data is done by states and local councils seldom sell and distribute data. Local councils liaise with states and in some cases there is very good collaboration in place, however, there is generally little collaboration between local councils (Table 3).

**Table 3.** Spatial data management arrangements within Australian jurisdictions

| | Federal | State | Local council |
|---|---|---|---|
| **Data Production** | PSMA, GA | State and local councils | Local council |
| **Policy development** | OSDM, ANZLIC | States and ANZLIC | State |
| **Access** | case by case basis, free to browse | for small clients through data resellers and for large clients State organizations | case by case basis |
| **Standards** | OSDM, ANZLIC, ICSM | State, ANZLIC | State |
| **Collaboration** | through PSMA and ANZLIC | With States through ANZLIC and PSMA, individually with local councils and through joint initiatives | individually with states, little collaboration with other councils through states |
| **Pricing** | OSDM, PSMA | State | State |
| **Spatial data integration** | in-house | users and third parties | users and third parties |

The diversity of key players in the spatial data area within different jurisdictional levels with different interests and priorities leads to a diversity and complexity in data coordination arrangements and policies. The diversity of policies and coordination approaches satisfy objectives and requirements of a particular organization.

The complexity of issues associated with spatial data integration and interoperability can not be addressed and facilitated, unless there is a well-structured and holistic platform to consider all effective components and issues of spatial data integration together.

For spatial services which utilize spatial data from different sources, it is important to access data at justifiable time and cost through the fastest channel. This is not possible unless the diversity of issues is managed through an enabling platform. Such a platform establishes interoperability at technical and non-technical levels and establishes effective interaction between different technical and non-technical components including policies, standards, collaboration and access. This platform can provide a comprehensive framework which assists spatial data stakeholders to develop and design required guidelines, policies and technical tools for effective spatial data integration and interoperability.

The development of SDIs can aid in providing a basis for establishment of an enabling platform. SDIs aim to facilitate the reuse, access, integration and sharing of data. In order to achieve this aim, SDIs potentially can provide the necessary technical and non-technical tools and policies together with guidelines.

# 4 SDI to Facilitate Effective Spatial Data Integration

SDIs are being developed by many countries throughout the world as an enabling platform to assist people to access, use and integrate spatial data effectively. It includes access networks, policies and standards of an enabling platform which facilitates the interaction of spatial data stakeholders with spatial data (Fig 2).

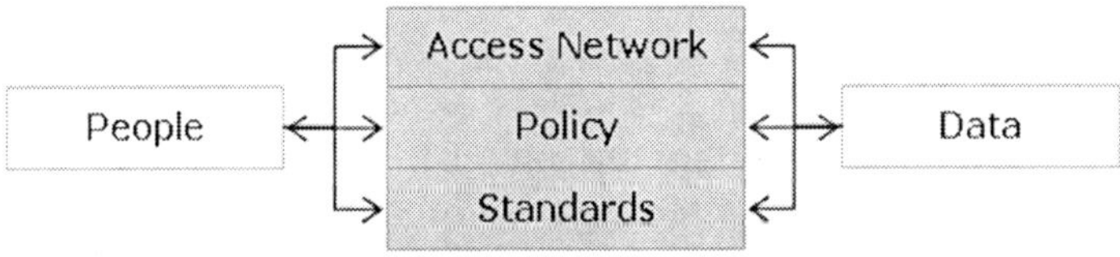


**Fig 2.** SDI components (Rajabifard et al 2001)

As Rajabifard et al (2001) have highlighted, SDIs are able to establish requisite arrangements and facilitate data integration. In this regard, necessary tools should be developed for stakeholders to interact with technical, policy and standardization tools to coordinate, integrate and use data more effectively. SDIs aim to address and coordinate the integration of multi-sourced data in a way which saves time and reduces costs; however for the stakeholders of spatial data, this aim has not been fully achieved. To facilitate data integration, issues and challenges need to be identified and addressed in the context of SDIs. These considerations can then assist to develop technical, policy, institutional and management tools for effective data integration.

In terms of Australia, Integrability of spatial datasets is one of four main streams in the development of the Australian National SDI (ANZLIC 2003). The perception of spatial data integration and its importance among states is different which leads to a lack of knowledge of spatial data Integrability issues in some states, however for other states it is a high priority. Therefore, spatial data integration has not yet been achieved fully at a national level in Australia. Special attention should be paid to data integration when developing technical mechanisms and tools taking in to account the legal, institutional, policy and social frameworks within an SDI initiative.

Data coordination and maintenance is one of the aims of SDIs. SDIs can develop guidelines for spatial data producers in order to facilitate the reuse and integration of spatial data. It consists of dictating standards, interoperability tools, semantic homogeneity among datasets, data quality and reference system, metadata guidelines, data models and attribution guidelines.

SDIs also need to take into consideration the Integrability of legal, institutional, social and policy frameworks. Without interoperable collaboration, spatial datasets can not be integrated and used to their maximum potential. Consistent pricing and privacy policies together with appropriate capacity building can ensure effective data integration. SDIs can also develop guidelines and tools to act as intermediators for many inconsistent systems. These guidelines and intermediators can establish effective links between inconsistent components of multi-source spatial data, services and policies.

With the advancement of Geo-WebServices (GWS), the access and availability of datasets which are targeted by SDIs is more facilitated (Nebert 2004). Geo-WebServices are significant technical tools which can also facilitate the integration of multi-sourced datasets. Geo-WebServices are developed based on open standards and provide tools, services and formats which comply with interoperability concepts. Geo-webservices can access and collect spatial datasets throughout the web and local databases. Based on this, a prototype geo-webservice has been developed to assess the level of readiness of datasets for the purpose of data integration.

## 5 Geo-WebService to Facilitate Spatial Data Integration

Geo-webservices accomplish spatial data interoperability at data and service levels by utilizing common standards and specifications (Gould 2001). Capitalizing on opportunities created by geo-webservices, multi-sourced spatial datasets and services can be accessed and utilized. Spatial data integration can also be facilitated with the aid of geo-webservices.

One of the major problems in an effective data integration service is the assessment of spatial data and its level of integratability. The assessment can be done with utilizing accessible, measurable and comparable measures including availability of metadata, format, datum, bounding box etc. In order to address other institutional, legal and social measures, a holistic study of the legal, institutional and social arrangements within different organizations and jurisdictions is required.

Many services gather data from distributed sources and assemble them within a single system for integration and analysis, but the lack of Integrability among datasets leads to many problems. Utilizing geo-webservice capabilities, a prototype system has been designed and developed as part of an ongoing research project on the integration of multi-sourced spatial data. This system is able to access Web Map Services (WMS) and Web Feature Services (WFS) as well as other common spatial formats, throughout the web and local databases. If other complementary information including availability of metadata, any restrictions on data or pricing and privacy policy are not extractable, this is requested from data provider. Data collected from different sources is then assessed against Integrability requirements based on a set of predefined criteria including the compatibility of format, datum, coordinate system, bounding box, availability of metadata, attribution (semantic interoperability) any restrictions on the use, manipulation and distribution of the data, and pricing and privacy policies associated with any particular data (Fig 3).

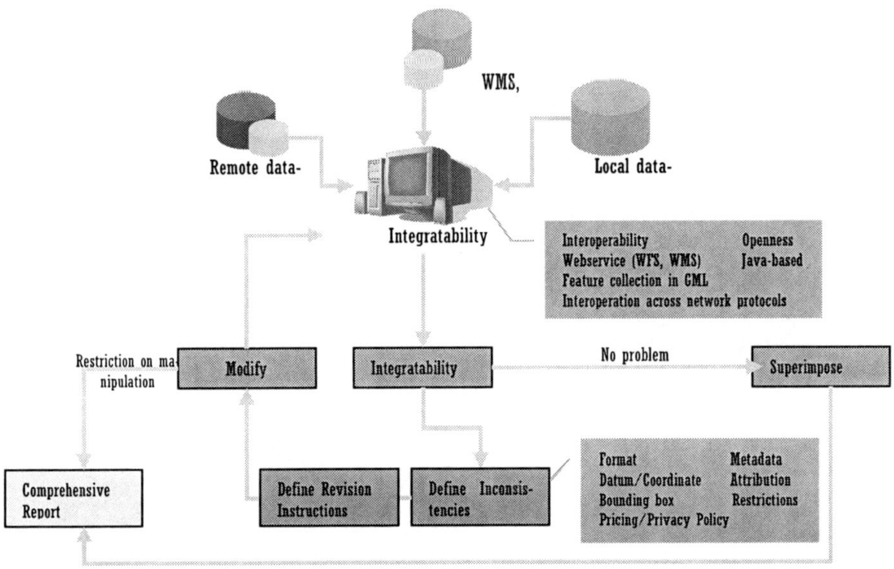


**Fig 3.** Prototype GWS flowchart

If there are no inconsistencies or any restrictions on access and use, which hinders integration recognized, datasets are superimposed into the system. Otherwise, the items of inconsistency are identified and a revision instruction is prepared to help the user amend data components. If there is no restriction on manipulation or use of the datasets, if the revision items have been met, the geo-webservice process is undertaken again to identify any possible problem. This process is repeated until all criteria are met and the user is able to see the data and any information that comes with the data.

The prototype has been tested utilizing datasets from Australian organizations including those provided by Federal and the State organizations including Geoscience Australia (GA) and states of NSW and Victoria and also some other datasets from different sources distributed worldwide including datasets from Geography Network Canada (2007). The result of the test showed that there were problems in the integration of datasets including geometrical mismatch, lack or incompleteness of metadata, attribution inconsistency and restrictions on data, even among datasets from a single jurisdiction.

Fig 4 is a snapshot of the developed geo-webservice which has collected data from different sources within Australia for a region near Melbourne based on above-mentioned assessments.

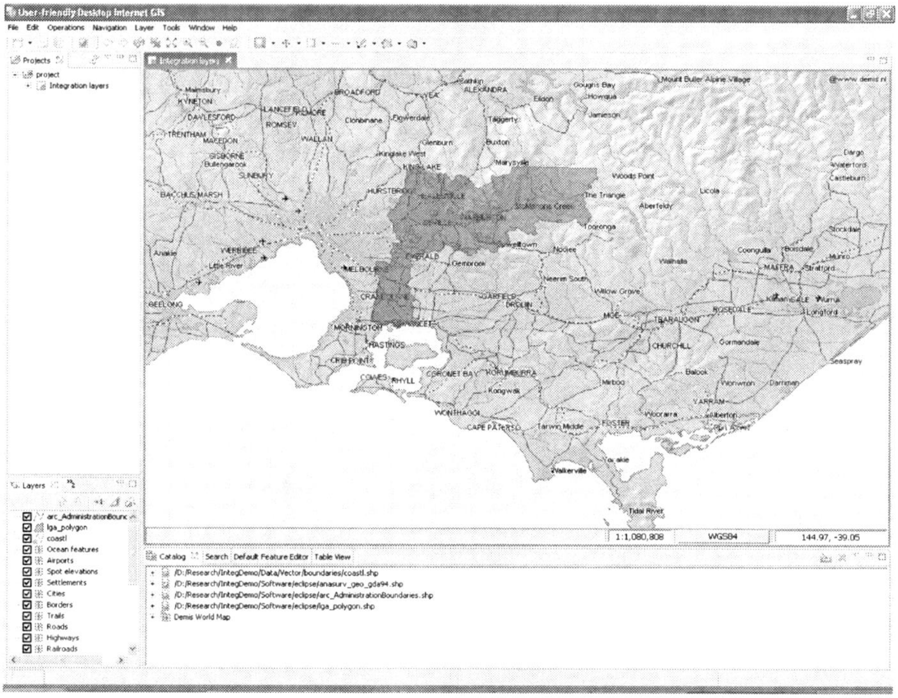

**Fig 4.** Prototype GWS snapshot-integrated data from distributed sources after Integrability assessment

The design and implementation of the Integrability geo-webservice and also the results of the test leads to some observations and recommendations as follows:

- Utilizing geo-webservices to asses the Integrability of multi-sourced datasets can saves time and money

- The Integrability test is an effective process which can decrease the workload associated with data integration
- Users and data providers need instructions for data integration
- Detailed metadata which contains data integration information including restrictions on data, privacy and pricing policies and attribution facilitates data integration
- Machine-readable metadata and complementary data highly facilitates data integration
- Data integration at attribution and data model level requires more investigation and sophisticated tools

At this stage the number of criteria is limited to the items mentioned above, but the prototype is under further development to adopt more criteria and also provide a separate data-custodian-centric instruction to help custodians of data to prepare datasets which comply appropriately with the integration criteria. The integration facilities of geo-webservice are limited to the superimposition of datasets, but further developments are needed to integrate attributes and also at more advanced levels to integrate data models.

The prototype will also be applied to the datasets from a number of countries in Asia-Pacific region through the channel of the Working Group 3 of the United Nations supported Permanent Committee on GIS Infrastructure for Asia and the Pacific (PCGIAP) to test the issues within countries with different characteristics. This will also identify more issues and considerations which should be taken into account for future developments.

The outcomes of the research including the tools developed can not only be customized based on the requirements of different jurisdictions, but also can assist spatial data policy makers to develop the necessary guidelines on how to prepare and amend data for effective data integration.

## 6 Conclusion

Spatial data integration is the most commonly performed task for many crucial spatial services including emergency management services. Despite the time and costs associated with data integration, due to the fragmentation of spatial custodians and inconsistency of approaches, effective spatial data integration has not been achieved in many cases. Many technical and non-technical issues result from these inconsistencies.

In terms of Australia as one of the leading countries in spatial data coordination, spatial data integration is still problematic and associated with many technical, institutional and policy, social and legal issues which hinders effective data integration at different jurisdictional levels. As a consequence, inconsistent collaboration models, pricing and access policies together with technical inconsistencies in standards, data model and metadata hinders spatial data integration and interoperability. To facilitate spatial data integration a holistic platform should be established which addresses technical and non-technical.

To fulfill this task SDIs can be utilized which aim to provide the necessary tools to facilitate spatial integration. SDI provides a platform of people with interaction to spatial data through technological components including access networks, standards and policies which can facilitate data integration and address associated technical and non-technical issues. SDIs can also provide guidelines for linking inconsistent data, services and data coordination policies. This framework can then be utilized by spatial data stakeholders to develop institutional arrangements, legal and policy tools and also social capacities to facilitate the integration of multi-sourced spatial data so that it is used to its maximum potential.

Geo-webservices provide effective technical tools to facilitate the access and integration of multi-sourced datasets. In this regard geo-webservices can by utilized not only to integrate multi-sourced spatial datasets but also to asses the Integrability of these datasets. The prototype GWS evaluates Integrability of multi-sourced spatial data against some criteria and provides instructions and guidelines to amend data based on criteria. This geo-webservice can be customized based on the requirements of different organizations and jurisdictions to meet specific criteria of that particular organization or jurisdiction. This tool can also help users to prepare datasets before integration and also can assist practitioners to develop required guidelines and specifications to be used by data users and providers to prepare data before use.

## References

Anzlic (2003). Implementing the Australian Spatial Data Infrastructure, ANZLIC-the Spatial Information Council.

Baker, A. J. and F. R. Young (2005). Digital Mapping Data Currency Through Sharing: A Practical Study. SSC2005 Spatial Intelligence, Innovation and Praxis, Melbourne, Australia.

Colless, R. (2005). Interoperability & Security in the Emergency Services Arena. http://www.anzlic.org.au/pubinfo/2413335134.html. 27th March

Geography Network Canada (2007), Geogaphy Network Canada data services. http://www.geographynetwork.ca/data/freedata.html. Access date: 20th March 2007

Gould, M. (2001). OGC: A Framework for Geospatial and Statistical Information Integration. Joint UNECE/Eurostat Work Session on Methodological Issues. Tallinn, Estonia.

Mohammadi, H., A. Rajabifard, A. Binns and I. P. Williamson (2006a). The Development of a Framework and Associated Tools for the Integration of Multi-Sourced Spatial Datasets. 17th UNRCC-AP. Bangkok, Thailand.

Mohammadi, H., A. Rajabifard, A. Binns and I. P. Willaimson (2006b). "Bridgind SDI Design Gaps to Facilitate Multi-source Data Integration." Coordinates 2(5): 26-29.

Nebert, D. D. (2004). Developing Spatial Data Infrastructure: The SDI Cookbook, Global Spatial Data Infrastructure (GSDI), Technical working group, Version 2, 25 January 2004.

Rajabifard, A. and I.P.Williamson (2001). Spatial Data Infrastructures: Concept, SDI Hierarchy and Future directions. GEOMATICS'80 Conference. Tehran, Iran.

Syafi'i, M. A. (2006). The Integration of Land and Marine Spatial Dataset as Part of Indonesian SDI Development. 17th UNRCC-AP. Bangkok, Thailand, UN.

# Information Services to Support Disaster and Risk Management in Alpine Areas

Alexander Almer[1], Thomas Schnabel[1], Klaus Granica[1], Manuela Hirschmugl[1], Johann Raggam[1], Michael van Dahl[2]

[1] JOANNEUM Research, Institute of Digital Image Processing, Wastiangasse 6, A-8010 Graz, Austria, alexander.almer@joanneum.at
[2] VCS Aktiengesellschaft, Bochum, Germany

**Abstract.** The concept for an operational service for natural disaster situations requires a scenario driven data access to different sensor information for all phases of a disaster management. This also includes the actual availability of image information of the earth surface concerning the specific requirements of each phase. From the temporal point of view, spaceborne data acquisition does not offer a sufficient data availability in order to support all different phases in specific crisis situations. Especially the event phase cannot be supported as required.

In this paper we describe a concept for on demand remote sensing image data acquisition and a rapid information flow within a crisis management system, which allows to support the decision making process for different crisis scenarios and user groups in charge. The investigation focuses on an airborne data acquisition platform as well as on the development of a multi platform geo-service framework to improve the risk management capacities in mountainous regions by realizing an integrated pre-operational service. The demonstrator includes the client applications for building up an overall crisis management system including mobile units, a mobile command centre, web based presentations and open interfaces to other systems. Additionally, it is described how Very High Resolution (VHR) remote sensing data could be used in this context. A landslide susceptibility mapping has been produced using airborne LIDAR data and QUICKBIRD satellite imagery.

**Keywords:** remote sensing, natural disaster, risk management, information services, GMOSS

## 1 Introduction

Analyses in the frame of the EU project ASSIST (Alpine Safety, Security & Information Services and Technologies) and within the EU network GMOSS (Global Monitoring for Security and Stability) have shown, that the actual availability and timeliness of spaceborne earth observation (EO) data is not sufficient for operational services which will cover all phases of natural disaster situations. At present, spaceborne EO data can only support risk prediction or management tasks in the regeneration phase, but not the concrete event phase of a disaster situation. Due to the extreme conditions in a high mountain environment concerning the availability of data and the limited bandwidth for data communication, disaster management is a challenging task.

One of the main objectives within the ASSIST project was to realize a rapid information flow within a crisis management system based on the development of a technical concept and the prototype realization for a geo-service framework which supports the usage of different data in all pre-crisis phases as well as concrete crisis situation. Therefore, emphasis was put onto data acquired from digital cameras mounted on airborne platforms (aircraft, helicopters, airships) with a pixel resolution of a few decimeters or even centimeters. They can be launched on demand and data recording and processing is feasible within a few hours as required. Automated processing lines further have to be available for geometric processing and change detection tasks. In addition to the data acquisition, the developments include an easy to use client application for using the data and building up an overall crisis management support system including a mobile command centre, web based presentations and open interfaces to other systems. Mobile solutions allow assisting rescue and security teams in terms of communication, navigation by visualizing the current position including the surrounded area. Furthermore, position tracking as well as presentation of geographically and thematically relevant information based on user and scenario driven customization of the mobile applications are additional advantages of a mobile solution. Next to the concrete crisis situation where detailed and up to date aerial images are required, also high resolution spaceborne data can be a valuable information source. As this kind of data provides resolutions up to a few decimeters, it can be used to support phases dealing with risk assessment, prevention and regeneration. A concrete field of application is the identification of parameters and/or indicators for natural hazards such as landslides and the derivation of susceptibility maps. For this task, QUICKBIRD data with a resolution of 60 cm have been chosen. The generation of landslide susceptibility maps relies mainly on information about land use/land cover, geology and geomorphology. Triggering factors such as meteorological parameters have not been taken into account in this study. In order to derive the above mentioned parameters with adequate accuracy, supervised classification and visual interpretation are combined with special emphasis on the use of automated tools. The aim is to segregate land cover classes possibly important for landslide occurrence with special focus on very small structured land cover classes. Furthermore, digital elevation models from different data sources are utilized to extract complementary geo-morphometric information. One further input for the modelling is a landslide inventory, which was generated by "pseudo-stereo" interpretation from QUICKBIRD imagery. Different models for estimating the landslide hazard probability will be used. These are focused on the probability of landslide occurrence but do not take into account the triggering temporal/meteorological factors at this phase.

Chapter 2 deals with the general architecture and the system components, followed by the requirements of the technical concept and the technical realization of the main system in chapter 3, where the different modules and interfaces are described in more detail. Chapter 4 presents an aerial platform as a system of rapid data acquisition and the applied data processing chain for the application within an event case. Furthermore, the results and their integration into the ASSIST system for a concrete crisis scenario are demonstrated. The usage of the data as well as interactions between the mobile command centre and the field staff are discussed. In chapter 5, the use of high resolution spaceborne data for the prevention and the preparedness phase in the special case of landslide susceptibility is demonstrated.

## 2 System Overview

The main focus within the project is the implementation of an integrated pre-operational service which provides a proof-of-concept for the integrated use of spaceborne as well as airborne remote sensing data. The technical demonstrator includes the client applications for building up an overall crisis management system. Mobile units, a mobile command centre, web based presentations and open interfaces to other services are the main components of a flexible node-based architecture. The overall system supports the most essential risk management aspects, e.g. acquisition and exchange, harmonization, visualization and distribution of the available data. Figure 1 gives a generalized graphical system overview showing the developed components of the multi platform geo-service framework.

The ASSIST geo-service framework has a node-based architecture and consists of the following system modules:

- ASSIST Service Node (ASN) – as a central data handling and management component including interfaces to external databases and security service centres.
- ASSIST Mobile Node (AMN) – as a browser based management application (mobile command centre) for visualizing geo-related data, supporting decision making processes and coordinating the field staff by using different communication possibilities.
- ASSIST End Device (AED) – as a PDA based solution combined with a multi channel communication box to support the field staff.

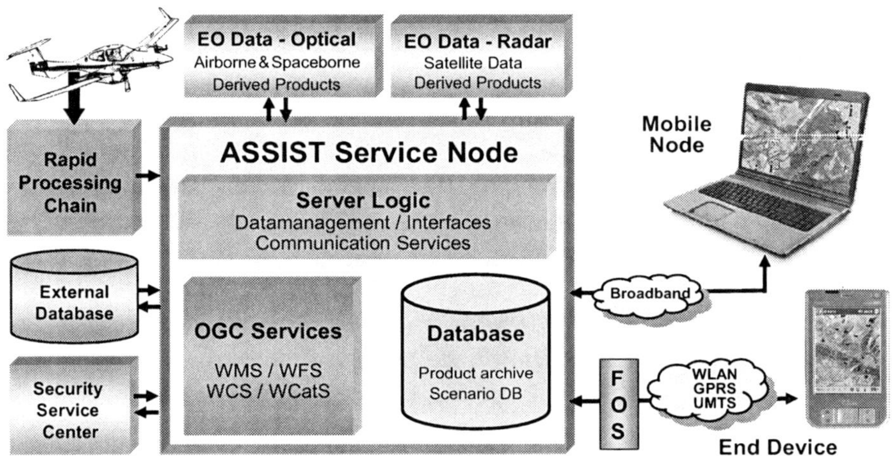


**Fig. 1.** Overview of the system components of the multi platform geo-service framework.

Furthermore, interfaces for external EO data support the usage of results from external processing nodes which use airborne and spaceborne remote sensing data as well as derived products for processing. The aerial platform offers the possibility of data acquisition on demand and is an important data source for the concrete event case.

These system components and their interactions between each other are described in chapter 3 in more detail.

# 3 Technical Realization of the Management Components

To meet the needs for supporting an overall crisis management, the technical concept has to fulfill the following requirements:

- Harmonization of the EO data exchange using OGC standards
- Enable an efficient centralized data/task management in combination with decentralized mobile units
- Use different communication channels depending on their availability in the operational area
- Support textual and geo-data visualization customized to the technical capabilities of the various platforms (mobile devices, web, etc.) and distribution channels
- An easy, user friendly and situation-dependent access to information for all involved parties
- Provide and integrate all services needed to build up a crisis management system which will support different crisis scenarios

This chapter gives a technical description of the ASN, AMN and AED, which build the main components of the ASSIST system.

## 3.1 ASSIST Service Node

The ASN is the central server node. It comprises the Product Archive, the OGC Services and the Scenario Database with its services. The Product Archive is used to integrate external data (products) into the ASN. It consists of the Product Meta Data Database, the Storage Area for the product files, the provider specific drop boxes and the Product Ingestion Daemon. Data providers put new products via FTP or web service into the ASN drop-box. The ingestion daemon automatically imports the products into the product archive. A product consists of product files and product meta-information. The data model for the meta-information is defined by an XML scheme. The ingestion daemon validates the product meta-information against the scheme. Figure 2 gives a graphical overview of the described main components of the ASN.

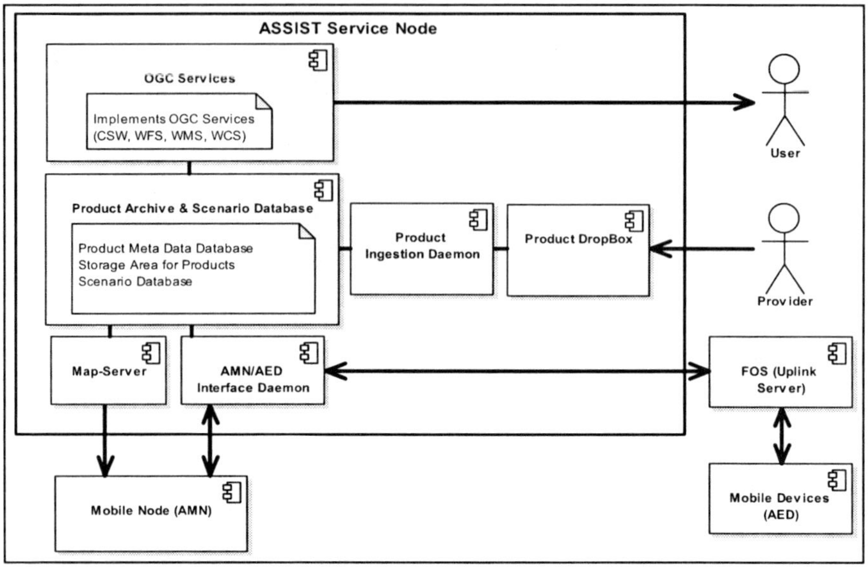


**Fig. 2.** ASN overview

A web-application is residing on top of the Product Archive, and implements OGC services like CSW, WFS, WMS, WCS (Open Geospatial Consortium 2007; WMS, WFS, WCS, CSW 2007). These are used to connect external OGC services as well as to visualize data on the AMN. The scenario database stores the scenario configuration like the kind of scenario, the scenario region, the products used for the scenario and the AEDs that are allowed within this scenario. In addition it stores all historical information from a scenario run. The scenario services are used within a scenario run, mainly to monitor and control a scenario run, e.g. the real-time service for exchanging data with the mobile end devices. The ASN also includes a communication service which is used to exchange data with the AED. Thereby, an external server called FOS (Field Operation Server) is used. It acts like a gateway to handle the link between the ASN and an AED because this is not realized as a direct communication channel.

### 3.2 ASSIST Mobile Node

The AMN is indented to be a decision support and mission control tool and enables to visualize, merge and analyze existing data to support the different mobile application use cases. This application is built as a slim client using AJAX and PHP-Mapscript which is based on the functionalities of the UMN Mapserver (PHP Mapscript 2007) to provide an interactive interface for the user (see Figure 7 - left). The implemented functionality covers the visualization of existing data (thematic maps, aerial images, etc.) as well as online tracking of the field staff, using a powerful but still easy to handle user interface. A structured and categorized menu allows efficient access to related themes and information. Features like storing and switching between favorite views enable an effective management in huge areas. Furthermore, inserting new geo-oriented data with additional information and distributing it by the server assures permanent up to date information (Granica et al. 2007). In addition, the client includes a communication module which allows the distribution of selected geo-oriented objects and the coordination of the field staff as well as presenting gathered up to date information from the mobile units. The AMN is an online client that communicates with the server (ASN) via HTTP over an existing TCP/IP connection. This allows multiple working groups to use the AMN simultaneously, using the same available data, which supports the coordination of various activities within a concrete crisis situation as well as within other phases of the crisis management circle.

### 3.3 ASSIST End Device

The AED is a Windows Mobile based solution to support the field staff. It consists of a PDA combined with a multi channel communication box (MCB), which allows to link to the ASN in different ways like WiFi, GSM or a direct satellite connection. The AED is built to support the field workers, providing comprehensive information about the area, the current state of critical objects and a linkage to the management centre to allow a better coordination of different groups and provide vital information for a decision making process. The functionalities include an online and buffered position tracking, visualization of vector- and raster data as well as additional information to the available objects. Furthermore it allows acquiring new vector data and sending it to the centralized server for further usage.

## 4 Technical Concept and Realization for an Event Case

### 4.1 Aerial Data Acquisition

In the present context an aerial data acquisition platform is to be implemented, which provides high flexibility concerning data acquisition, including highly overlapping images to support 3D surface reconstruction and the feasibility of a quasi true ortho image and mosaic generation. The images are taken by a digital camera which is mounted on a manned or unmanned air vehicle such as a helicopter or plane. The data is transferred to a ground station where it is used as input for a Rapid Processing Chain. For disaster monitoring near 'real time' image processing is required, implying geo-referencing without using ground control points for optimization and validation purposes. Further requirements refer to low cost, easy operating, low weight, automatic as well as hand-held operating possibility, sufficiently high image resolution and quality of the GPS and IMU instruments.

First platform prototype: The first platform prototype was realized by a high resolution digital camera (12 megapixels), which was further connected to a L1/L2 GPS phase receiver and operated from board of a helicopter (Raggam et al. 2006). A related data acquisition experiment was made in the context of landslide mapping after a period of intensive rainfall at the end of August 2005. The procedures and results are briefly summarized for demonstration purposes in chapter 6. This first prototype was not equipped with an IMU (inertial measuring unit), and hence an acceptable geo-location accuracy could only be achieved by means of GCPs and photogrammetric adjustment procedures applied to the image orientation parameters.

Current platform configuration: As a consequence, the platform setup was extended by a low cost IMU, which was mounted on the camera, thus providing approximate values of the camera's exterior orientation (see Table 1). The IMU accuracy values as given by the manufacturer are supposed to assure pointing accuracies of about 1 percent of the flying height above ground. Moreover, the capacity of the 12 megapixels camera was considered to be low for larger area coverage at higher pixel resolution. Therefore the camera was replaced by a Hasselblad H2D with 39 megapixels. This setup is shown in Figure 3.

Future platform conception: In a future conception, the GPS equipment of the aerial data acquisition platform shall be further improved, and the platform shall be equipped with a higher accuracy IMU. This shall assure a pointing accuracy in the range of 1 to 2 decimeters for flying heights of about 500 meters above ground, and near real time data processing at a highly acceptable accuracy level shall definitely become feasible.

**Table 1.** Airborne data acquisition platform concepts

| | First prototype | Current conception | Future conception |
|---|---|---|---|
| Camera system | consumer camera | high end consumer camera | high end consumer camera |
| Camera resolution | 12 Mp | 39 Mp | 39 Mp |
| GPS system | L1/L2 Phase receiver | L1/L1 Phase receiver | L1/L2 Phase receiver + EGNOS |
| IMU system | - | Xsens | Novatel |
| IMU accuracy | - | low ~0.5°/1.0° | high ~0.015°/0.03° |
| Stabilization | - | - | yes |
| Image processing | post processing | post processing | near real time |

**Fig. 3.** Hardware components of the current airborne acquisition platform, as operated from board of a helicopter (left), comprising a Hasselblad camera, an IMU and GPS attached to it.

## 4.2 Aerial Data Processing

As for data acquisition and data processing, the following processing scenarios and throughput goals are envisaged:

A. Rapid mapping (near real time): Ortho-rectification (2D mapping) relying on exterior orientation as provided via the GPS and IMU recordings to get rapidly geo-referenced image data, providing an overview on hazard events in order to launch e.g. first rescue and relief operations. The required accuracy is low and the acceptable location error is around a few meters. The image data can be processed sequentially (ortho-rectified, mosaicked) as acquired, and should be made available within a few seconds.

B. Enhanced mapping: Intrinsic optimization of image data set through utilization of (automatically detected and measured) tie-points. These are used to refine relative orientation of images in order to make them consistent in a relative sense, although absolute geo-positioning accuracy may still be low as mentioned above. Optimization of relative orientation is possible only after all images (all sets of tie-points) are available, and processed products are supposed to become available in a time frame of several minutes up to a few hours.

C. Precision mapping: Optimization of exterior orientation using ground control points (GCPs). These may be acquired either automatically, e.g. from GCP chip matching (Raggam et al. 2003), or interactively, typically via GPS –measurements in the field. Product availability then may be assured only within a few hours up to a few days but at the advantage of a very high location accuracy.

## 4.3 Demonstration of Results

A first realistic application for the aerial platform in combination with the mobile components of the ASSIST system was in the summer of 2005 where several Austrian areas were affected by intensive rainfall and subsequent damage. For a severely affected area (see Figure 4), a landslide mapping experiment was launched. During a helicopter overflight, more than 200 images were captured using the first prototype of the aerial data acquisition platform. Images were taken from a height of 400 meters above ground at a pixel size of about 15cm and with an overlap around 70%. Figure 5 shows a close-up of one image, covering areas affected by landslides.

**Fig. 4.** Hazard events in summer 2005

**Fig. 5.** Acquired image, showing areas affected by landslides

Concerning geometric post-processing of the image data the subjects of ortho-rectification, data mosaicking and surface mapping were investigated. The latter aspect is illustrated by Figure 6 showing profiles drawn over a mudslide area before (left top) and after (right top) the event. The profiles indicate a decline of the terrain by 1 to 3 meters, thereby giving an idea on the amount of mass movements. A more detailed description of the procedures and results achieved for this mapping experiment is given in Raggam et al. (2006).

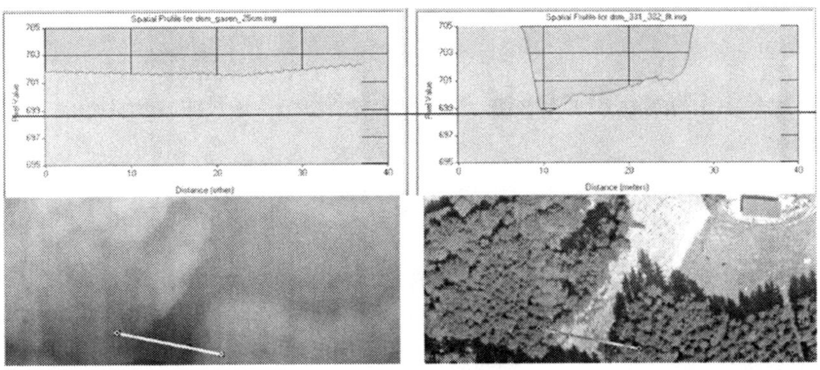

**Fig. 6.** Ground model profile before (left top) and surface model profile after (right top) event. Left bottom: Surface model (gray value coded); Right bottom: ortho image.

An overlap of an analyzed profile and ortho-image was integrated into the ASN as a new product. Subsequently, the scenario configuration tool enabled to add this information for further usage in the related scenario. From there on, it is available on the Mobile Node. Figure 7 shows the AMN as well as the AED, both displaying a result of the data processing chain. The user of the application has now the possibility to get additional data for the shown objects and can also send it to one or more specific teams out in the field. This allows the field staff to get up-to-date information about the defined area and access to all new data. The communication module allows the field staff to exchange acquired data with the ASN as well as with other AEDs. This fulfils the demand for extensive but still structured information and interaction of the involved parties.

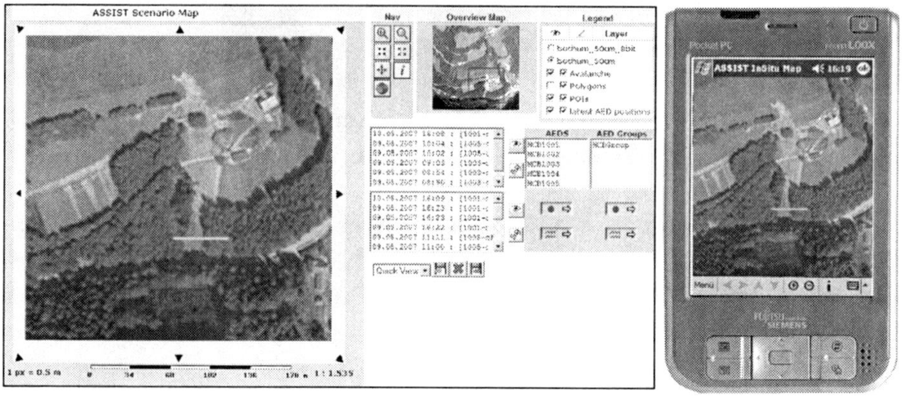


**Fig. 7.** AMN (left) and AED (PDA application, right), displaying a result of the data processing.

## 5 Optical Satellite Data Analysis for the Prevention Phase

### 5.1 Requirements and Main Goals

As already mentioned in the introduction, one focus in this investigation is on the identification of parameters and/or indicators for natural hazards with special emphasis on landslides and the derivation of susceptibility maps. In order to satisfy the high accuracy demands for this task, QUICKBIRD data on one hand and LIDAR data on the other hand have been chosen as data sources.

Until recently only medium resolution remote sensing data or traditional airborne images have been at hand for such investigations. As the generation of landslide susceptibility maps relies mainly on information about land use/land cover, geology and geomorphology, it is obvious from this point of view, that the quality of these parameters should satisfy the expectations in terms of finer data resolution and high level processing.LIDAR data offer new opportunities in this context and enables the generation of a highly precise DTM (Digital Terrain Model) even below forest surface.

Although the models applied do not take into account the triggering temporal/meteorological factors in this phase of the investigations, the mentioned new data sources are expected to improve the results of the susceptibility mapping significantly. The analysis of the available and derived parameters should lead into the selection of the most important variables, which has to flow into the computation of the envisaged hazard maps.

## 5.2 Methodology

In the test site area manifold surface types are present due to the difference in elevation of almost 2400 m. In order to handle this aspect, the applied methodology has to be elaborated in a flexible way involving different approaches for the derivation of the assigned parameters from geologic maps, QUICKBIRD images and DTMs, especially high resolution ones from LIDAR data.

A prerequisite for any further processing is the accurate ortho-rectification. This task was performed using DEM-based geocoding available in the Remote Sensing software package Graz (RSG). The geocoded multi-spectral and panchromatic QUICKBIRD scenes were then merged using a modified Brovey transform (Hirschmugl et al. 2005) in order to obtain a high resolution multispectral image. Subsequently, the resulting pan-sharpened image was topographically normalized to compensate the strong illumination effects. This was performed using an algorithm called 'incidence normalization' developed at the Institute of Digital Image Processing of JOANNEUM RESEARCH (not yet published). Based on these preprocessing steps the classification on spectral and textural features has been performed, geomorphometric features were calculated from the DTM and finally, a landslide susceptibility analysis was done.

The textural feature classification is performed using an algorithm that calculates certain statistical values based on mean or variance within the sectors surrounding a pixel (Gallaun et al. 2007). The radius and the number of sectors (or wedges) can be specified by the user. The texture measure proofed to be a valuable tool for the differentiation of spectrally similar classes such as fine and coarse debris.

Different geomorphometric features were tested in this study: aside from slope and aspect, the drainage network was calculated and curvature was analyzed. Based on both the land cover information from the classification and the geomorphometric features from the DTM, the susceptibility mapping is performed. The methodology of univariate statistical models for susceptibility mapping is based on the assumption that landslides are likely to occur under the same environmental conditions as those under which they occurred in the recent past. In order to 'train' the model, past landslides were mapped by visual interpretation of simulated stereo Quickbird data. The simulation of stereo condition from mono-temporal QUICKBIRD data showed to be congruent to the quality of landslide inventory from standard stereo aerial photographs.

The main steps for the whole processing chain encompass
1. Creation of a Landslide Inventory Map,
2. Geocoding (Raggam et al. 1991),
3. Pansharpening (Hirschmugl et al. 2005),
4. Topographic Normalisation (Colby, 1991),
5. Derivation of Morphometric Terrain Parameters,
6. Spectral Based Classification (Schardt et al. 1998),
7. Texture Based Classification,
8. Landslide Susceptibility Analysis (Van Westen, 1993).

### 5.3 Demonstration of the Results

The results of these investigations are evaluated according to the potential of the new VHR optical remote sensing data for their use in hazard mapping and also the derived maps for their relevance to the ASSIST project.

Within the project a test site has been selected in a high mountainous terrain in the western part of Tyrol/Austria. The inhabitants of this region have regularly experienced the impacts of different natural hazards, e.g. landslides, debris flows, floods or snow avalanches. In August 2005 a catastrophic flood caused heavy damage on infrastructure and settlements disrupting the important Arlberg railway route for more than three months.

In the present investigations two different DEMs are available, i.e. a 25m DEM and a high resolution LiDAR DEM with 1m resolution. The differences are obvious, as the high resolution DEM shows much more details enabling a more precise analysis of the surface (see Figure 8). The shaded relief of the LIDAR data represents well all forest roads, ravines, small ridges and undercutting of slopes, while the 25m DEM only roughly shows the slopes and very coarse geomorphometric features. Fresh undercutting of unstable slopes can clearly be interpreted by a trained expert in the 1m DEM, while it can hardly be discriminated from old, stable slopes in the 25m model. Also the area of the landslide itself, which was triggered by these undercutting processes and has severely damaged the road, is not recognizable in lower resolution DEMs.

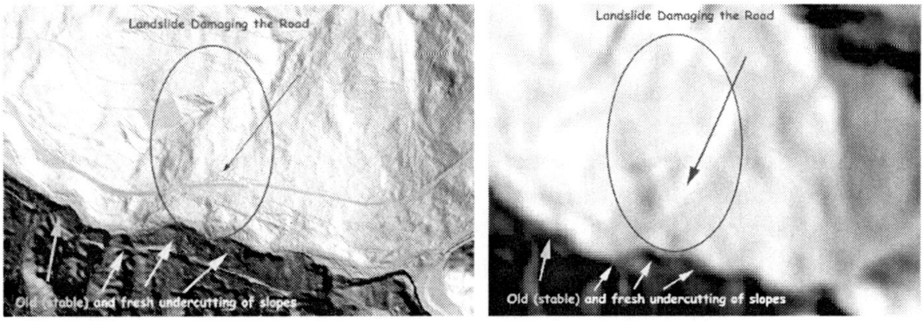


**Fig. 8.** Left - DEM 1m Resolution; Right - DEM 25m Resolution.

Not all of the required indicators, which are inherent in the QUICKBIRD image, can be derived by automatic processing, such as rockglaciers, elongated ponds or wet areas. For the derivation of these classes, visual interpretation is still an adequate procedure, including the complementary information from 'pseudo-stereo' QUICKBIRD imagery.

For the derivation of land cover information, the QUICKBIRD images show a significant advantage in terms of larger coverage and higher radiometric stability in contrast to aerial images. This proved to be of high benefit for the development of automatic tools. The classification process itself is based in a pre-phase on a supervised classification using Maximum Likelihood, and in a second phase followed by the computing of texture information from the panchromatic image. The content derived from this texture layer can then be used to refine the classification result, e.g. the differentiation of fine debris versus coarse debris, in a post-processing step.

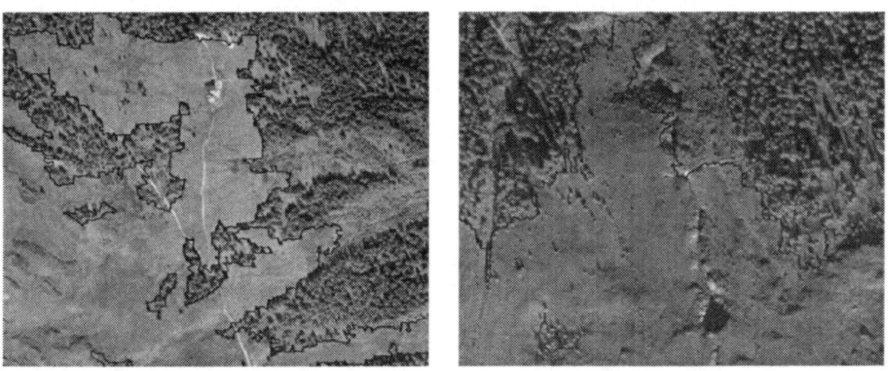

**Fig. 9.** Detailed Views of the automatically derived Upper Forest Border.

Another important application by deriving texture features is the determination of the upper forest border, which is essential to quantify the forest area and to differentiate between different "non-vegetation" areas within the image. For instance, non-vegetation areas may be settlements and streets in the valley, whereas the same spectral response from above the forest border line shows talus scree and eroded areas. In Figure 9 the successful derivation of the upper forest border is displayed. Although is has to be stated, that some errors remain in the shadowed areas, an estimated accuracy out of visual interpretation of about 90% could be achieved. A detailed quantitative verification has not yet been performed.

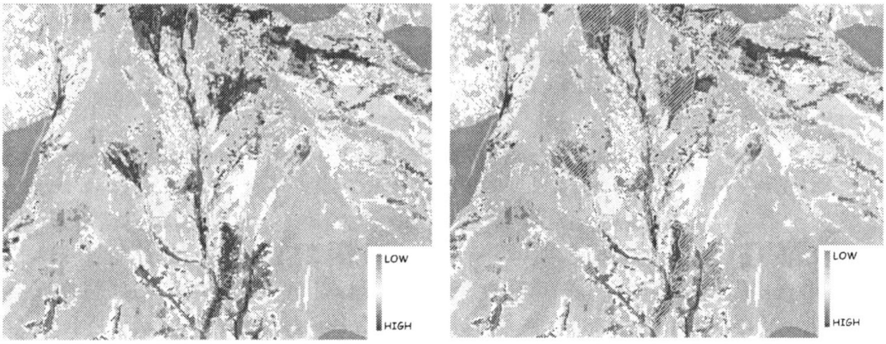


**Fig. 10.** Landslide Susceptibility Map (right: superimposed with historical slides).

After the identification and derivation of valuable input parameters the weights of evidence modeling (Bonham-Carter et al. 1990), as explained in Van Westen (1993) was employed for the susceptibility mapping. It is important to mention, that during the susceptibility analysis, the parameter 'aspect' turned out to be strongly biased and therefore had to be excluded. The reason was found in the geologic situation of the test site. The area is located at the border between carbonatic rock in the north (= south-facing slopes) and crystalline rock in the south (= north-facing slopes). Since landslides are much more frequent in the carbonate area, a high relevance is given to the parameter aspect, although the determinant factor of the landslide distribution is the specific geologic situation.

The calculations have been executed on superficial translational slides only, as this type is the dominating landslide type within the test area. The resulting map (see Figure 10) shows the distribution of the different landslide probability classes, from low to high. It matches well with the historical landslide areas (Figure 10 right) and shows the disposition of the terrain under the same conditions. These results are one part of the information that can be integrated into the ASN and further visualized on the AMN and the AED as already described in the previous chapter.

## 6 Conclusion and Outlook

In this paper we have described the concept and the implementation of a crisis management system. Additionally, we showed two different methods of obtaining information for this system from remote sensing applications.

The overall crisis management system allows the assistance of a mobile command centre and also the support of security field teams in terms of communication, navigation and spatial data visualizing. The modular concept of the management system enables a user and scenario driven customization of the mobile command centre as well as the mobile applications. The detailed definition of the scenario profiles for the different disaster situations will be elaborated within the last period of the ASSIST project.

The first method to integrate remote sensing is designed to rapidly deliver data for the event case. Airborne remote sensing image data is generated on demand and a rapid information flow is designed. This includes the steps of data processing, data management and distribution as well as data accessibility for first responder teams using mobile solutions. A demonstrator was already developed to show the applicability of this approach.

The second method to integrate remote sensing data in the management system is related to the prevention and preparedness phases of the disaster management circle. In this case, the time is not a crucial component, but the quality of the data. In this study, landslide susceptibility maps were generated based on optical satellite data and LiDAR terrain data.

All results from both methods were integrated into a WEBGIS solution with a demonstrator system to complete the building up of an overall crisis management support system (http://www.assist-gmes.org/).

## Acknowledgment

We thank the co-ordinator of the ASSIST project Michael Henke (VCS), Germany, ASI, Landeck (Alpine Safety and Security) and the "Landesregierung Tirol" for their support. We would also like to thank Patrick Luley, Roland Wack, Heinz Gutjahr and Michael Wurm from Joanneum Research for their essential developments within this project.

## References

Bonham-Carter, G.F., Agterberg, F.P., Wright, D.F. (1990): Weights of Evidence modelling: a new approach to map mineral potential. Geological Survey of Canada Paper 8-9. Agterberg, F.P. and Bonham-Carter, G.F. (eds.). Ottawa, Ca, pp. 171-183.

Colby J.D. (1991): Topographic normalisation in rugged terrain. Photogrammetric Engineering and Remote Sensing, Vol. 57, No. 5, pp. 531-537.

Carrara, A., Cardinali, M. & Guzzetti, F. (1992). Uncertainty in assessing landslide hazard and risk.- ITC Journal 1992-2, pp.172-183, Enschede 1992.

H. Gallaun, M. Schardt, and S. Linser (2007). Remote sensing based forest map of austria and derived environmental indicators. In Proceedings of the ForestSAT 2007 Scientific Workshop, Montpellier, France, 2007. published on CD.

Granica, K., Almer, A., Hirschmugl, M., Proske, H., Wurm, M., Schnabel, Th., Kenyi, L.W. & Schardt, M. 2007: Generation and WebGIS representation of landslide susceptibility maps using VHR satellite data. To be published in: Proceedings of IGARSS2007, Barcelona, July, 23rd – 27th 2007.

Hirschmugl, M., Gallaun, H., Perko, R. & Schardt, M. (2005): "Pansharpening" – Methoden für digitale, sehr hoch auflösende Fernerkundungsdaten. Strobl/Blaschke/Griesebner (Hrsg.): Angewandte Geoinformatik 2005. Proc. of 17th AGIT Symposium (06. – 08. 07. 2005), Salzburg. Verlag Wichmann, Heidelberg, pp. 270 – 276.

Open Geospatial Consortium 2007: http://portal.opengeospatial.org/files/?artifact_id=16075 Last date accessed: 6/2007

PHP Mapscript 2007: http://www.maptools.org/php_mapscript/ Last date accessed: 6/2007

Raggam H., Almer A., Strobl D. & Buchroithner M.F. (1991): RSG - State-of-the-Art Geometric Treatment of Remote Sensing Data. In Proc. of 11'th EARSeL Symposium: Europe: From Sea Level to Alpine Peaks, from Iceland to the Urals, pp. 111-120, Graz, Austria, July 3-5 1991.

Raggam H. and M.D. Villanueva Fernandez (2003): Approaches to automate image geocoding and registration. "IEEE International Geoscience and Remote Sensing Symposium (IGARSS)", Toulouse, 21-25 July 2003, published on CD

Raggam H., Wack R., and Gutjahr K., 2006: Assessment of the Impact of Floods using Image Data acquired from a Helicopter. 26th Earsel Symposium, "New Developments and Challenges in Remote Sensing", Warsaw, May 29th – June 1st, 2006

Schardt, M.; Gallaun, H. & Häusler, T. (1998): Monitoring of Environmental Parameters in the Alpine Regions by Means of Satellite Remote Sensing. Proceedings of the International ISPRS-Symposium on "Resource and Environmenal Monitoring, Local, Regional, Global, Commission VII, September 7-4, 1998, Budapest, Hungary

Van Westen, C.J., van Asch, T.W.J., Soeters, R. (2006): Landslide Hazard and Risk Zonation-why is it still so difficult? Bull. Eng. Geol. Env. 65: 167-184.

Van Westen, C.J. (1993): GISSIZ – Geographic Information Systems in Slope Instability Zonation.- ITC Publication Number 15, Vol. 1, Enschede 1993.

WMS, WFS, WCS, CSW 2007: http://portal.opengeospatial.org/files/?artifact_id=16075 Last date accessed: 6/2007

# User Performance in Interaction with Web-GIS: A Semi-Automated Methodology Using Log-Files and Streaming-Tools

Jens Ingensand, François Golay

Laboratory of Geographical Information Systems, Swiss Federal Institute of Technology, Lausanne (EPFL) Institute of Urban and Regional Planning & Design – Geomatics Lausanne, Switzerland
jens.ingensand, francois.golay @ epfl.ch

**Abstract.** This paper describes a framework of methods for the measurement and evaluation of users' performance in interaction with a web-GIS. The framework involves testing of a system with real-world users, streaming of the user's screen during the evaluation and the analysis of web-server log files after the evaluation. We have developed and tested this method within a project called RIV together with real-world users that are using a web-GIS. We found out that users have different strategies when interacting with a web-GIS that offers different manners of interaction. The findings that we present in this paper are useful for developers and designers of web-GIS and can help improving such systems.

**Keywords:** web-GIS, performance, usability, evaluations

## 1 Introduction

The use of spatial data and systems to manipulate these data used to be highly restricted to experts in the domain. In recent years the development and use of web-GIS has been changing this fact as an increasing number of people get access to spatial information and to interactive tools to visualize and to manipulate this information. The fact that non-expert GIS users get access to geospatial tools and data has also dramatically increased the necessity of their fitness to user's needs and abilities, and therefore for usability testing. (Preece et al. 2002)

During the 1990 some researchers have focused on the usability of standard desktop GIS, e.g (Traynor and Williams 1995). However, usability testing of web-GIS is still a rather blank field.

In most definitions of usability the terms effectiveness, efficiency and satisfaction (ISO 9241 1994) and speed of performance, time to learn, retention over time and rate of errors by users and subjective satisfaction (Shneiderman 1998) and efficiency learnability, memorability, errors/safety and satisfaction (Nielsen 1993) are the main components of usability, yet some researchers (Hunter et al. 2002) tried to refine usability and claimed that there are many more components than just these three main categories.

We believe that measuring a user's performance can help identify many of these components, as we claim that a user's performance is not only about measuring the absolute time to complete a specific task, but also about analyzing what exactly a user is doing with a system and what problems occur during the interaction.

In this paper, we want to describe how the performance of a user interacting within a web-GIS can be measured and evaluated. First we want to give a brief survey of how web-GIS work from a conceptual and technological point of view in order to set forth how our methodology framework works. We then describe our methods for analyzing performance and demonstrate how the methodologies work in a real-world case study. We then show how the measures that have been collected during the evaluation-session can be used and interpreted. Finally we present our conclusions and ideas for further research and evaluations.

## 2 Characteristics of Web-GIS

All web-GIS have an architecture in common with one or more servers and several clients that are accessing the server/s via a web-interface. This architecture implies that all web-GIS are submitted to using network connections and that all web-GIS are using a web-browser (such as Internet Explorer or Firefox) for navigating the maps.

Web-GIS typically access only small amounts of the whole data that is available in the system at the time, which implies that clients send queries to the server/s with the information that is needed. The server responds with the data that is required and the client displays the server output.

From the point of view of data we can distinguish systems with a different server – client communication:

- raster-based systems; the server replies with an image on a client's request; the request can consist of a position in the image;
- vector-based systems: the server replies with vector data on a client's request
- hybrid systems (vector + raster data combined); the request can consist of a selected vector element or of a set of elements.

Another important characteristic of web-GIS is the fact that web-GIS usually use some kind of "extension" to normal homepages in order to dynamically display geographic content and to let the user navigate the data. Today several different technologies can be combined to form a whole system:

On the client side:

- Plugin-based clients (using plugins such Flash, Java-applets, SVG, etc) Examples for this kind of application are Mappy (Flash), Map24 (Java–applet) and the Swiss televisions' election maps (SVG)
- HTML-Javascript based clients (such as Google maps)
- Scripting-based clients; e.g the Swiss canton of Vaud's official mapping platform, Geoplanet.

A technology framework that has been widely used in recent years is AJAX, a javascript-based technology which only charges some parts of the interface if needed and thus minimizes the bandwidth use of the application.

On the server side all systems must use a web-server (such as Apache or IIS) to respond to the client's requests. This web-server is the interface between the client's requests and a server that is adapted to producing the requested spatial to the client.

## 3 Means for Measuring a User's Performance

In order to measure the performance of a user interacting with a web-GIS we can take advantage of the facts that:

- any web-GIS uses a network-connection that can be used to stream the screen of a user-interface through the network and thus capturing a whole interaction-session
- any web-GIS uses a web-server that produces a log-file that contains the query that the client has sent to the server

Using a network-connection for streaming a computer's interface is a method that is widely used for server-administration – the administrator can remotely connect to the server and access its entire interface from a remote computer. Examples for such streaming-applications are Windows' Remote connection or the open-source framework VNC.

A web-server's log-file is commonly used by computer administrators to analyze who accessed the server and what was requested. Most web-servers such as Apache do log all activity out of the box.

## 4 The Evaluation of a Web-GIS with Real-World Users

At the GIS-Lab at the Swiss Federal Institute of Technology in Lausanne we have developed a web-GIS for all actors involved in wine-growing business in the area. This system called RIV has been developed in close interaction with the end-users, which involved interviews, the development of prototypes and the evaluation of these prototypes with the end-user. The project and the development process have been further described in Ingensand (2005), Ingensand et al. (2006) and Ingensand and Golay (2007).

In order to evaluate the system we had invited 20 users. All users were winegrowers and thus potential end-users of the system. The evaluation-sessions involved a first questionnaire to gather the user's personal data (age, experience, etc) a hands-on evaluation and a second questionnaire where we asked the user his opinion about specific aspects of the interface and the whole system. Before the hands-on evaluation there were no explanations given to the users how the interface might be used. All users saw the system for the first time and were forced to figure out how it worked in order to solve 10 given tasks. The evaluation-expert only helped in case the user was unable to solve the tasks after trying for at least three minutes.

We wanted to determine what occurred during the hands-on evaluation itself, but also to find an ideal way to find evidence how the system was used in practice, therefore we tried to find a methodology that captured as much as possible of the user's physical interaction with the system but also what the user thought and said. We therefore wanted to take the following measures

- The user itself, what he said and what he showed with gestures
- The user's interaction with the system, in that way that it was possible to reconstruct each evaluation afterwards and to analyze it in depth.

For the first measure we put a video-camera in the room, filming the user sitting in front of the computer.

One point of departure for measuring the user's interaction's with the system was Aoidh and Bertolotto's (Aoidh and Bertolotto 2007) methodology of analyzing the spatial location of the user's mouse interactions where the mouse' spatial location is considered as representative for the user's interest in a specific feature. However as RIV is a system with several menus, tools and features it was difficult to apply the same methodology to our evaluations. Another incitement was Tulis' (Tulis et al. 2002) comparison of lab and remote testing where a specifically instrumented browser was used for capturing the user's interactions with the interaction of a web-site.

Our system uses Apache as web server which logs each user activity into a log-file. Fig. 1 shows an example of what the web-server writes in one log file each time the user clicks on the map, on any of the navigation tools or on the layer-selection:

70.52.205.158 [1] - - [31/May/2006:15:53:46 +0200] [2] "GET [3] /riv.php?bbox=561553.0320588425,137880.742787655;561553.0320588425,137880.742787655&tool=zoomin&layers=cartes,orthophoto,cadastre,parcelles&_= [4] HTTP/1.1" 200 [5] 20044 [6]

[1] The IP-address: Who tried to access the web-server
[2] The exact time when the access occurred
[3] The method that was used to access the server (GET or POST)
[4] The URL that the client requested
[5] If the query was successful or not (200 means successful)
[6] The amount of bytes that were sent back to the client.

**Fig 1.** The output of RIV's web-server

Parameters [1], [2], [3], [5] and [6] are non-system-specific (every system using the Apache web-server produces such a record in the log-file). Parameter [4] is specific for RIV. In our case the URL that was requested by the client shows us:

- a bbox-parameter, that shows what region the map on the user's screen is displaying
- a tool-parameter (which navigation-tool the user was using)
- a layers-parameter (what layers were requested to be displayed on the map)

In order to analyze the log-file we created a tool that parses the log-file and puts it into a database which every column showing a parameter. Thereafter we created a tool that extracts the information from the database and visualizes the whole interaction-session that one specific user did with the system. (Fig. 2) The different parameters were translated into a more human-readable format, filtering unnecessary elements and emphasizing important elements:

- the time when the interaction occurred (with an absolute timestamp and relative timestamp to analyze the log-file afterwards in synchronization with the recorded screen and video)
- what tools the user was using (e.g. zoom in, recentering)
- what layers the user was requesting
- if there were gaps of more than 10 seconds in between the different queries

For streaming the user's desktop we installed a VNC-server on the evaluation-computer that can be used for remotely controlling the computer. The signal of such a VNC-server is able to send the remote computer's screen to a client, but is also accepting user input from the client computer's VNC-client. For our evaluation we used a tool, installed on a second computer, that streams this signal directly to a video-file.

Before we used our log-file visualization-tool for measuring the user's performance we verified it's functionality with the screen captures that had been recorded during each session.

| | |
|---|---|
| 2:57/09:32:14 | Zoom in |
| 3:06/09:32:23 | Zoom in |
| 3:19/09:32:36 | Scalebar 1:25000 |
| 3:31/09:32:48 | Scalebar 1:5000 |
| 3:41/09:32:58 | Scalebar 1:1000 |
| 13 | |
| 3:56/09:33:13 | Scalebar 1:2000 |
| 30 | |
| 4:28/09:33:45 | Scalebar 1:3000 |

**Fig. 2.** Output of the log-file visualization tool.

We then utilized the log-file visualization tool as an index for all evaluations that were recorded. It helped us to detect:

- How much time it took to complete a given task
- How many attempts where necessary to solve the task
- What errors the user made during the evaluation (e.g. if he got lost in a menu that did not offer the functionality that was required to solve a task

- What navigation tools the user was using
- If the user had different strategies to solve a given task (e.g. to navigate to a specific location)

Moreover we used the gap-detection feature together with the videos (the user's screen + the user itself) recorded in order to determine what was happing when the user was hesitating at a specific moment.

## 5 Evaluating Measurements

In our evaluation we used the tools that what just presented to answer to the following research-questions:

- are there different user strategies to navigate a web-GIS?
- which strategies result in better performance?

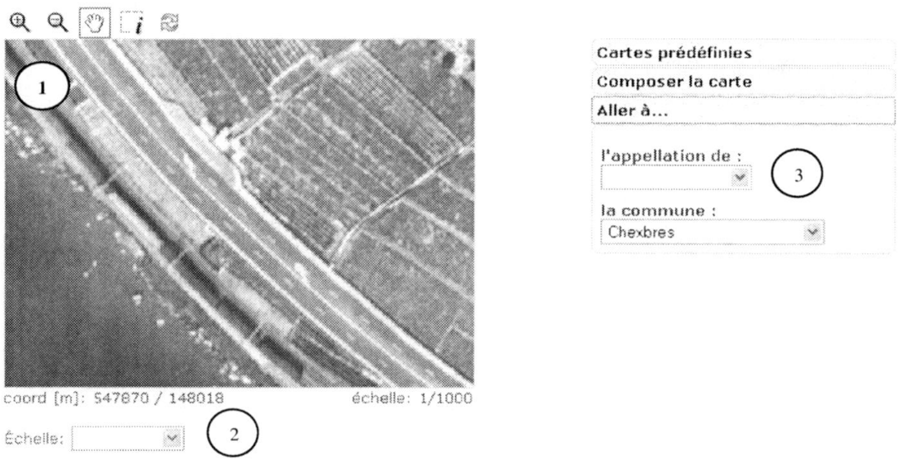


1: Zoom in, Zoom out and Pan; the selected tool is highlighted
2: The scalebar with a choice of 16 scales
3: Scrollbars with direct access to the growing region and villages

**Fig 3.** RIV's navigation tools.

In RIV the user had the choice of five different navigation-tools (Fig. 3). Three navigation tools (zoom in, zoom out and pan (to move the map)) where the selected tool is marked by a red frame, the scalebar with 16 scales and a menu that lets the user choose growing regions and villages.

In order to analyze the user's strategies in the first task we analyzed the following measures with our tools:

- how many "navigation-clicks" the user made
- what clicks the user made

At first we noticed some differences in the frequency of use of the different tools by counting the number of clicks (see Fig. 4):

- One out of 20 users tried out all tools during the first task
- Six users used four different tools
- Seven users used three different tools and
- Six users managed to navigate to the right spot (where the parcel had to be digitized) with only two different navigation tools.
- Eight users clearly used the pair zoomin- zoomout for changing the scale (however some tried out other manners as well).
- Five users used the scalebar as least as often as the zoom-in zoom-out pair.

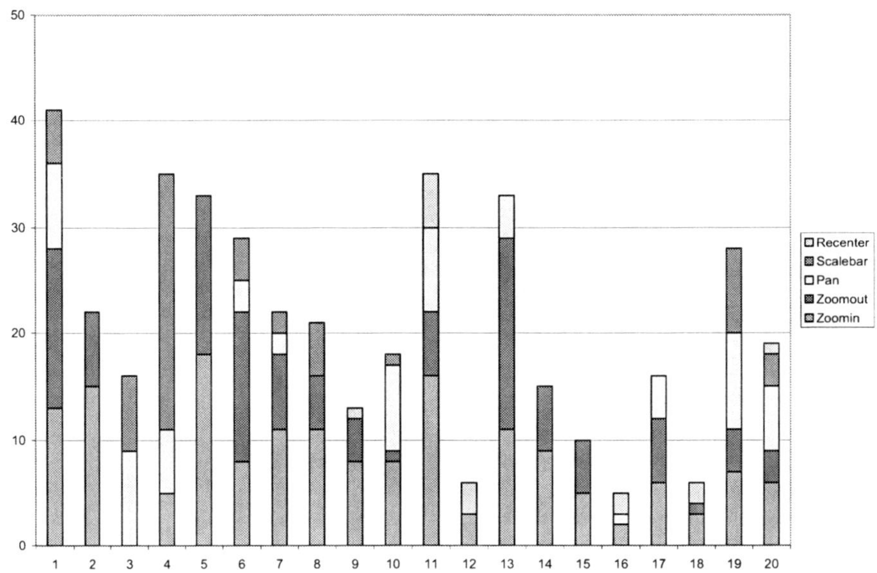


**Fig. 4.** Number of clicks with the navigation-tools to navigate to a parcel

Furthermore we noticed that users who only use few navigation-tools also need few navigation-clicks to complete the first task. On the other hand, users who used many different navigation-tools also made many clicks.

In ten cases we noticed that users tried to click on the zoom-tools and expected the system to zoom in through our gap-detection-feature. Eight of those ten users found out later that it is necessary to click on the map in order to zoom in or zoom out and two users went over to other navigation-tools (such as the scalebar and the recenter-tool)

In order to analyze the user's performance during the task to navigate the maps, we analyzed how much time it took to navigate to a specific place. We considered that map-navigation is a continuous process where the user is:

- considering the map and trying to put it into relation with the real world
- finding out a strategy in order to change the state of the map (e.g. the map is not a the right scale or is not showing the right place)
- applying the strategy (zooming, recentering)
- re-considering the map and so forth

However measuring navigation-time was difficult in our evaluation due to the following reasons:

- our evaluations were using verbal protocols (Ericson and Simon, 1993) – the user was encouraged to talk aloud while interacting with the system, thus gaps in the interaction with the system were quite frequent.
- a click on the map or a click on another navigation-tool (scalebar and recenter-menu) resulted in a query that was sent to the server. The server had to process a new map and send it to the client. This time was approximately 1-2 seconds.

Due to these reasons we decided to measure navigation-time as follows:

- A navigation-flow (many navigation-clicks in a row) has less than 10 seconds in between the actions - we counted the time from the beginning of the flow until the end of the flow
- Single clicks with ten seconds before and after the click were counted as three seconds

For all users we accumulated navigation-flows and single clicks to one measure. We also counted the gap-time separately.

As a result we found out that there were big differences in the user's performance. Some users managed to solve the task in 12 seconds, while others needed almost 6 minutes. Moreover we found out that users who only used one combination of few navigation tools, e.g. zoom-in and zoom-out or scalebar + pan or recenter-tool + pan-tool needed much fewer clicks and overall time than users who used three or more navigation tools. However this fact is certainly also related to the system's response time on clicking and due to the fact that each click produces a new map-state which requires the user to re-align himself and to figure out the next steps. Considering the gaps in between the "navigation-flows" we also noticed differences – 75% of the test-users needed more "gap-time" than "pure" navigation-time, while 25% of the user's needed less. During the gaps much of the user's time was spent on re-aligning himself after zooming to a very low scale.

## 6 Conclusions and Further Research

Within our project, we have found an evaluation-method for measuring performance that can be easily deployed and used. We believe that this method can be used for most web-GIS that are using different technologies, both on the client-side as well as on the server-side. We think that it is important to consider not only the log-file, but also the user's interface. The log file gives information about the number of interactions, the kind of interactions and the absolute and relative time. The user's screen can give explanations why the user interacted in a certain manner. Furthermore the log-file's correctness and accuracy can be verified by considering the user's screen.

Despite the fact that our web-GIS' interface was limited compared to a standard GIS' interface, we could see that users performed very differently. Users who only used few navigation-tools performed best, compared to users who navigated with more than three different tools. While we considered what the user was doing in between the navigation clicks, we found out that users seem to have certain expectations on the system's reaction on their action – in many cases this expectation was not fulfilled and the user waited for the system to react. Moreover many users spent a considerable amount of time in re-aligning themselves after zooming to a very low scale.

In a further step of our project it will be a goal to verify if the performance that we have measured during our evaluations does have a connection to the user's satisfaction and the user's previous experience with similar systems and if specific interface features (e.g. features that have been adapted from standard-GIS or features that are mostly used within a web-context) cause a higher load for the user. Moreover we want to test and refine our evaluation methodology with further tests involving different systems, users and evaluation-tasks.

Another long-term perspective will be to evaluate the effectiveness of the interface, in terms of its ability to help resolve the tasks at hand.

## References

Aoidh, E.M., Bertolotto, M.: Improving Spatial Data Usability By Capturing User Interactions. Proceedings of AGILE (2007), Aalborg, Denmark

Ericsson, K., & Simon, H. (1993). Protocol Analysis: Verbal Reports as Data, 2nd ed., Boston: MIT Press.

Hunter, G. J., Wachowicz, M, Bregt, A.: (2002) Understanding Spatial Data Usability. Data Science Journal, 2, 79-89

Ingensand, J. (2005): Developing web-GIS applications acording to HCI guidelines: the viti-vaud project in: Hypermedia: Concepts and Systems, Springer, Zurich.

Ingensand, J. Caloz, R, Pythod, K. (2006): Creating an Interactive Network for Wine-cultivation: Proceedings of the Nordic Geographers Meeting, Lund

Ingensand, J. Golay, F. (2007) Evaluating collaborative web-GIS. To be published in Cybergeo, European Journal of Geography, Paris

ISO 9241 (1994): Ergonomic requirements for office work with visual display terminals (VDT's). Part 11. Guidance on usability.

Nielsen, J. (1993), Usability Engineering, Academic Press, London

Preece, J., Rogers, Y. & Sharp, H. (2002) Interaction Design: Beyond Human-Computer Interaction. New York, NY: John Wiley & Sons.

Traynor, C. and Williams, M (1995). : Why are Geographic Information Systems Hard to Use? CHI 1995 Proceedings. ACM Press.

Shneiderman, B. (1998), Designing the User Interface, Addison-Wesley Publishing Company, USA

Tullis, T.S., Fleischman, S. McNulty, M. Cianchette, C. Bergel, M. (2002): An Empirical Comparison of Lab and Remote Usability Testing of Web Sites. Usability Professionals Association Conference, Orlando, USA

# Index

Printed in the United Kingdom
by Lightning Source UK Ltd.
135725UK00006BA/170/P

9 783540 789451